T0269046

LONDON MATHEMATICAL SOCIETY LECTURE NOTE SERIES

Managing Editor: Professor J.W.S. Cassels, Department of Pure Mathematics and Mathematical Statistics,
University of Cambridge, 16 Mill Lane, Cambridge CB2 1SB, England

The titles below are available from booksellers, or, in case of difficulty, from Cambridge University Press.

46	*p*-adic Analysis: a short course on recent work, N. KOBLITZ
59	Applicable differential geometry, M. CRAMPIN & F.A.E. PIRANI
66	Several complex variables and complex manifolds II, M.J. FIELD
86	Topological topics, I.M. JAMES (ed)
87	Surveys in set theory, A.R.D. MATHIAS (ed)
88	FPF ring theory, C. FAITH & S. PAGE
89	An F-space sampler, N.J. KALTON, N.T. PECK & J.W. ROBERTS
90	Polytopes and symmetry, S.A. ROBERTSON
92	Representation of rings over skew fields, A.H. SCHOFIELD
93	Aspects of topology, I.M. JAMES & E.H. KRONHEIMER (eds)
96	Diophantine equations over function fields, R.C. MASON
97	Varieties of constructive mathematics, D.S. BRIDGES & F. RICHMAN
98	Localization in Noetherian rings, A.V. JATEGAONKAR
99	Methods of differential geometry in algebraic topology, M. KAROUBI & C. LERUSTE
100	Stopping time techniques for analysts and probabilists, L. EGGHE
104	Elliptic structures on 3-manifolds, C.B. THOMAS
105	A local spectral theory for closed operators, I. ERDELYI & WANG SHENGWANG
107	Compactification of Siegel moduli schemes, C.-L. CHAI
109	Diophantine analysis, J. LOXTON & A. VAN DER POORTEN (eds)
113	Lectures on the asymptotic theory of ideals, D. REES
114	Lectures on Bochner-Riesz means, K.M. DAVIS & Y.-C. CHANG
116	Representations of algebras, P.J. WEBB (ed)
119	Triangulated categories in the representation theory of finite-dimensional algebras, D. HAPPEL
121	Proceedings of *Groups - St Andrews 1985*, E. ROBERTSON & C. CAMPBELL (eds)
128	Descriptive set theory and the structure of sets of uniqueness, A.S. KECHRIS & A. LOUVEAU
130	Model theory and modules, M. PREST
131	Algebraic, extremal & metric combinatorics, M.-M. DEZA, P. FRANKL & I.G. ROSENBERG (eds)
132	Whitehead groups of finite groups, ROBERT OLIVER
133	Linear algebraic monoids, MOHAN S. PUTCHA
134	Number theory and dynamical systems, M. DODSON & J. VICKERS (eds)
137	Analysis at Urbana, I, E. BERKSON, T. PECK, & J. UHL (eds)
138	Analysis at Urbana, II, E. BERKSON, T. PECK, & J. UHL (eds)
139	Advances in homotopy theory, S. SALAMON, B. STEER & W. SUTHERLAND (eds)
140	Geometric aspects of Banach spaces, E.M. PEINADOR & A. RODES (eds)
141	Surveys in combinatorics 1989, J. SIEMONS (ed)
144	Introduction to uniform spaces, I.M. JAMES
146	Cohen-Macaulay modules over Cohen-Macaulay rings, Y. YOSHINO
148	Helices and vector bundles, A.N. RUDAKOV *et al*
149	Solitons, nonlinear evolution equations and inverse scattering, M. ABLOWITZ & P. CLARKSON
150	Geometry of low-dimensional manifolds 1, S. DONALDSON & C.B. THOMAS (eds)
151	Geometry of low-dimensional manifolds 2, S. DONALDSON & C.B. THOMAS (eds)
152	Oligomorphic permutation groups, P. CAMERON
153	L-functions and arithmetic, J. COATES & M.J. TAYLOR (eds)
155	Classification theories of polarized varieties, TAKAO FUJITA
156	Twistors in mathematics and physics, T.N. BAILEY & R.J. BASTON (eds)
158	Geometry of Banach spaces, P.F.X. MÜLLER & W. SCHACHERMAYER (eds)
159	Groups St Andrews 1989 volume 1, C.M. CAMPBELL & E.F. ROBERTSON (eds)
160	Groups St Andrews 1989 volume 2, C.M. CAMPBELL & E.F. ROBERTSON (eds)
161	Lectures on block theory, BURKHARD KÜLSHAMMER
162	Harmonic analysis and representation theory, A. FIGA-TALAMANCA & C. NEBBIA
163	Topics in varieties of group representations, S.M. VOVSI
164	Quasi-symmetric designs, M.S. SHRIKANDE & S.S. SANE
166	Surveys in combinatorics, 1991, A.D. KEEDWELL (ed)
168	Representations of algebras, H. TACHIKAWA & S. BRENNER (eds)
169	Boolean function complexity, M.S. PATERSON (ed)
170	Manifolds with singularities and the Adams-Novikov spectral sequence, B. BOTVINNIK
171	Squares, A.R. RAJWADE
172	Algebraic varieties, GEORGE R. KEMPF
173	Discrete groups and geometry, W.J. HARVEY & C. MACLACHLAN (eds)
174	Lectures on mechanics, J.E. MARSDEN
175	Adams memorial symposium on algebraic topology 1, N. RAY & G. WALKER (eds)
176	Adams memorial symposium on algebraic topology 2, N. RAY & G. WALKER (eds)
177	Applications of categories in computer science, M. FOURMAN, P. JOHNSTONE & A. PITTS (eds)
178	Lower K- and L-theory, A. RANICKI
179	Complex projective geometry, G. ELLINGSRUD *et al*
180	Lectures on ergodic theory and Pesin theory on compact manifolds, M. POLLICOTT

181 Geometric group theory I, G.A. NIBLO & M.A. ROLLER (eds)
182 Geometric group theory II, G.A. NIBLO & M.A. ROLLER (eds)
183 Shintani zeta functions, A. YUKIE
184 Arithmetical functions, W. SCHWARZ & J. SPILKER
185 Representations of solvable groups, O. MANZ & T.R. WOLF
186 Complexity: knots, colourings and counting, D.J.A. WELSH
187 Surveys in combinatorics, 1993, K. WALKER (ed)
188 Local analysis for the odd order theorem, H. BENDER & G. GLAUBERMAN
189 Locally presentable and accessible categories, J. ADAMEK & J. ROSICKY
190 Polynomial invariants of finite groups, D.J. BENSON
191 Finite geometry and combinatorics, F. DE CLERCK et al
192 Symplectic geometry, D. SALAMON (ed)
194 Independent random variables and rearrangement invariant spaces, M. BRAVERMAN
195 Arithmetic of blowup algebras, WOLMER VASCONCELOS
196 Microlocal analysis for differential operators, A. GRIGIS & J. SJÖSTRAND
197 Two-dimensional homotopy and combinatorial group theory, C. HOG-ANGELONI,
    W. METZLER & A.J. SIERADSKI (eds)
198 The algebraic characterization of geometric 4-manifolds, J.A. HILLMAN
199 Invariant potential theory in the unit ball of C^n, MANFRED STOLL
200 The Grothendieck theory of dessins d'enfant, L. SCHNEPS (ed)
201 Singularities, JEAN-PAUL BRASSELET (ed)
202 The technique of pseudodifferential operators, H.O. CORDES
203 Hochschild cohomology of von Neumann algebras, A. SINCLAIR & R. SMITH
204 Combinatorial and geometric group theory, A.J. DUNCAN, N.D. GILBERT & J. HOWIE (eds)
205 Ergodic theory and its connections with harmonic analysis, K. PETERSEN & I. SALAMA (eds)
206 An introduction to noncommutative differential geometry and its physical applications, J. MADORE
207 Groups of Lie type and their geometries, W.M. KANTOR & L. DI MARTINO (eds)
208 Vector bundles in algebraic geometry, N.J. HITCHIN, P. NEWSTEAD & W.M. OXBURY (eds)
209 Arithmetic of diagonal hypersurfaces over finite fields, F.Q. GOUVÊA & N. YUI
210 Hilbert C*-modules, E.C. LANCE
211 Groups 93 Galway / St Andrews I, C.M. CAMPBELL et al (eds)
212 Groups 93 Galway / St Andrews II, C.M. CAMPBELL et al (eds)
214 Generalised Euler-Jacobi inversion formula and asymptotics beyond all orders, V. KOWALENKO,
    N.E. FRANKEL, M.L. GLASSER & T. TAUCHER
215 Number theory 1992–93, S. DAVID (ed)
216 Stochastic partial differential equations, A. ETHERIDGE (ed)
217 Quadratic forms with applications to algebraic geometry and topology, A. PFISTER
218 Surveys in combinatorics, 1995, PETER ROWLINSON (ed)
220 Algebraic set theory, A. JOYAL & I. MOERDIJK
221 Harmonic approximation, S.J. GARDINER
222 Advances in linear logic, J.-Y. GIRARD, Y. LAFONT & L. REGNIER (eds)
223 Analytic semigroups and semilinear initial boundary value problems, KAZUAKI TAIRA
224 Computability, enumerability, unsolvability, S.B. COOPER, T.A. SLAMAN & S.S. WAINER (eds)
225 A mathematical introduction to string theory, S. ALBEVERIO, J. JOST, S. PAYCHA, S. SCARLATTI
226 Novikov conjectures, index theorems and rigidity I, S. FERRY, A. RANICKI & J. ROSENBERG (eds)
227 Novikov conjectures, index theorems and rigidity II, S. FERRY, A. RANICKI & J. ROSENBERG (eds)
228 Ergodic theory of Z^d actions, M. POLLICOTT & K. SCHMIDT (eds)
229 Ergodicity for infinite dimensional systems, G. DA PRATO & J. ZABCZYK
230 Prolegomena to a middlebrow arithmetic of curves of genus 2, J.W.S. CASSELS & E.V. FLYNN
231 Semigroup theory and its applications, K.H. HOFMANN & M.W. MISLOVE (eds)
232 The descriptive set theory of Polish group actions, H. BECKER & A.S. KECHRIS
233 Finite fields and applications, S. COHEN & H. NIEDERREITER (eds)
234 Introduction to subfactors, V. JONES & V.S. SUNDER
235 Number theory 1993–94, S. DAVID (ed)
236 The James forest, H. FETTER & B. GAMBOA DE BUEN
237 Sieve methods, exponential sums, and their applications in number theory, G.R.H. GREAVES,
    G. HARMAN & M.N. HUXLEY (eds)
238 Representation theory and algebraic geometry, A. MARTSINKOVSKY & G. TODOROV (eds)
239 Clifford algebras and spinors, P. LOUNESTO
240 Stable groups, FRANK O. WAGNER
241 Surveys in combinatorics, 1997, R.A. BAILEY (ed)
242 Geometric Galois actions I, L. SCHNEPS & P. LOCHAK (eds)
243 Geometric Galois actions II, L. SCHNEPS & P. LOCHAK (eds)
244 Model theory of groups and automorphism groups, D. EVANS (ed)
245 Geometry, combinatorial designs and related structures, J.W.P. HIRSCHFELD, S.S. MAGLIVERAS
    & M.J. DE RESMINI (eds)
246 p-Automorphisms of finite p-groups, E.I. KHUKHRO
247 Analytic number theory, Y. MOTOHASHI (ed)
248 Tame topology and o-minimal structures, LOU VAN DEN DRIES
249 The atlas of finite groups: ten years on, ROBERT CURTIS & ROBERT WILSON (eds)
250 Characters and blocks of finite groups, G. NAVARRO
251 Gröbner bases and applications, B. BUCHBERGER & F. WINKLER (eds)
252 Geometry and cohomology in group theory, P. KROPHOLLER, G. NIBLO, R. STÖHR (eds)

London Mathematical Society Lecture Note Series. 249

The Atlas of Finite Groups: Ten Years on

Edited by

Robert Curtis and Robert Wilson
University of Birmingham

PUBLISHED BY THE PRESS SYNDICATE OF THE UNIVERSITY OF CAMBRIDGE
The Pitt Building, Trumpington Street, Cambridge CB2 1RP, United Kingdom

CAMBRIDGE UNIVERSITY PRESS
The Edinburgh Building, Cambridge, CB2 2RU, United Kingdom
40 West 20th Street, New York, NY 10011-4211, USA
10 Stamford Road, Oakleigh, Melbourne 3166, Australia

First published 1998

A catalogue record for this book is available from the British Library

ISBN   0 521 57587 7 paperback

Transferred to digital printing 2003

Contents

Introduction   vii

Addresses of registered participants   ix

Addresses of non-participating authors   xiii

Programme of lectures   xiv

Conference photograph and key   xvi

C. M. Campbell, G. Havas, S. A. Linton and E. F. Robertson
Symmetric presentations and orthogonal groups   1

F. Celler and C. R. Leedham-Green
A constructive recognition algorithm for the special linear group   11

J. H. Conway and C. S. Simons
Relations in M_{666} 27

R. T. Curtis
A survey of symmetric generation of sporadic simple groups   39

G. Hiss
Harish-Chandra theory, q-Schur algebras, and decomposition matrices
for finite classical groups   58

D. F. Holt
The Meataxe as a tool in computational group theory   74

J. F. Humphreys
Branching rules for modular projective representations of the
symmetric groups   82

G. A. Jones
Characters and surfaces: a survey   90

W. Kimmerle
On the characterization of finite groups by characters   119

A. S. Kondratiev
Finite linear groups of small degree   139

W. Lempken, C. Parker and P. Rowley
Minimal parabolic systems for the symmetric and alternating groups   149

M. W. Liebeck and A. Shalev
Probabilistic methods in the generation of finite simple groups   163

K. Lux and M. Wiegelmann
Condensing tensor product modules   174

V. D. Mazurov and V. I. Zenkov
Intersections of Sylow subgroups in finite groups   191

vi

S. P. Norton
Anatomy of the Monster. I   198

R. A. Parker
An integral 'Meat-axe'   215

W. Plesken
Finite rational matrix groups: a survey   229

P. Rowley
Chamber graphs of sporadic group geometries   249

R. A. Wilson
An Atlas of sporadic group representations   261

F. Zara
Presentations of reductive Fischer groups   274

J. H. Conway, R. T. Curtis and R. A. Wilson
A brief history of the ATLAS   288

Introduction

This book contains the proceedings of a conference on group theory and its applications held in Birmingham, July 10th–13th, 1995, to mark the tenth anniversary of the publication of the 'ATLAS of finite groups'. The theme of the conference was to survey developments in the subject during the intervening ten years, and in particular those that were facilitated or inspired in some way by the ATLAS itself. The conference was supported by a grant from the Engineering and Physical Sciences Research Council, and underwritten by the London Mathematical Society.

The conference brought together all five authors of the ATLAS, for the first time in many years (if not the first time ever), and a number of distinguished speakers whose talks are listed below. The twenty papers in this volume mostly represent expanded versions of some of these talks. We were also able to celebrate not only the birthday of the ATLAS of Finite Groups, but the birth of a new ATLAS, the 'ATLAS of Brauer characters', whose publication was brought forward to enable the first 30 copies to be sold at the conference. If this is volume 2 of the ATLAS series, perhaps we can expect volume 3 in another ten years.

Although the articles in this book do not fall easily into well-defined categories, we can roughly divide them into those concerned with *presentations of groups*, those dealing with *representations and characters*, followed by *computational methods*, and various aspects of *subgroups and geometries*, and finally *applications and generation*.

In the theory and practice of presentations, we have papers by *Conway and Simons* on presentations of the Monster, *Campbell et al.* and *Curtis* discussing two different types of symmetric presentations, and *Zara* with series of presentations for some Fischer 3-transposition groups.

In representation theory, we begin with two surveys, one by *Hiss* on recent advances in the representation theory of classical groups in non-defining characteristic, the other by *Kimmerle* on the Zassenhaus conjectures on integral group rings. Then we have an article by *Humphreys* on modular representations of double covers of symmetric groups, and *Wilson*'s description of a computerized ATLAS of group representations.

This area overlaps substantially with the area of computational methods, where *Parker* describes new algorithms making up a complete Meataxe system for integral representations. A different approach to integral representations is explored in *Plesken*'s paper, which describes a computational classification of lattices in dimensions up to 31. *Lux and Wiegelmann* describe an extension of the Meataxe, using a condensation technique to analyse the structure of tensor product modules, and *Holt* discusses Meataxe techniques in general. Somewhat different computational ideas are developed by *Celler and Leedham-Green* in their paper describing an explicit isomorphism test for projective special linear groups.

The papers on subgroups and geometries range widely, from *Norton*'s article on subgroups of the Monster, which makes generally accessible for the first time a wealth of detailed information, and *Kondratiev*'s paper on subgroups of general linear groups of degree up to 27, via the paper of *Mazurov and Zenkov* on intersections of Sylow subgroups, to geometric aspects of Sylow subgroups discussed by *Lempken et al.* and *Rowley*.

Finally, *Jones*'s comprehensive survey on applications of character theory to surfaces includes some discussion of ways of generating simple groups, a topic which is investigated from a probabilistic point of view by *Liebeck and Shalev*. We round off the book with a short historical note, outlining the development of the ATLAS project over many years.

In addition to the serious business of the conference, we had a birthday cake made in the shape of an ATLAS, complete with red icing and all the lettering from the cover of the real ATLAS, and ten candles. The evening boat trip on the canal to the Gas Street Basin proved popular, as did the reception in the Barber Institute (with guided tour of the art gallery by Prof. Richard Verdi included), and the conference dinner in the botanical gardens.

Addresses of registered participants

M. Ashworth, School of Mathematics and Statistics, The University of Birmingham, Edgbaston, Birmingham B15 2TT.

Dr. R. W. Baddeley, Department of Mathematics and Computer Science, University of Leicester, Leicester LE1 7RH.

Prof. R. A. Bailey, School of Mathematical Sciences, Queen Mary and Westfield Colleges, London University, Mile End Road, London E1 4NS.

B. Baumeister, Department of Mathematics, Imperial College, Huxley Building, 180 Queen's Gate, London SW7 2BX.

Dr. A. Borovik, Department of Mathematics, UMIST, P.O. Box 88, Manchester M60 1QD.

J. N. Bray, School of Mathematics and Statistics, The University of Birmingham, Edgbaston, Birmingham B15 2TT.

Dr. T. Breuer, Lehrstuhl D für Mathematik, RWTH Aachen, Templergraben 64, D-52062 Aachen, Germany.

Prof. R. M. Bryant, Department of Mathematics, UMIST, P.O. Box 88, Manchester M60 1QD.

Prof. P. J. Cameron, School of Mathematical Sciences, Queen Mary and Westfield Colleges, London University, Mile End Road, London E1 4NS.

Dr. C. M. Campbell, Mathematics Institute, St Andrews University, North Haugh, St Andrews, Fife KY16 9SS, Scotland.

Prof. R. W. Carter, Mathematics Institute, Warwick University, Coventry CV4 7AL.

Dr. Frank Celler, Lehrstuhl D für Mathematik, RWTH Aachen, Templergraben 64, D-52062 Aachen, Germany.

Mr. A. D. Chaplin, Department of Mathematics, UMIST, P.O. Box 88, Manchester M60 1QD.

Dr. I. M Chiswell, School of Mathematical Sciences, Queen Mary and Westfield Colleges, London University, Mile End Road, London E1 4NS.

Prof. J. H. Conway, Department of Mathematics, University of Princeton, Fine Hall, Princeton, NJ 08544, USA.

Dr. P. Covey-Crump, GCHQ, Priors Road, Cheltenham, Gloucester, GL1 1NC.

Dr. R. T. Curtis, School of Mathematics and Statistics, The University of Birmingham, Edgbaston, Birmingham B15 2TT.

Dr. H. Cuypers, Department of Mathematics, Technical University Eindhoven, P.O. Box 513, 5600 MB Eindhoven, Netherlands.

Dr. M. Edjvet, Department of Mathematics, University of Nottingham, University Park, Nottingham NG7 2RD.

A. Feig, Fachbereich Mathematik der Universität Mainz, Saarstrasse 21, D-55099 Mainz, Germany.

Dr. P. J. Flavell, School of Mathematics and Statistics, The University of Birmingham, Edgbaston, Birmingham B15 2TT.

Dr. A. D. Gardiner, School of Mathematics and Statistics, The University of Birmingham, Edgbaston, Birmingham B15 2TT.

D. A. Gewurz, Via Muzio Attendolo, 13, 00176 Roma, Italy.

Dr. R. Green, Mathematical Institute, 24–29 St Giles, Oxford, OX1 3LB.

Prof. R. K. Guy, University of Calgary, Calgary, Alberta, Canada T2N 1N4.

A. Hanysz, DPMMS, 16 Mill Lane, Cambridge CB2 1SB.

Dr. S. Heiss, Martin-Luther-Universität, Fachbereich Mathematik und Informatik, D-06099 Halle–Saale, Germany.

Prof. Dr. C. Hering, Mathematisches Institut, Universität Tübingen, Auf der Morgenstelle 10, D-72076 Tübingen, Germany.

Prof. Dr. G. Hiss, Lehrstuhl D für Mathematik, RWTH Aachen, Templergraben 64, D-52062 Aachen, Germany.

Dr. A. H. M. Hoare, School of Mathematics and Statistics, The University of Birmingham, Edgbaston, Birmingham B15 2TT.

Dr. D. F. Holt, Institute of Mathematics, University of Warwick, Coventry CV4 7AL.

Dr. J. Hrabě de Angelis, Fachbereich Mathematik der Universität Mainz, Saarstrasse 21, D-55099 Mainz, Germany.

S. Hudson, School of Mathematics and Statistics, The University of Birmingham, Edgbaston, Birmingham B15 2TT.

Dr. J. F. Humphreys, Department of Pure Mathematics, Liverpool University, P.O. Box 147, Liverpool L69 3BX.

Dr. N. Inglis, Department of Mathematics and Statistics, Lancaster University, Fylde College, Lancaster LA1 4YF.

Dr. A. A. Ivanov, Department of Mathematics, Imperial College, Huxley Building, 180 Queen's Gate, London SW7 2BX.

Prof. G. D. James, Department of Mathematics, Imperial College, Huxley Building, 180 Queen's Gate, London SW7 2BX.

Dr. C. Jansen, Lehrstuhl D für Mathematik, RWTH Aachen, Templergraben 64, D-52062 Aachen, Germany.

Dr. G. A. Jones, Department of Mathematics, Southampton University, Highfield, Southampton SO17 1BJ.

Dr. R. W. Kaye, School of Mathematics and Statistics, The University of Birmingham, Edgbaston, Birmingham B15 2TT.

Dr. W. Kimmerle, Math. Institut B der Universität Stuttgart, Pfaffenwaldring 57, D-70569 Stuttgart, Germany.

Prof. A. S. Kondratiev, Institute of Mathematics and Mechanics, Ural Branch of Russian Academy of Sciences, S. Kovalevskaya Street 16, 620219 Ekaterinburg, Russia.

Dr. P. H. Kropholler, School of Mathematical Sciences, Queen Mary and Westfield Colleges, London University, Mile End Road, London E1 4NS.

Dr. C. R. Leedham-Green, School of Mathematical Sciences, Queen Mary and Westfield Colleges, London University, Mile End Road, London E1 4NS.

Dr. W. Lempken, Institute for Experimental Mathematics, University of Essen, Ellernstrasse 29, D-45326 Essen, Germany.

Prof. M. W. Liebeck, Department of Mathematics, Imperial College, Huxley Building, 180 Queen's Gate, London SW7 2BX.

Dr. S. A. Linton, Mathematics Institute, St Andrews University, North Haugh, St Andrews, Fife KY16 9SS, Scotland.

Dr. K. Lux, Department of Mathematics, University of Arizona, Tucson, Arizona 85721, USA.

M. Marjoram, Department of Mathematics, University College Dublin, Bellfield, Dublin 4, Ireland.

Dr. R. Marsh, Department of Mathematics, University of Glasgow, University Gardens, Glasgow G12 8QW.

Dr. A. W. Mason, Department of Mathematics, University of Glasgow, University Gardens, Glasgow G12 8QW.

Dr. A. Mathas, Department of Mathematics, Imperial College, Huxley Building, 180 Queen's Gate, London SW7 2BX.

Prof. V. D. Mazurov, Institute of Mathematics, Novosibirsk 630090, Russia.

Dr. T. McDonough, Department of Mathematics, University of Wales, Aberystwyth SY23 3BZ.

Prof. J. McKay, Department of Computer Science, SGWC, Concordia University, Montréal, QC, Canada H3G 1M8.

N. Mudamburi, School of Mathematics and Statistics, The University of Birmingham, Edgbaston, Birmingham B15 2TT.

B. Mühlherr, Mathematisches Institut, Universität Tübingen, Auf der Morgenstelle 10, D-72076 Tübingen, Germany.

Dr. J. Müller, IWR Universität Heidelberg, Im Neuenheimer Feld 368, Heidelberg, Germany.

J. Muncke, Denzenberg Strasse 41, 72074 Tübingen, Germany.

Dr. P. M. Neumann, Queen's College, Oxford, OX1 4AW.

Dr. S. P. Norton, DPMMS, 16, Mill Lane, Cambridge CB2 1SB.

Dr. C. W. Parker, School of Mathematics and Statistics, The University of Birmingham, Edgbaston, Birmingham B15 2TT.

R. A. Parker, 'Springfields', Frog Lane, Shepton Mallet, Somerset, BA4 4PP

Dr. D. V. Pasechnik, RIACA/CAN, 419 Kruislaan, 1098 VA Amsterdam, The Netherlands.

S. Perlo-Freeman, Mathematics Institute, Warwick University, Coventry

CV4 7AL.

Dr. G. Pfeiffer, Mathematics Institute, St Andrews University, North Haugh, St Andrews, Fife KY16 9SS, Scotland.

B. J. Philp, School of Mathematics and Statistics, The University of Birmingham, Edgbaston, Birmingham B15 2TT.

L. Pirnie, Department of Mathematics, UMIST, P.O. Box 88, Manchester M60 1QD.

Prof. Dr. W. Plesken, Lehrstuhl B für Mathematik, RWTH Aachen, Templergraben 64, D-52062 Aachen, Germany.

Prof. C. Praeger, Department of Mathematics, University of Western Australia, Nedlands, WA 6907, Australia.

Prof. G. Robinson, Department of Mathematics and Computer Science, University of Leicester, Leicester LE1 7RH.

S. J. F. Rogers, School of Mathematics and Statistics, The University of Birmingham, Edgbaston, Birmingham B15 2TT.

Dr. C. A. Rowley, Open University, 527 Finchley Road, London NW3 7BG.

Dr. P. J. Rowley, Department of Mathematics, UMIST, P.O. Box 88, Manchester M60 1QD.

Dr. J. Saxl, DPMMS, 16, Mill Lane, Cambridge CB2 1SB.

Dr. I. J. Siemons, School of Mathematics, University of East Anglia, University Plain, Norwich NR4 7TJ.

Dr. G. C. Smith, School of Mathematical Sciences, Bath University, Claverton Down, Bath BA2 7AY.

Dr. L. H. Soicher, School of Mathematical Sciences, Queen Mary and Westfield Colleges, London University, Mile End Road, London E1 4NS.

F. Spiezia, School of Mathematics, University of East Anglia, University Plain, Norwich NR4 7TJ.

Dr. C. Stack, Department of Mathematics, The University of Reading, Whiteknights, P.O. Box 220, Reading RG6 6AX.

S. A. Stanley, School of Mathematics and Statistics, The University of Birmingham, Edgbaston, Birmingham B15 2TT.

H. Sulaiman, Department of Mathematics, UMIST, P.O. Box 88, Manchester M60 1QD.

Prof. C. Tamburini, Dipartimento di Matematica, Università Cattolica, Via Trieste 17, 25121 Brescia, Italy.

Dr. R. M. Thomas, Department of Mathematics and Computer Science, University of Leicester, Leicester LE1 7RH.

J. Tompson, Oxford University Press, Walton Street, Oxford.

Dr. J. van Bon, School of Mathematical Sciences, Queen Mary and Westfield Colleges, London University, Mile End Road, London E1 4NS.

Dr. M. Vaughan-Lee, Christ Church, Oxford, OX1 1DP.

Dr. L. A. Walker, Department of Mathematics and Statistics, University of Central Lancashire, Preston PR1 2TQ.
P. G. Walsh, School of Mathematics and Statistics, The University of Birmingham, Edgbaston, Birmingham B15 2TT.
Dr. D. L. Wilkens, School of Mathematics and Statistics, The University of Birmingham, Edgbaston, Birmingham B15 2TT.
Prof. J. S. Wilson, School of Mathematics and Statistics, The University of Birmingham, Edgbaston, Birmingham B15 2TT.
Dr. R. A. Wilson, School of Mathematics and Statistics, The University of Birmingham, Edgbaston, Birmingham B15 2TT.
J. York, School of Mathematics and Statistics, The University of Birmingham, Edgbaston, Birmingham B15 2TT.
Prof. F. Zara, 34 bis, av. Victor-Hugo, 92340 Bourg-la Reine, France.
T. E. Zieschang, University of Saarland, Geb. 36, Graduiertenkolleg, Postfach 1150, D-66041 Saarbrücken, Germany.

Addresses of non-participating authors

G. Havas, Department of Computer Science, University of Queensland, Brisbane, 4072 Australia.
E. F. Robertson, Mathematics Institute, St Andrews University, North Haugh, St Andrews, Fife KY16 9SS, Scotland.
A. Shalev, Institute of Mathematics, The Hebrew University, Jerusalem 91904, Israel.
C. S. Simons, Department of Mathematics, University of Princeton, Fine Hall, Princeton, NJ 08544, USA.
M. Wiegelmann, Lehrstuhl D für Mathematik, RWTH Aachen, Templergraben 64, D-52062 Aachen, Germany.
V. I. Zenkov, Institute of Mathematics and Mechanics, S. Kovalevskaja Street 16, Ekaterinburg, 620066, Russia.

Programme of lectures

C. M. Campbell, Symmetric presentations

R. W. Carter, Paths and crystals

J. H. Conway, The Atlas: past, present and future

R. T. Curtis, Monomial modular representations and symmetric generation of sporadic simple groups

G. Hiss, Representations of finite groups of Lie type in non-defining characteristic

D. F. Holt, The Meataxe as a tool in computational group theory

J. Hrabě de Angelis, A new presentation of the Held group

J. F. Humphreys, Branching rules for modular projective representations of the symmetric groups

G. D. James, Problems queuing up

G. Jones, Applications to the study of coverings of surfaces

W. Kimmerle, Character tables and Zassenhaus' conjecture for $\mathbb{Z}G$

A. S. Kondratiev, Modular representations of finite quasisimple groups

C. R. Leedham-Green, Recognising tensor product representations

W. Lempken, On p'-semiregular groups

M. W. Liebeck, Simple groups and probabilistic methods

K. Lux, Condensation methods in modular representation theory

V. Mazurov, Intersections of Sylow subgroups in finite groups

J. McKay, Some thoughts on replicable functions and the Monster

P. M. Neumann, Goldbach's conjecture and soluble permutation groups

S. P. Norton, Properties of the Monster algebra

R. A. Parker, Modular characters

W. Plesken, Finite rational matrix groups

C. E. Praeger, The ATLAS: a tool for finite permutation groups

P. Rowley, Chambers

J. Saxl, Factorizations of groups

L. H. Soicher, The classification of multiplicity-free and of distance-transitive primitive representations of the sporadics

J. G. Thompson, The sporadic simple groups as Galois groups

R. A. Wilson, Constructing the Monster?

F. Zara, Presentations of classical Fischer groups

Key to conference photograph

1. Hans Cuypers
2. Rick Thomas
3. Robert Baddeley
4. Tony Gardiner
5. Geoff Robinson
6. Gareth Jones
7. Chris Parker
8. Götz Pfeiffer
9. Roger Carter
10. Paul Martin
11. Alec Mason
12. Stephen Stanley
13. Julian York
14. Sasha Ivanov
15. Peter Neumann
16. John Thompson
17. Chris Rowley
18. Colin Campbell
19. Peter Kropholler
20. Gordon James
21. Thilo Zieschang
22. Wilhelm Plesken
23. Peter Cameron
24. Jürgen Müller
25. Leonard Soicher
26. John van Bon
27. Daniele Gewurz
28. Bernhard Mühlherr
29. John Bray
30. Roger Bryant
31. Steve Linton
32. Stefan Heiss
33. Frank Celler
34. A mystery customer
35. Charles Leedham-Green
36. Anatolii Kondratiev
37. Frederica Spiezia
38. Michelle Ashworth
39. Jo Maio
40. Murray Macbeath
41. John McKay
42. Dima Pasechnik
43. Wolfgang Lempken
44. Howard Hoare
45. Peter Rowley
46. Johannes Siemons
47. Jörg Hrabě de Angelis
48. François Zara
49. Wolfgang Kimmerle
50. Derek Holt
51. Christoph Hering
52. Lewis Pirnie
53. Thomas Breuer
54. Sue Bennett
55. Chiara Tamburini
56. Samuel Perlo-Freeman
57. Robert Marsh
58. Gerhard Hiss
59. Pavel Zalesskii
60. Andrew Mathas
61. Andrew Chaplin
62. Nick Inglis
63. Viktor Mazurov
64. Michael Collins
65. Richard Green
66. Richard Kaye
67. Tetsuo Deguchi
68. Barbara Baumeister
69. Angelika Feig
70. Louise Walker
71. Julia Tompson
72. Klaus Lux
73. Richard Parker
74. Rob Curtis
75. John Conway
76. Simon Norton
77. Rob Wilson
78. Christoph Jansen
79. John Wilson
80. Hajar Sulaiman
81. Cheryl Praeger
82. Rosemary Bailey

Symmetric presentations and orthogonal groups

Colin Campbell, George Havas, Stephen Linton and Edmund Robertson

Abstract

In this paper we examine series of finite presentations which are invariant under the full symmetric group acting on the set of generators. Evidence from computational experiments reveals a remarkable tendency for the groups in these series to be closely related to the orthogonal groups. We examine cases of finite groups in such series and look in detail at an infinite group with such a presentation. We prove some theoretical results about 3-generator symmetric presentations and make a number of conjectures regarding n-generator symmetric presentations.

1 Introduction

Symmetric presentations have been much studied over a long period; see for example [1, 9, 10, 14, 23], which focus on symmetric relations. An alternative approach, directed towards symmetric generating sets, is taken by Curtis [11, 12, 13].

Suppose G is a finite 2-generator group with generators x_1, x_2 and an automorphism θ with $x_1\theta = x_2$ and $\theta^2 = 1$. Then G has a *symmetric presentation*

$$\mathcal{P} = \{x_1, x_2 \mid w_i(x_1, x_2) = w_i(x_2, x_1) = 1, \ i = 1, 2, \ldots, m\}.$$

Note that any finite non-abelian simple group has such a symmetric presentation. For, if G is a finite non-abelian simple group, then $G = \langle a, b \rangle$ where $a^2 = 1$ [19]. Now consider $H = \langle b, b^a \rangle \leq G$. Either $G = H$ or $|G : H| = 2$. But G is simple, so $G = H = \langle b, b^a \rangle$ and hence G has a symmetric presentation.

Recently Miklos Abert (unpublished) has generalized the results of [14] and conjectures that, given any non-trivial group G, then G has a presentation which, when symmetrized, presents a non-trivial image of G. The above comment on symmetric presentations for simple groups is a consequence of the results of Abert.

1

Next, for any group G with symmetric presentation $\mathcal{P}$ and any $n \geq 2$, we define the n-generator symmetric presentation

$$\mathcal{P}(n) = \{x_1, x_2, \ldots, x_n \mid w_i(x_j, x_k) = 1,\ i = 1, 2, \ldots, m,\ 1 \leq j \neq k \leq n\}.$$

For example, if $\mathcal{P} = \{x_1, x_2 \mid x_1^3 = x_2^3 = (x_1 x_2)^2 = 1\}$ then $\langle \mathcal{P} \rangle = A_4$ and $\langle \mathcal{P}(n) \rangle = A_{n+2}$, the alternating group of degree $n + 2$, see [7]. This example was generalized by Sidki [27] (with a relevant correction in [28]) who considered $\mathcal{Q}_t(n)$, where $\mathcal{Q}_t = \{x_1, x_2 \mid x_1^t = x_2^t = (x_1^i x_2^i)^2 = 1,\ 1 \leq i \leq (t-1)/2,\ t \text{ odd}\}$. Sidki showed that the groups $\langle \mathcal{Q}_t(n) \rangle$ are related to orthogonal groups. In this paper we show that orthogonal groups arise as the groups $\langle \mathcal{P}(n) \rangle$ for other symmetric presentations $\mathcal{P}$. Moreover, the nature of $\langle \mathcal{P}(n) \rangle$ varies with increasing n in very diverse ways depending not only on the group $\langle \mathcal{P} \rangle$ but also on the presentation $\mathcal{P}$.

We achieve many of our results by systematic and substantial use of implementations of algorithms. Access to group-theoretic algorithms is provided via the computer algebra systems Cayley [6], GAP [25] and MAGMA [2]; the packages Quotpic [18] and the ANU p-Quotient Program [16, 21]; and various stand-alone programs, see for example [17]. A fundamental tool for computing with finitely presented groups is coset enumeration. We use descendants of the procedure described in [15] which are available in MAGMA and in stand-alone programs. We use Maple for general symbolic calculation. We use the ATLAS [8] as our source of structural information and notation for simple groups.

In section 2 we examine the groups $\langle \mathcal{P}(3) \rangle$ where $\mathcal{P}(2)$ is one of a family of presentations for $L_2(p)$. Section 3 examines 4-generator versions of these groups and shows that both finite and infinite groups can occur. A series of symmetric presentations can collapse to the trivial group as n increases and section 4 investigates two different cases, each starting from a symmetrical presentation of $L_3(3)$. The final section includes a number of conjectures, based on further computational evidence.

2   A family of presentations

We consider the 2-generator symmetric presentation

$$\mathcal{S}_{p,c} = \{x_1, x_2 \mid x_1^p = x_2^p = (x_1^i x_2^{c/i})^2 = 1,\ 1 \leq i \leq p - 1\}$$

where p is prime, $c \neq 0$ and division is performed in $GF(p)$.

Beetham [1] showed that $\langle \mathcal{S}_{p,1} \rangle \cong L_2(p)$ for p prime and $p \geq 3$. Sidki [26] showed that it suffices to take $i = 1, 2, 4$ in $\mathcal{S}_{p,1}$. It is easy to see that, for any c, $1 \leq c \leq p - 1$, $\langle \mathcal{S}_{p,c} \rangle \cong L_2(p)$ for p prime and $p \geq 3$. The main result of this section is the analysis of $\langle \mathcal{S}_{p,c}(3) \rangle$ for general prime $p \geq 5$. (For $p = 3$ we obtain alternating groups, as shown in [7].)

Theorem 1 *Let $G = \langle S_{p,c}(3) \rangle$ (p prime, $p \geq 5$ and $c \neq 0$) with presentation*

$$\{x, y, z \mid x^p = y^p = z^p = (x^\alpha y^{c/\alpha})^2 = (y^\alpha z^{c/\alpha})^2 = (z^\alpha x^{c/\alpha})^2 = 1, \forall \alpha \in GF(p)^*\}.$$

If $\sqrt{c/2} \in GF(p)$, then $G \cong O_4^+(p) \cong 2(L_2(p) \times L_2(p))$.

Proof: The proof is by way of general hand calculations of the style of those in [4]. The details of the proof involve frequent use of the following lemma.

Lemma 1 *If $u, v \in G$, $u^p = v^p = 1$, for p an odd prime, and $(u^\alpha v^{2/\alpha})^2 = t \in Z(G)$ for $\alpha \in \{1, 2\}$, then $u^{-1}vu = vuv^{-1}$.*

Proof of Lemma 1:

$$
\begin{aligned}
v^2 u^2 &= v^2 u^2 \\
\implies v(vu^2) &= (v^2 u)u \\
\implies vu^{-2}v^{-1}t &= u^{-1}v^{-2}ut
\end{aligned}
$$

Cancel the t's and raise to the power $-1/2 \pmod p$ to get the result. $\square$

The proof of the theorem has the following simple structure. For $\sqrt{c/2} \in GF(p)$ we show (1) that $2(L_2(p) \times L_2(p))$ is a homomorphic image of G and (2) that G is a homomorphic image of $2(L_2(p) \times L_2(p))$. We use the following presentation for $H = 2(L_2(p) \times L_2(p))$:

$\{x_1, x_2, y_1, y_2, t \mid x_i^p = y_i^p = 1, [x_1, x_2] = [x_1, y_2] = [y_1, x_2] = [y_1, y_2] = 1,$
$(x_i^\alpha y_i^{c/\alpha})^2 = t, \ t^2 = 1, \ [t, x_i] = [t, y_i] = 1; \ i = 1, 2; \ \alpha \in GF(p)^*\}.$

Observe that it suffices to prove (1) and (2) for $c = 2$. This follows from a consideration of the Tietze transformation of G given by $x \to x^{\sqrt{2/c}}$, $y \to y^{\sqrt{2/c}}$, $z \to z^{\sqrt{2/c}}$ (replacing α by $\sqrt{c/2}\alpha$) and analogous transformations for H.

The details of both steps of the proof are straightforward but tedious. Here we give only the homomorphisms and omit the explicit calculations.

(1) Define $x\phi = x_1 y_2$, $y\phi = y_1 x_2$, $z\phi = x_1 y_1 x_1^{-1} x_2 y_2 x_2^{-1}$.

We claim that $x\phi$, $y\phi$ and $z\phi$ satisfy the relations of G and that the map ϕ is onto.

(2) Define $x_1\psi = (yxz)^2$, $y_1\psi = (zyx)^2$, $x_2\psi = (xyz)^2$, $y_2\psi = (zxy)^2$.

Again we claim that the images of the generators of H satisfy the relations of H and that the map ψ is onto. $\square$

Theorem 2 *Let $G = \langle S_{p,c}(3) \rangle$ (p prime, $p \geq 5$ and $c \neq 0$).*
If $\sqrt{c/2} \notin GF(p)$, then $O_4^-(p) \cong L_2(p^2)$ is a homomorphic image of G.

Proof: This time we provide mappings to matrix generators. Define
$$x\phi = \begin{pmatrix} 1 & 1 \\ 0 & 1 \end{pmatrix}, \ y\phi = \begin{pmatrix} 1 & 0 \\ -2/c & 1 \end{pmatrix}, \ z\phi = \begin{pmatrix} 1+\sqrt{2/c} & 1 \\ -2/c & 1-\sqrt{2/c} \end{pmatrix}.$$

These matrices generate $SL_2(p^2)$ since the images of x and y generate the maximal subgroup $SL_2(p)$ and the image of z is outside this, as $\sqrt{2/c} \notin GF(p)$. Then their images factored by the centre generate $L_2(p^2)$.

Conjecture 1 *Let* $G = \langle S_{p,c}(3) \rangle$ *(p prime, $p \geq 5$ and $c \neq 0$). If $\sqrt{c/2} \notin GF(p)$, then $G \cong O_4^-(p) \cong L_2(p^2)$.*

A similar argument to that in the proof of Theorem 1 implies that we only need to prove this for one such c.

To complete this proof it would be sufficient to find a homomorphism from $L_2(p^2)$ onto G. We have not succeeded in doing so, using either matrix generators for $L_2(p^2)$ or a presentation. We have tried to use various presentations from [5] and [29].

We have verified the conjecture for $p \leq 43$ using MAGMA.

3 Four-generator presentations

Definition 1 *Let $X_{\mathcal{P}}(n)$ be the set $\{x_1, \ldots, x_n\}$ of generators in $\langle \mathcal{P}(n) \rangle$. When $\mathcal{P}$ is understood, we simply write $X(n)$.*

The following easy lemma is useful here.

Lemma 2 *If $\langle \mathcal{P}(n-1) \rangle$ is a simple group then either*

(i) $\langle \mathcal{P}(n) \rangle$ is the trivial group, or

(ii) any $n-1$ elements of $X(n)$ generate a subgroup isomorphic to $\langle \mathcal{P}(n-1) \rangle$.

Theorem 3 *Let $G = \langle S_{5,1}(4) \rangle$. Then $G \cong O_5(5)$.*

Proof: Coset enumeration over $H = \langle x_1, x_2, x_3 \rangle$ reveals that this subgroup has index 600 in G. By Theorem 1 and Lemma 2, this shows that $|G| = |O_5(5)|$.

Using the MAGMA system we confirm that the permutations giving the action of G on the cosets of H generate a group isomorphic to $O_5(5)$, which is thus a homomorphic image of G. Together with the value of $|G|$ obtained above, this concludes the proof. □

In contrast, however, we have

Theorem 4 *$\langle S_{5,2}(4) \rangle$ is an infinite group. It has a homomorphism onto $O_5(4)$, and the kernel of this map has a homomorphism onto an infinite 2-group.*

Proof: Using the low index subgroup algorithm in Cayley we quickly find a subgroup $\langle x_1,\ x_2,\ x_3,\ x_4 x_1 x_2^{-1} x_3^{-1} x_2^{-1} x_1^{-1} x_4^{-1},\ x_4 x_2^{-1} x_3 x_4 x_2^{-1} x_1^{-1} x_4^{-1} \rangle$ of index 136. Using MAGMA, we show that the permutation action on the cosets of this subgroup generates the group $O_5(4)$ of order 979 200.

Using Quotpic, we construct the homomorphism onto this group, considered as a permutation group on 85 points, and obtain a presentation for the pre-image H of the point stabilizer, $2^6:(A_5 \times 3)$. Then, using the low index subgroups algorithm inside Quotpic, we find a subgroup K of index 4 in H, whose core, $\bigcap_{g \in G} K^g$, has index $2^8 |O_5(4)|$ in G.

The presentation of K obtained by Quotpic is then simplified by Tietze transformations (also in Quotpic) to obtain a presentation with 3 generators and 17 relations, of total length 265. We then restart Quotpic with this presentation as input.

The abelian quotient of K is cyclic of order 3, and the derived subgroup has a single homomorphism onto A_5 which is found by Quotpic, leading to a presentation of a subgroup L of index 180 in K.

We simplify this presentation, first using the Tietze transformation program in Quotpic and then (to further improve the presentation) the one in GAP, to a presentation with 61 generators, 624 relations and total length 16244. The ANU p-Quotient Program, applied to this presentation, quickly reveals that the largest exponent-2 class-2 quotient of this group has central factors of orders 2^{61} and 2^{1529}. A theorem of M. F. Newman [20] (adjusted for 2-groups as in [22]) then implies that this group must have an infinite 2-quotient.

Remarks: The Cayley run which found the index 136 subgroup also found an index 272 subgroup. A presentation of the index 136 subgroup of G was also obtained and simplified using stand-alone programs to one with 2 generators and 21 relations of total length 253. Subgroups of this group, of index 24, 32 and 64, were then found by low index subgroup methods or random coincidences [2, Example H11E20].

4 Presentations for $L_3(3)$

We consider two symmetric presentations for $L_3(3)$ and observe how they behave differently both from the examples already considered and from each other.

First consider $\mathcal{P}_1 = \{x_1, x_2 \mid x_1^{13} = x_2^{13} = x_1^4 x_2 x_1 x_2^{-3} = x_2^4 x_1 x_2 x_1^{-3} = 1\}$. The group $\langle \mathcal{P}_1 \rangle$ is $L_3(3)$, see [3]. $\langle \mathcal{P}_1(3) \rangle$ is trivial, and in fact the same results hold for the corresponding finitely presented semigroup. Proofs of both results are via coset enumerations. For the semigroup given by the presentation $\{x_1, x_2 \mid x_1^{14} = x_1,\ x_2^{14} = x_2,\ x_1^4 x_2 x_1 = x_2^3,\ x_2^4 x_1 x_2 = x_1^3\}$ the procedure described in [24] shows that the semigroup is actually the same as the group with that presentation, and the same holds for the analogous 3-generator presentation.

We now consider a member of the parametrized family of symmetric presentations: $\mathcal{T}(p; i_1, i_2; i_3, i_4; \ldots) =$
$\{x_1, x_2 \mid x_1^p = x_2^p = (x_1^{i_1} x_2^{i_2})^2 = (x_2^{i_1} x_1^{i_2})^2 = (x_1^{i_3} x_2^{i_4})^2 = (x_2^{i_3} x_1^{i_4})^2 = \cdots = 1\}$.
Let $\mathcal{P}_2 = \mathcal{T}(13; 1, 2; 3, 3; 6, 6)$. Then the group $\langle \mathcal{P}_2 \rangle$ is $L_3(3)$, by coset enumeration.

However, in this case, $\langle \mathcal{P}_2(3) \rangle$ is easily seen to be $L_3(3) \times L_3(3)$, but a direct coset enumeration of $\mathcal{P}_2(4)$ over the trivial subgroup fails. We can however show:

Theorem 5 *The group* $\langle \mathcal{P}_2(4) \rangle$ *is trivial.*

Proof: A coset enumeration shows that the index $|\langle \mathcal{P}_2(4) \rangle : H|$, where H is the subgroup generated by three of the symmetric generators, is 1.

Then let G be $\langle \mathcal{P}_2(4) \rangle$, which must be isomorphic to a quotient group of $L_3(3) \times L_3(3)$. If G is not trivial, then G admits a group S_4 of automorphisms acting naturally and faithfully on the generators $X_{\mathcal{P}_2}(4)$. Since the outer automorphism group of $L_3(3) \times L_3(3)$ is dihedral of order 8, it is easy to see that there must be a group $A \cong A_4$ of inner automorphisms of G, also acting naturally on $X(4)$.

Since this is a transitive action the generators are all conjugate in G. The point stabilizer in this action is cyclic of order 3, implying that the centralizer of a generator in G contains a cyclic subgroup of order 39. This implies that $G \not\cong L_3(3)$ since $L_3(3)$ contains no elements of order 39.

Finally, suppose that $G \cong L_3(3) \times L_3(3)$. Elements of order 13 in G have centralizer either $13 \times L_3(3)$ (when we say they are of non-diagonal type) or 13^2 (when we say they are of diagonal type). It is easy to see that each $x \in X(4)$ must be of diagonal type, or else they could not be conjugate and generate the whole group G. This contradicts the existence of an element of order 39 in $C_G(x)$. □

Remarks: The coset enumeration $|\langle \mathcal{P}_2(4) \rangle : \langle x_1, x_2, x_3 \rangle|$ needs a maximum and a total of between ten and twenty thousand cosets using coset table oriented methods. Experience suggests that to enumerate over $\langle x_1, x_2 \rangle$ would require a maximum of about 5000 times as many cosets. This would lead to a much easier proof from a theoretical point of view, but would require a much harder coset enumeration.

Even though both presentations, $\mathcal{P}_1$ and $\mathcal{P}_2$, are based on pairs of generators of order 13 they can be distinguished by observing that in $\mathcal{P}_1$ the generators are in different conjugacy classes while in $\mathcal{P}_2$ they are in the same class.

5   Concluding remarks

For $G_n = \langle S_{5,1}(n) \rangle$, the results of sections 2 and 4 show that:

$$G_2 \cong L_2(5) \cong O_3(5)$$
$$G_3 \cong L_2(25) \cong O_4^-(5)$$
$$G_4 \cong S_4(5) \cong O_5(5)$$

Using the computational tools described earlier, we can show that

$$G_5 \cong 2L_4(5) \cong 2O_6^+(5)$$
$$G_6 \cong 5^6{:}G_5$$
$$G_7 \cong 2O_8^+(5)$$
$$G_8 \cong O_9(5)$$

Based on this information, we conjecture that

$$G_n \cong O_{n+1}(5), n \equiv 0, 2, 4(8)$$
$$G_n \cong O_{n+1}^-(5), n \equiv 1, 3(8)$$
$$G_n \cong 2O_{n+1}^+(5), n \equiv 5, 7(8)$$
$$G_n \cong 5^n{:}G_{n-1}, n \equiv 6(8)$$

Similarly, for $H_n = \langle S_{5,2}(n) \rangle$, the results of sections 2 and 3 show that $H_2 \cong L_2(5) \cong O_3(5)$, $H_3 \cong 2(L_2(5) \times I_2(5)) \cong O_4^+(5)$ and H_4 is infinite. Note that we can equally well say that $H_2 \cong L_2(4) \cong O_3(4)$ and $H_3 \cong 2(L_2(4) \times L_2(4)) \cong O_4^+(4)$. Furthermore H_4 has a quotient isomorphic to $S_4(4) \cong O_5(4)$. We conjecture that H_n is infinite for $n \geq 4$. Notice that the results of section 4 show that, in certain cases, the group can collapse as n increases, so there is no obvious proof of this last conjecture.

For other values of p similar results hold for $\langle S_{p,c}(n) \rangle$ although for $5 \leq p \leq 23$ both of the series appear to give finite groups for all n, except when $p = 5$ and $\sqrt{c/2} \in GF(5)$ or $p = 7$ and $\sqrt{c/2} \notin GF(7)$.

For $n = 2$, $\langle S_{17,1}(n) \rangle$ and $\langle T(9; 1, 2; 4, 4)(n) \rangle$ are both $L_2(17)$ and their series agree for $2 \leq n \leq 4$, despite having generators with different orders; likewise for $\langle S_{17,3}(n) \rangle$ and $\langle T(9; 1, 1; 3, -4)(n) \rangle$. It is probable that, were we able to compute these series for larger n, we would find that the four series are all distinct. We can exhibit four distinct series of finite groups containing orthogonal groups over $GF(8)$.

Presentation	n				
	2	3	4	5	6
$T(9; 1, -2; 3, 4)$	$O_3(8)$	$O_4^-(8)$	$O_5(8)$	$O_6^+(8)$	$O_7(8)$
$T(9; 1, 2; 3, -4)$	$O_3(8)$	$O_4^+(8)$	$O_5(8)$	$O_6^+(8)$	$O_7(8)$
$T(9; 1, 2; (a^2b^{-2}ab^{-2})^2 = 1)$	$O_3(8)$	$O_4^+(8)$	$O_5(8)$	$O_6^-(8)$	$O_7(8)$
$Q_7(n)$	$2^6{:}O_2^-(8)$	$O_4^+(8)$	$O_5(8)$	$O_6^-(8)$	$2^{18}{:}O_6^-(8)$

We have computed many other series of groups with symmetric presentations, all containing groups closely related to orthogonal groups. The results are available from the authors.

ACKNOWLEDGEMENTS

The authors wish to acknowledge the support from European Community Grant ERBCHRXCT930418, which helped to make possible the visit of the second author to the University of St. Andrews during which this work was undertaken. The second author was also partially supported by the Australian Research Council.

References

[1] M. J. Beetham, A set of generators and relations for the group $PSL(2, q)$, q odd, *J. London Math. Soc.* **3** (1971), 554–557.

[2] W. Bosma and J. Cannon, *Handbook of* MAGMA *functions*, Computer Algebra Group, University of Sydney, 1995.

[3] C. M. Campbell and E. F. Robertson, Some problems in group presentations, *J. Korean Math. Soc.* **19** (1983), 123–128.

[4] C. M. Campbell, E. F. Robertson and P. D. Williams, Efficient presentations of the groups $PSL(2, p) \times PSL(2, p)$, p prime, *J. London Math. Soc. (2)* **41** (1989), 69–77.

[5] C. M. Campbell, E. F. Robertson and P. D. Williams, On presentations of $PSL_2(p^n)$, *J. Australian Math. Soc.* **48** (1990), 333–346.

[6] J. J. Cannon, An introduction to the group theory language, Cayley, in *Computational group theory (ed. M. D. Atkinson)*, pp. 145–183. Academic Press, London/New York, 1984.

[7] R. D. Carmichael, Abstract definitions of the symmetric and alternating groups and certain other permutation groups, *Quart. J. Pure Appl. Math.* **49** (1923), 226–283.

[8] J. H. Conway, R. T. Curtis, S. P. Norton, R. A. Parker and R. A. Wilson, ATLAS *of finite groups*, Clarendon Press, Oxford, 1985.

[9] H. S. M. Coxeter, Symmetrical definitions for the binary polyhedral groups, *Proc. Sympos. Pure Math.* **1** (1959), 64–87.

[10] H. S. M. Coxeter and W. O. J. Moser, *Generators and relations for discrete groups*, 4th edition, Springer, Berlin, 1979.

[11] R. T. Curtis, Symmetric presentations. I. Introduction, with particular reference to the Mathieu groups M_{12} and M_{24}, in *Groups, combinatorics and geometry (eds. M. W. Liebeck and J. Saxl)*, pp. 380–396. LMS Lecture Note Series 165, Cambridge University Press, 1992.

[12] R. T. Curtis, Symmetric presentations. II. The Janko group J_1, *J. London Math. Soc. (2)* **47** (1993), 294–308.

[13] R. T. Curtis, A survey of symmetric generation of sporadic simple groups, *these proceedings*, pp. 26–44.

[14] W. Emerson, Groups defined by permutations of a single word, *Proc. Amer. Math. Soc.* **21** (1969), 386–390.

[15] G. Havas, Coset enumeration strategies, in *Proceedings of the 1991 international symposium on symbolic and algebraic computation (ed. S. M. Watt)*, pp. 191–199. ACM Press, New York, 1991.

[16] G. Havas and M. F. Newman, Application of computers to questions like those of Burnside, in *Burnside groups (ed. J. L. Mennicke)*, pp. 211–230. Lecture Notes in Math. 806, Springer-Verlag, Berlin, 1980.

[17] G. Havas and E. F. Robertson, Application of computational tools for finitely presented groups, in *Computational support for discrete mathematics (eds. N. Dean and G. E. Shannon)*, pp. 29–39. DIMACS Ser. Discrete Math. Theoret. Comput. Sci., Vol. 15, 1994.

[18] D. F. Holt and S. Rees, A graphics system for displaying finite quotients of finitely presented groups, in *Groups and computation (eds. L. Finkelstein and W. M. Kantor)*, pp. 113–126. DIMACS Ser. Discrete Math. Theoret. Comput. Sci., Vol. 11, 1993.

[19] G. Malle, J. Saxl and T. Weigel, Generation of classical groups, *Geom. Dedicata* **49** (1994), 85–116.

[20] M. F. Newman, Proving a group infinite, *Arch. Math.* **54** (1990), 209–211.

[21] M. F. Newman and E. A. O'Brien, Application of computers to questions like those of Burnside, II, *Internat. J. Algebra Comput.* **6** (1996), 593–605.

[22] D. B. Nikolova and E. F. Robertson, One more infinite Fibonacci group, *C. R. Acad. Bulgare Sci.* **46** (1993), 13–15.

[23] E. F. Robertson and C. M. Campbell, Symmetric presentations, in *Group theory (eds. K. N. Cheng and Y. K. Leong)*, pp. 497–506. Walter de Gruyter, Berlin/New York, 1989.

[24] E. F. Robertson and Y. Ünlü, On semigroup presentations, *Proc. Edinburgh Math. Soc.* **36** (1992), 55–68.

[25] M. Schönert *et al.*, GAP—*Groups, Algorithms and Programming*, Lehrstuhl D für Mathematik, Rheinisch-Westfälische Technische Hochschule, Aachen, 1995.

[26] S. Sidki, $HK \cap KH$ in groups, Trabalho de Matemática 96, Universidade de Brasilia, 1975.

[27] S. Sidki, A generalization of the alternating groups—a question on finiteness and representation, *J. Algebra* **75** (1982), 324–372.

[28] S. Sidki, SL_2 over group rings of cyclic groups, *J. Algebra* **134** (1990), 60–79.

[29] P. D. Williams, *Presentations of linear groups*, Ph. D. thesis, University of St. Andrews, 1982.

A constructive recognition algorithm for the special linear group

F. Celler

and

C. R. Leedham-Green

Abstract

In the first part of this note we present an algorithm to recognise constructively the special linear group. In the second part we give timings and examples.

1   Introduction

It seems possible, using Aschbacher's celebrated analysis of subgroups of classical groups [5], to develop algorithms that will answer basic questions about the group G generated by a subset X of $GL(d, q)$, for modest values of d and q, as is already possible for permutation groups. The best strategy may involve trying to recognise very large subgroups of $GL(d, q)$ by special techniques.

In the case of permutation groups, special techniques are used to recognise the alternating and symmetric groups. This is done by making a random search for elements of a certain cycle type. If such elements are found in a primitive group, the group is known to contain the alternating group. If no such elements are found after a sufficiently long search, one proceeds with the expectation that one is dealing with a smaller group. For linear groups, the corresponding question is to determine whether or not the group in question contains a classical group.

It is possible to recognise the classical groups in a *non-constructive* way as described in [6], [7], and [2]. This still leaves the further problem of exhibiting an explicit isomorphism, that is to say, given that the group $G = \langle X \rangle$ contains a classical group, how can one express a given element A of G as a word in X? We call an algorithm to solve such a problem a *constructive* recognition algorithm.

The natural idea is to find a suitable generating set Y for G and expressions for the elements of Y as words in X, such that we have an algorithm expressing A as a word in Y. We present an algorithm which allows one to do this in the case of the special linear group; here Y will be a suitable set of transvections as it is well known that $SL(d, q)$ is generated by transvections. Using Gaussian elimination it is possible to rewrite an element of $SL(d, q)$ as a product of elementary matrices, which in turn can be written as a product of transvections. Hence, our goal is to find the transvections required in a Gaussian elimination as words in X, and then use the Gaussian elimination to rewrite an element of G as a word in these transvections.

The words in X that define the transvections may be very long. It is therefore sensible to give these words as 'straight line programs'. That is to say, to define a word w in X we define a sequence of words $w_1, w_2, \ldots, w_n$ in X, where each w_i is either of the form $x^{\pm 1}$ for $x \in X$, or of the form $w_j^{\pm 1} w_k^{\pm 1}$ for $j, k < i$ or w_j^{-1} for $j < i$. This may reduce the number of multiplications required to define w dramatically. For example if $w = x^n$, then w can be defined in this way by $O(\log n)$ multiplications.

One of the main applications of the constructive recognition algorithm will be in the following setting. In investigating a matrix group H along the lines of Aschbacher's classification, one ends up with either a classical group, an almost simple group, or a reduction to a smaller group; in the latter case, we get a homomorphism φ of H into S, where S is cyclic, a permutation group, or some matrix group of smaller dimension or over a smaller field. If $\varphi(H)$ is a matrix group containing the special linear group in its natural representation, we can use the constructive algorithm to produce elements of the kernel of φ.

We assume that the algorithm is applied when we have already proved, using the much faster non-constructive recognition algorithm, that $\langle X \rangle$ does contain $SL(d, q)$. As an alternative, the program could report failure if some randomized component of the algorithm failed to complete in the expected time. The effect of proving in advance that G does contain $SL(d, q)$ is to change the algorithm from a Monte Carlo algorithm, when failure would give statistical evidence that G does not contain $SL(d, q)$, to a Las Vegas algorithm, where the randomized features of the algorithm merely make the runtime uncertain.

We also assume $d > 1$ in the following.

The algorithm described in this paper is randomized; however explicit isomorphisms constructed will always be correct. Such algorithms are called Las Vegas algorithms; the output is either correct or the algorithm reports failure (with prescribed probability $\leq \varepsilon$).

To say that a Las Vegas algorithm has (Las Vegas) complexity f, where f is some function of the input, means that, for some ϵ, $0 < \epsilon < 1$, it will, with probability at least $1 - \epsilon$, terminate in time less than kf for some constant k independent of the input. The algorithm will then terminate with probability $1 - \epsilon^2$, in time less than $2kf$, as one could simply run the program twice. So ϵ can be squared by doubling k, and hence the algorithm has complexity f for any ϵ, $0 < \epsilon < 1$, however small. Hence running the algorithm until success has expected complexity $O(f)$. The stated running times and complexities in the following sections are always to be understood as expected running times and complexities.

2   Constructing a transvection

A transvection is an element of $GL(d, q)$ acting trivially on a hyperplane and on its quotient space; we shall call this hyperplane the *centralized subspace* of the transvection. A transvection is conjugate to $\mathrm{diag}(\mathsf{J}_2(1), I_{d-2})$, the block matrix with $\mathsf{J}_2(1)$ and I_{d-2} on the diagonal, where $\mathsf{J}_2(1)$ is a Jordan block of dimension 2 and eigenvalue 1 and I_{d-2} is the identity of $GL(d-2, q)$. Although only a small proportion of the elements of $GL(d, q)$ are transvections, we show in this section that the transvections are relatively easy to find provided q is not too large.

Let F_q be the finite field with q elements, $q = p^k$, where p is prime.

In order to find a transvection, we look for an element conjugate to $\mathrm{diag}(\mathsf{J}_2(\alpha), R)$, the block matrix with $\mathsf{J}_2(\alpha)$ and R on the diagonal; $\mathsf{J}_2(\alpha)$ is a Jordan block of dimension 2 and eigenvalue $\alpha \neq 0$, R is an element of $GL(d-2, q)$, such that α is not an eigenvalue of R and R is semisimple. As R is semisimple, p does not divide the order $o(R)$ of R. Raising $\mathrm{diag}(\mathsf{J}_2(\alpha), R)$ to the least common multiple m of the orders of R and α yields a transvection; R^m is trivial, $\mathsf{J}_2(\alpha)^m = \begin{pmatrix} 1 & m\alpha^{-1} \\ 0 & 1 \end{pmatrix}$ and $m\alpha^{-1} \neq 0$ because p does not divide $o(R)$ or $o(\alpha)$. In order to find random elements in $\langle X \rangle = GL(d, q)$ we use the algorithm described in [3], which requires, after a preprocessing phase, one matrix multiplication per random element; but in addition we also keep track of the expressions for the random elements in the given generators X.

Note that it is possible to obtain a suitable multiple of $o(R)$ from the degrees of the factors of the minimal polynomial of R, but as we want short expressions in the given generators for our transvection, we use the precise order of R, see [1].

We now estimate the proportion of elements in $GL(d, q)$ of this form. The first lemma counts the elements in $GL(d, q)$ having a given eigenvalue, the second lemma counts the semisimple elements.

Lemma 2.1 *Let α be an element of $F_q{}^*$ and let M_α be the set of elements in $GL(d, q)$ having eigenvalue α. Then*

$$\frac{1}{q-1} - \left(\frac{1}{q-1}\right)^2 \leq \frac{|M_\alpha|}{|GL(d,q)|} \leq \frac{1}{q-1}.$$

Proof: In order to get the stated upper bound we use a counting argument similar to that in [4]: for each non-zero vector v choose a vector space complement W_v such that $F_q^d = \langle v \rangle \oplus W_v$. Let $\mathcal{E}$ be the set of triples $\{(v, \beta, \tau) \mid v \in F_q^d, \ \beta \in GL(W_v), \ \tau \in \mathrm{Hom}(W_v, \langle v \rangle)\}$. Now if $A \in GL(d, q)$ has an eigenvalue α, and we choose a corresponding eigenvector v, then the pair (A, v) corresponds to exactly one triple in $\mathcal{E}$. On the one hand

$$|\mathcal{E}| = (q^d - 1) \cdot |GL(d-1, q)| \cdot q^{d-1} = (q^d - 1) q^{d-1} \prod_{i=0}^{d-2} (q^{d-1} - q^i) = |GL(d, q)|.$$

On the other hand each A with eigenvalue α has at least $q - 1$ eigenvectors. Therefore the number of matrices in $GL(d, q)$ with eigenvalue α is at most $|GL(d, q)|/(q - 1)$.

    In order to get the stated lower bound we count the matrices whose characteristic polynomials have α as root with multiplicity one; that is, the matrices conjugate to $\mathrm{diag}(\alpha, R)$ for $R \in \mathcal{R}_\alpha \subset GL(d-1, q)$ where $\mathcal{R}_\alpha$ is the set of matrices which do not have eigenvalue α. Let $\mathcal{C}_\alpha$ be a set of representatives of the conjugacy classes in $\mathcal{R}_\alpha$ under the action of $GL(d-1, q)$. Then $\sum_{R \in \mathcal{C}_\alpha}[GL(d, q) : C_{GL(d,q)}(\mathrm{diag}(\alpha, R))]$ matrices in $GL(d, q)$ are conjugate to $\mathrm{diag}(\alpha, R)$. The centralizer of $\mathrm{diag}(\alpha, R)$ is isomorphic to $F_q^* \times C_{GL(d-1,q)}(R)$; so that the above sum gives

$$
\begin{aligned}
|M_\alpha| &\geq \sum_{R \in \mathcal{C}_\alpha} \frac{|GL(d, q)|}{(q-1) \cdot |C_{GL(d-1,q)}(R)|} \\
&= \frac{|GL(d, q)|}{(q-1) \cdot |GL(d-1, q)|} \sum_{R \in \mathcal{C}_\alpha} [GL(d-1, q) : C_{GL(d-1,q)}(R)] \\
&= \frac{|GL(d, q)|}{q-1} \cdot \frac{|\mathcal{R}_\alpha|}{|GL(d-1, q)|} \\
&\geq \frac{1}{q-1} \cdot \left(1 - \frac{1}{q-1}\right) \cdot |GL(d, q)|,
\end{aligned}
$$

since from the first part of the proof above we already know that at least $(q-2)|GL(d-1, q)|/(q-1)$ matrices do not have eigenvalue α, and therefore $|\mathcal{R}_\alpha| \geq (q-2)|GL(d-1, q)|/(q-1)$.   □

Lemma 2.2 *Let S be the set of semisimple, regular elements in $GL(d, q)$; that is to say, the set of elements whose characteristic polynomial is square-free. Then*

$$\left(\frac{q-1}{q}\right)^2 |GL(d, q)| \leq |S|.$$

Proof: Let $\mathcal{P}$ be the set of all square-free polynomials of degree d with leading coefficient one and non-zero constant term. The set S is a union of

$GL(d, q)$-conjugacy classes. We can choose as a representative for each class the companion matrix C_f of a polynomial f in $\mathcal{P}$.

Let $f = \prod_{i=1}^{r} f_i \in \mathcal{P}$, where f_i is irreducible of degree d_i. Then the centralizer of C_f is isomorphic to $C_{GL(d_1,q)}(C_{f_1}) \times \cdots \times C_{GL(d_r,q)}(C_{f_r})$, so that $|C_{GL(d,q)}(C_f)| = \prod_i (q^{d_i} - 1) \leq q^d$. Therefore we get

$$
\begin{aligned}
|S| &= \sum_{f \in \mathcal{P}} [GL(d, q) : C_{GL(d,q)}(C_f)] \\
&\geq \sum_{f \in \mathcal{P}} \frac{|GL(d, q)|}{q^d} \\
&= \frac{|\mathcal{P}|}{q^d} |GL(d, q)|.
\end{aligned}
$$

Now let $\mathcal{N}$ be the set of all polynomials of degree d with leading coefficient one and non-zero constant term that do have a non-trivial square factor. In order to get an upper bound on $|\mathcal{N}|$ we use the following counting argument. As $f \in \mathcal{N}$ is not square-free it contains a non-trivial factor h of degree $k \leq d/2$ at least twice. Counting all possibilities for k, the factor h, and the co-factor f/h^2, we count each polynomial with repeated factors at least once.

$$
\begin{aligned}
|\mathcal{N}| &\leq \sum_{k=1}^{\lfloor d/2 \rfloor} \left(q^{k-1}(q - 1) \right) \cdot \left(q^{d-2k-1}(q - 1) \right) \\
&= \frac{(q - 1)^2}{q^2} \sum_{k=1}^{\lfloor d/2 \rfloor} q^{d-k} \\
&= \frac{(q - 1)}{q^2} (q^d - q^{\lceil d/2 \rceil}) \\
&< \frac{(q - 1)}{q^2} q^d.
\end{aligned}
$$

Hence we have $|\mathcal{P}| = (q - 1)q^{d-1} - |\mathcal{N}| > (q - 1)q^{d-1}(1 - \frac{1}{q})$ and therefore $|S| > \left(\frac{q-1}{q} \right)^2 \cdot |GL(d, q)|$. $\qquad \square$

Lemma 2.3 *Let $q > 3$. Let N_α be the set of elements A of $GL(d, q)$ such that both the characteristic polynomial $c(x)$ and the minimal polynomial of A have α as a root of multiplicity two and $c(x)/(x - \alpha)^2$ is square-free. Then*

$$
\frac{1}{q} \left(\left(\frac{q-1}{q} \right)^2 - \frac{1}{q-1} \right) \leq \frac{|\cup_{\alpha \in F_q^*} N_\alpha|}{|GL(d, q)|}.
$$

Proof: Let A be an element of N_α. Then A is conjugate to $\mathrm{diag}(\mathsf{J}_2(\alpha), R)$, where $R \in GL(d-2, q)$, α is not an eigenvalue of R and R is semisimple. The set $\mathcal{R}_\alpha$ of all such R is a union of $GL(d-2, q)$-conjugacy classes; let $\mathcal{C}_\alpha$ be a system of representatives for these classes. As α is not an eigenvalue of R the centralizer of $\mathrm{diag}(\mathsf{J}_2(\alpha), R)$ is isomorphic to $C_{GL(2,q)}(\mathsf{J}_2(\alpha)) \times C_{GL(d-2,q)}(R)$, therefore its order is $q \cdot (q-1) \cdot |C_{GL(d-2,q)}(R)|$. Taking the sum of the indices $[GL(d, q) : C(\mathrm{diag}(\mathsf{J}_2(\alpha), R))]$ for all $R \in \mathcal{C}_\alpha$, we get

$$|N_\alpha| = \frac{1}{q(q-1)} \cdot |GL(d, q)| \cdot |\mathcal{R}_\alpha| / |GL(d-2, q)|.$$

According to Lemma 2.2 at least $\frac{(q-1)^2}{q^2} \cdot |GL(d-2, q)|$ matrices are semisimple, and according to Lemma 2.1 at most 1 in $q-1$ matrices have eigenvalue α; therefore $|\mathcal{R}_\alpha| / |GL(d-2, q)|$ is at least $\frac{(q-1)^2}{q^2} - \frac{1}{q-1}$. As R is semisimple it does not contain any other 2-dimensional Jordan block, so $N_\alpha \cap N_\beta = \emptyset$ for $\alpha \neq \beta$. Therefore the total proportion of matrices of the required form is at least $\frac{1}{q} \cdot (\frac{(q-1)^2}{q^2} - \frac{1}{q-1})$. $\square$

We can now analyse an algorithm for finding a transvection if $\langle X \rangle = GL(d, q)$. The general case $SL(d, q) \leq \langle X \rangle$ is discussed at the end of this section.

Theorem 2.4 *It is possible to find a transvection in $GL(d, q) = \langle X \rangle$ as a word in X using Las Vegas $O(qd^3)$ finite field operations if $O(q)$ random elements as words in X are given.*

Proof: If $d = 2$ then 1 in $q + 1$ matrices in $GL(d, q)$ are conjugate to $J_2(\alpha)$ for some α.

Now assume $d > 2$ and $q > 2$. Lemma 2.3 shows that the proportion of suitable elements is at least $1/(5q)$. Therefore the probability of failure is less than e^{-1} if we look at $5q$ random elements.

Now assume $d > 2$ and q arbitrary. Counting the matrices conjugate to $\mathrm{diag}(J_2(\alpha), R)$, where the characteristic polynomial of R is square-free, in the same way as in Lemma 2.3, shows that there are at least $\frac{(q-1)^2}{q^5} \cdot |GL(d, q)|$ such matrices. Hence the proportion $\frac{(q-1)^2}{q^5}$ of suitable elements is independent of d for $q = 2$ and 3.

It follows that the proportion of suitable elements is at least $1/32$ for $q = 2$, at least $4/243$ for $q = 3$, and at least $1/(5q)$ for $q > 3$. Hence the probability of failure will be less then e^{-1} for any q if we choose $21q$ elements.

Computing each characteristic polynomial $c(x)$ requires $O(d^3)$ finite field operations [1]. By looking at the greatest common divisor of $c(x)$ and its derivative it is possible to check if $c(x)$ has a root of multiplicity 2 and square-free co-factor without quantifying over the field. Computing the gcd requires

$O(d^2)$ finite field operations. We do not need to compute the minimal polynomial because the condition on the minimal polynomial given in Lemma 2.3 can be checked by looking at the dimension of the eigenspace for the eigenvalue α. This again requires $O(d^3)$ finite field operations. Therefore we can check if we have found a suitable element using $O(d^3)$ finite field operations.

We have to look at $21q$ elements to get a probability of failure of less than e^{-1}; computing and checking one element requires $O(d^3)$ finite field operations. Having found a suitable element $\mathrm{diag}(J_2(\alpha), R)$ we can compute the order of R using $O(d^3 \log q \log t)$ finite field operations, where t is the maximal number of prime factors in $q^i - 1$ for $i \leq d$ (see [1]). □

Remark: After a preprocessing phase, finding a random element using the method described in [3] requires one matrix multiplication.

Remark: Note that we do not have to raise the element to the power $\mathrm{lcm}(o(R), o(\alpha))$ explicitly, which would require $O(d^4 \log q)$ finite field operations. However, although we need only $O(q)$ matrix multiplications to find the transvection, evaluating the corresponding straight line program requires $O(q + d \log q)$ multiplications because of this last powering step.

We now consider the general case when $SL(d, q) \leq \langle X \rangle \leq GL(d, q)$.

If $d > 3$, the determinant will simply partition the set of suitable elements into subsets of approximate equal cardinality. If $d > 1$ the chance that a random element has an eigenvalue α is still about one in $q - 1$ even if we impose a condition on the determinant of R.

If $d = 3$ and $G = SL(3, q)$, however, the determinant condition forces $R \in GL(1, q)$ to be (α^{-2}). Hence, if F_q^* has elements of order 3 we get a slightly worse chance, with the exception of the rogue group $SL(3, 4)$ which has no elements of the required form.

To catch these exceptions we use the following variation of the algorithm. We try to find a random element A with minimal polynomial m_A and characteristic polynomial c_A, such that m_A has a root α of multiplicity two, $m_A/(x - \alpha)^2$ is square-free and c_A has α as root of multiplicity two or three. If q is even and we find an element where α is a root of multiplicity three of m_A, we try A^2 instead.

3   Constructing a transvection basis

We assume in this section that $\langle X \rangle$ contains $SL(d, q)$.

We now describe an algorithm to find a set of transvections that generate $SL(d, q)$. We first find a set of transvections that centralize a common subspace M of co-dimension one. We then look for conjugating elements g_j such that $M \cap \bigcap_j M g_j$ is trivial.

Assume that we have found a transvection using Theorem 2.4. Hence we know an element $A \in \langle X \rangle$, where A has order pm, $t_1 = A^m$ is a transvection, and A is conjugate to $\mathrm{diag}(\mathrm{J}_2(\alpha), R)$, where R is semisimple and regular.

Let M be the subspace of co-dimension 1 that is centralized by t_1, and let b be a vector not in M. For a transvection t centralising M define $\pi(t)$ to be $b - b^t \in M$. Let $M = S \oplus T$ be the A-invariant decomposition of M into a one-dimensional subspace S and a $(d-2)$-dimensional subspace T.

Our goal is to find a set of $(d-1)k$ transvections $\{t_i\}$ with centralized subspace M, such that $\{\pi(t_i)\}$ is a basis for $M < F_q{}^d$ over the *prime field* of F_q. Using these transvections it is possible to write any transvection t centralising M as a word in the transvections t_i; if $\pi(t) = \sum \alpha_i \pi(t_i)$ for $0 \le \alpha_i < p$ then $t = \prod t_i{}^{\alpha_i}$.

We find such a set of transvections in two stages. In the first stage we find another transvection by taking conjugates of t_1 with random elements of $\langle X \rangle$ until we find a transvection that also normalizes M. In the second stage we conjugate our transvections with A. We then iterate until we find a basis.

The next lemma estimates the chances of finding a conjuagte of t_1 that will normalize M or fix b.

Lemma 3.1 *Let t be a transvection, M its centralized subspace and b a vector outside M. Let g be a random element of $GL(d, q)$.*
(1) The probability that t^g normalizes M is at least 1 in $q + 1$.
(2) The probability that t^g fixes b is at least 1 in $q + 1$.

Proof: It is clear that t^g normalizes M if and only if t^g either centralizes M or $\pi(t)g$ lies in M.

If $d = 2$ and t^g normalizes M it must also centralize this one-dimensional subspace. We have a probability of 1 in $q + 1$ that Mg and M are equal as we can assume that g is a random invertible matrix, and so Mg must be a random, one-dimensional subspace.

If $d > 2$ the probability that t^g centralizes M is one in $(q^d - 1)/(q - 1)$ and we ignore this possibility for the analysis. Now g is a random element of $GL(d, q)$, so $\pi(t)g$ is a random, non-zero vector. The probability that a random, non-zero vector of $F_q{}^d$ lies in a given hyperplane is $q^{d-1} - 1$ in $q^d - 1$.

Part (2) is proved in the same way.   □

Using a transvection normalising M we can now construct a conjugate of t centralising M.

Lemma 3.2 *Let t be transvection with centralized subspace M. It is possible to find a conjugate t' of t centralising the same subspace M using Las Vegas $O(d^3 q)$ finite field operations if $O(q)$ random elements of $GL(d, q)$ are given.*

Proof: Lemma 3.1 states that a conjugate s of t has a chance of at least 1 in $q + 1$ of normalising M.

If $d = 2$ than a transvection normalising M already centralizes M and we can choose $t' = s$.

If $d > 2$, then either s centralizes M and we choose $t' = s$ or it does not. In the latter case we choose $t' = t^s$ which does centralize M.   □

For the second stage we assume that we have found another transvection t_2 fixing M such that the T-component of $\pi(t_2)$ is non-trivial. The conjugate $t_2{}^A$ still centralizes M and A acts as R on the T-component of $\pi(t_2)$.

We now form $t_3 = t_2{}^A, t_4 = t_3{}^A, \ldots$, until some linear relation occurs between the $\pi(t_i)$, $1 < i$. We are looking for a basis over the prime field of F_q for T; if a linear relation occurs before we have such a basis, we add in a new transvection constructed in the same way as t_2. Iterating this process, we get a basis for T. We shall see later that we expect to construct this basis with $O(k)$ iterations.

Remark: Using [3] for finding random elements, we need $O(q)$ multiplications to evaluate the straight line program for A, $O(d \log q)$ additional multiplications to evaluate the straight line program for t_1. We expect to need $O(k)$ iterations, therefore evaluating the straight line programs for all t_i simultaneously requires $O(qk + d \log q)$ multiplications.

After we have found the $\{t_i\}$, our next goal is to find $d - 1$ elements $g_j \in \langle X \rangle$ such that $M \cap \bigcap_j Mg_j = \{0\}$.

Taking random elements we expect to find such a set $\{g_j\}$ after $O(d)$ tries. However, as we assume that constructing a random element already requires $O(d^3)$ finite field operations, this would require $O(d^4)$ finite field operations altogether. So instead of taking $O(d)$ random elements, we start with just one random element g_1. We now form $g_2 = g_1 A$, $g_3 = g_2 A$, and so on. If we do not get $M \cap \bigcap_j Mg_j = \{0\}$, we add in a new random element. As R is semisimple and regular, we can choose a basis such that A is sparse. Hence multiplication by A can be done in $O(d^2)$ finite field operations.

We now prove that, with positive probability independent of d and q, two random vectors of the underlying vector space will generate the vector space as an A-module.

Lemma 3.3 *Let R be a semisimple, regular element of $GL(d, q)$. Then the underlying row space $F_q{}^d$ is generated as an R-module by Las Vegas $O(1)$ random vectors.*

Proof: We can assume that R is $\mathrm{diag}(R_1, ..., R_l)$, where R_i has an irreducible characteristic polynomial. Let $F_q{}^d = \oplus_{i=1}^l V_i$ be the corresponding decomposition of $F_q{}^d$ into irreducible A-modules V_i. Then each non-trivial vector in V_i generates V_i as R-module.

Let $a(j)$ be the number of j-dimensional subspaces occurring in the above decomposition. First of all we assume that there are only blocks of dimension

j. Taking two random non-trivial vectors we have a chance of one in $1/q^{2j}$ that both vectors have a trivial V_i component for a fixed i. Hence we have a chance of $(1 - 1/q^{2j})^{a(j)}$ to generate the whole vector space with two random vectors. It is clear that $a(j) < q^j$. Hence

$$
\begin{aligned}
(1 - 1/q^{2j})^{a(j)} &\geq (1 - 1/q^{2j})^{q^j} \\
&\geq (1 - q^2)^{q^{-j}} \\
&\geq 4^{-q^{-j}}.
\end{aligned}
$$

Now, if there are blocks of various dimensions, we have a chance of success of at least $\prod_{j=1}^d 4^{-q^{-j}} \geq 4^{-1/(q-1)}$. $\qquad\qquad\square$

4 Recognising SL constructively

Putting the algorithms of sections 2 and 3 together we have now proved

Theorem 4.1 *Given $\langle X \rangle = GL(d,q)$, we can construct straight line programs for a set of $(d-1)k$ transvections $\{t_i\}$ with common centralized subspace M, such that $\{\pi(t_i)\}$ is linearly independent over the prime field of F_q using Las Vegas $O(qkd^3)$ finite field operations provided that $O(qk)$ random elements as words in X are given.*

Using an additional Las Vegas $O(d^3)$ finite field operations we can construct a set of elements $\{g_j\}$ as straight line programs such that $\bigcap_j Mg_j = \{0\}$.

Using the transvections $\{t_i\}$ and the elements $\{g_j\}$ we can now write any element of $G = \langle X \rangle$ as a word in X. This word is represented as a straight line program. If $G > SL(d,q)$, it is a triviality to find an element Z of G such that $G = \langle Z, SL(d,q) \rangle$, and to give Z as a word in X.

As $\bigcap Mg_j$ is trivial, we can choose a basis $\mathcal{B} = \{b_1, \ldots, b_d\}$ for F_q^d such that $b_j \notin Mg_j$ and $b_j \in Mg_{j'}$ for $j' \neq j$. From now on assume that $\mathcal{B}$ is the standard basis of F_q^d.

If we have a transvection t centralising M and $\pi(t) = \sum \alpha_i \pi(t_i)$ then we know that $t = \prod t_i^{\alpha_i}$. As we know straight line programs for the t_i we can produce a straight line program for t. If we have a transvection t centralising Mg_j, then $t^{g_j^{-1}}$ centralizes M, and we can therefore produce a straight line program for t using the straight line programs for the $\{t_i\}$ and g_j as before.

In order to write an element A of G as a word in X, we proceed as follows. First set $A' = Z^n A$, where n is the integer of smallest modulus such that $\det(A') = 1$. The algorithm then uses a Gaussian elimination, that is biased to prefer column operations, to write A' as a product of transvections T_k, where each transvection T_k centralizes M_{i_k} for some i_k. By solving a system of linear equations the algorithm is now able to write each T_k as a product of $t_{i_k g_j}$. This enables one to write A as a straight line program in $X \cup Z$.

5 Variations

5.1 Computing a transvection, q large

If $q > d + 1 > 3$, the following algorithm can be used to find a transvection in G.

First we try to find an element A that is almost irreducible, that is, that acts irreducibly on a subspace of co-dimension 1. About 1 in d of the elements of G have this property if G contains $SL(d, q)$, see [6]. After raising A to the power $o(A)/\gcd(o(A), q - 1)$, this power B will either be a scalar matrix, in which case we try again, or it will be a diagonal matrix with two eigenvalues, say α of multiplicity 1 and β of multiplicity $d - 1$.

If G is $SL(d, q)$, we know that $\alpha \cdot \beta^{d-1} = 1$; since q is larger than $d + 1$, the worst case is $q = 2d - 1$, in which case $\beta^{d-1} = \pm\beta^{-1}$. If $\beta^{d-1} = \beta^{-1}$ then B is a scalar matrix and we have to try again.

When we find such a B, we conjugate B by a random element of G. Since q is large, the eigenspaces of B and C with eigenvalue β will almost certainly be distinct. These spaces then intersect in a subspace W of co-dimension 2 which is invariant under the action of B and C. Let $\overline{B}$ and $\overline{C}$ be the images of B and C acting on V/W, and $\overline{G}$ the group they generate. We shall now search for a transvection in $\overline{G}$ using the algorithm described in section 2 hoping that $\overline{G}$ contains $SL(2, q)$ and try to pull it back. In an attempt to avoid the situation where $SL(d, q) \not\leq \overline{G}$ we demand that the squares of $\overline{B}, \overline{C}, \overline{BC}$ do not commute. Hence $\overline{G}$ is neither abelian nor imprimitive.

Assume that we have found a transvection in $GL(2, q)$. Its pre-image S in $GL(d, q)$, after using a suitable basis transformation, is of the form

$$S = \begin{pmatrix} 1 & \delta & t_1 \\ & 1 & t_2 \\ & & \beta I_{d-2} \end{pmatrix},$$

where $\delta \neq 0$ and t_i are row vectors of length $d - 2$.

First assume that $\beta \neq 1$ and let n be the order of β. Using the same basis as above we see that

$$S^n = \begin{pmatrix} 1 & n\delta & (\sum_{i=0}^{n-1} \begin{pmatrix} 1 & \delta \\ 0 & 1 \end{pmatrix}^i \beta^{n-1-i}) \begin{pmatrix} t_1 \\ t_2 \end{pmatrix} \\ 0 & 1 & \\ & & I \end{pmatrix}.$$

Note that $n\delta \neq 0$ because n and p are coprime. Now we need to look more closely at the upper right corner. Expanding the sum gives

$$\sum_{i=0}^{n-1} \begin{pmatrix} 1 & \delta \\ 0 & 1 \end{pmatrix}^i \beta^{n-1-i} = \sum_{i=0}^{n-1} \begin{pmatrix} \beta^{n-1-i} & i\delta\beta^{n-1-i} \\ 0 & \beta^{n-1-i} \end{pmatrix}.$$

Let $x = \sum_{i=0}^{n-1} \beta^i$; then $x\beta = x$ because n is the order of β. As we assume $\beta \neq 1$, x must be zero, and therefore $\sum_{i=0}^{n-1} \begin{pmatrix} 1 & \delta \\ 0 & 1 \end{pmatrix}^i \beta^{n-1-i}$ is a matrix with zeros on and below the diagonal. We get $S^n = \begin{pmatrix} 1 & n\delta & \\ 0 & 1 & \\ & & I \end{pmatrix}$, which is a transvection.

Now assume that $\beta = 1$ and q is even. In this case it follows that

$$S^2 = \begin{pmatrix} 1 & 0 & \delta t_2 \\ 0 & 1 & 0 \\ 0 & 0 & I \end{pmatrix}.$$

If t_2 is trivial then $S = \begin{pmatrix} 1 & \delta & t_1 \\ 0 & 1 & 0 \\ 0 & 0 & I \end{pmatrix}$ is already a transvection or $t_2 \neq 0$ and S^2 is a transvection.

If $\beta = 1$ but q is odd then the order of S is p and S is a transvection if and only if $t_2 = 0$. So the only bad case is if we find an element S with $\beta = 1$, $t_2 \neq 0$ and q is odd. We avoid this case from the beginning by using only diagonal elements with $\beta \neq 1$ if $p \neq 2$. As β is the power of an element from a field with q^{d-1} elements lying in the subfield with q elements we have a chance of one in $\varphi(l)$ of getting a bad β, where l is the π-part of $q^{d-1} - 1$, for π the set of primes occurring in $q - 1$, and φ is the Euler phi function.

5.2 Computing a transvection basis, k large

If k is large the following can reduce the iteration required in section 3. We use the same notation as in section 3.

After we have found the second transvection t_2, we form $t_3 = t_2^A, \ldots$, until some linear relation occurs as in section 3. But now instead of iterating we construct an element B that leaves the subspace M invariant and centralizes b. Conjugating our transvection t_i with B will most likely extend our basis.

We look for a conjugate t_1^g of t_1 that fixes b. If a non-trivial linear combination $\sum \alpha_i \pi(t_i)g$ lies in M then the corresponding transvection $B = (\prod t_i^{\alpha_i})^g$ fixes b and normalizes M. We now conjugate the given t_i with B.

5.3   Black box recognition

It would not be hard to produce an analogous algorithm that works in a black box group. That is to say, we assume that G is an arbitrary group isomorphic to a group lying between $SL(d,q)$ and $GL(d,q)$. We assume, of course, that we have efficient algorithms to multiply and compare elements of G. We also assume that we have an efficient algorithm to compute the order of any element of G, and that, given an elementary abelian p-subgroup W of G, where p is the charactistic of the field over which G is defined, we can efficiently determine linear relations in W. These conditions would be satisfied if, for example, G was given by some faithful representation over a finite field of charactistic p. It would be possible to adapt our algorithm to work in G, thus getting an isomorphism of some subgroup of G into $GL(d,q)$. The question then arises of computing the image of X in $GL(d,q)$, thus proving that the whole of G is embedded in $GL(d,q)$. We are investigating an elaboration of these ideas into an efficient algorithm.

6   Timings and examples

We have implemented the algorithm in GAP [8] and this implementation is distributed with GAP.

We investigate the group G generated by

$$
\begin{pmatrix} 5 & 13 & 4 & 3 \\ 2 & 16 & 7 & 0 \\ & & 1 & 3 \\ & & 16 & 15 \end{pmatrix} \text{ and } \begin{pmatrix} 1 & 14 & 14 & 14 \\ 1 & 15 & 10 & 11 \\ & & 3 & 4 \\ & & & 1 \end{pmatrix}
$$

over the field $GF(17)$. Let the matrices act from the right, so the upper left hand corner describes the action on a two–dimensional quotient space. First we investigate the group acting on this quotient space.

```
gap> x1:=[[5,13,4,3],[2,16,7,0],[0,0,1,3],[0,0,16,15]]*Z(17)^0;;
gap> x2:=[[1,14,15,14],[1,15,10,11],[0,0,3,4],[0,0,0,1]]*Z(17)^0;;
gap> y1 := [ [ 5, 13 ], [ 2, 16 ] ] * Z(17)^0;;
gap> y2 := [ [ 1, 14 ], [ 1, 15 ] ] * Z(17)^0;;
gap> cr := CRecognizeSL( Group(y1,y2), [y1,y2] );
#I  <G> is GL( 2, 17 )
<< constructive SL recognition record >>
```

This show that G acts as $GL(2,17)$ on the quotient space; rewriting two "random" elements we get two elements in the kernel. We now investigate the action of the kernel on the invariant subspace.

```
gap> w1 := Rewrite( cr, y2 );
```

```
(t2_1)^4*(t1_1)^3*(t2_1)^7
gap> DisplayMat( (x2) / Value( w1, [x1,x2] ) );
  1  .  7  2
  .  1 14  4
  .  .  5  2
  .  .  8  6
gap> w2 := Rewrite( cr, y1^2*y2 );
z^2*(t2_1)^4*(t1_1)^3*(t2_1)^7
gap> DisplayMat( (x1^2*x2) / Value( w2, [x1,x2] ) );
  1  .  8  .
  .  1 16  .
  .  .  5 10
  .  .  5  6
gap> z1 := [ [ 5, 2 ], [ 8, 6 ] ] * Z(17)^0;;
gap> z2 := [ [ 5, 10 ], [ 5, 6 ] ] * Z(17)^0;;
gap> cr := CRecognizeSL( Group(z1,z2), [z1,z2] );
#I  <G> is GL( 2, 17 )
<< constructive SL recognition record >>
```

Hence the kernel acts as $GL(2,17)$ on the invariant subspace. Rewriting a "random" element in the kernel shows that G is $17^4.GL(2,17).GL(2,17)$ because $GL(2,17)$ acts from both sides on the upper right corner; therefore we either get 17^4 or the trivial subspace in this corner.

```
gap> DisplayMat( (xx1) / Value( Rewrite( cr, z1 ), [xx1,xx2] ) );
  1  . 14  8
  .  1 11 16
  .  .  1  .
  .  .  .  1
```

The timings in Fig. 1 were obtained by running the program on an Intel Pentium P5, 133 Mhz, running FreeBSD 2.1.0. They are an average of 100 runs of running the algorithm using 2 random matrices generating $SL(d,q)$. Note that GAP computes in finite fields using a Zech logarithm table; therefore the time required for one finite field multiplication does not depend on q.

ACKNOWLEDGEMENTS

We are grateful to M. Geck, K. Lux, W. Nickel, and E. A. O'Brien for helpful conversations.

References

[1] Frank Celler and C. R. Leedham-Green, Calculating the order of an in-

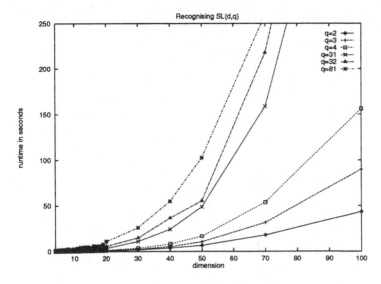

Figure 1: Timings

vertible matrix, DIMACS proceedings, to appear.

[2] Frank Celler and C. R. Leedham-Green, A non-constructive recognition algorithm for the special linear and other classical groups, in *Groups and computation II (eds. L. Finkelstein and W. M. Kantor)*. DIMACS series, vol. 28, Amer. Math. Soc., 1997.

[3] Frank Celler, Charles R. Leedham-Green, Scott H. Murray, Alice C. Nie-meyer, and E. A. O'Brien, Generating random elements of a finite group, *Comm. Algebra* **23** (1995), 4931–4948.

[4] Derek F. Holt and Sarah Rees, Testing modules for irreducibility, *J. Austral. Math. Soc. Series A* **57** (1994), 1–16.

[5] Peter Kleidman and Martin Liebeck, *The subgroup structure of the finite classical groups*, London Math. Soc. Lecture Note Series 129, Cambridge University Press, 1990.

[6] Peter M. Neumann and Cheryl E. Praeger, A recognition algorithm for special linear groups, *Proc. London Math. Soc. (3)* **65** (1992), 555–603.

[7] Alice C. Niemeyer and Cheryl E. Praeger, A recognition algorithm for classical groups over finite fields, *Proc. London Math. Soc.*, to appear.

[8] Martin Schönert *et al.*, GAP—*Groups, Algorithms and Programming*, Lehrstuhl D für Mathematik, RWTH Aachen, 1994.

Relations in $\mathbb{M}_{666}$

John H. Conway
and
Christopher S. Simons

Abstract

We sketch geometrical proofs of the equivalence of various relations which hold in the Bimonster.

1 Introduction

The Ivanov–Norton theorem [4, 5] asserts that the "Bimonster", or wreathed square $(\mathbf{M} \times \mathbf{M})\!:\!2$ of the Monster group $\mathbf{M}$, is the abstract group defined by the Coxeter relations of the $\mathbb{M}_{666}$ diagram (see Fig. 1) together with a single additional relation, initially taken as

$$(ab_1c_1ab_2c_2ab_3c_3)^{10} = 1 \qquad \text{(S)}$$

The purpose of this paper is to provide purely geometrical proofs of the equivalence between various alternatives to this so-called "spider" relation (S). The original proofs of these equivalences, mostly due to L. Soicher [2], involved machine coset enumerations. We found those given here during the preparation of Simons's Ph. D. dissertation [6].

There is considerable point in providing machine-free proofs, because the coset enumerations that would be required to complete this work into a new proof of the Ivanov–Norton theorem are much too large ever to be done by machine. However there is no similar bar to extending our geometric arguments, and a purely geometrical proof of the Ivanov–Norton theorem would almost certainly yield a new and simple way to compute inside the Monster.

We shall use $c\mathbb{M}_{666}$ for the Coxeter group defined by the graph in Fig. 1, and $c\mathbb{M}_{pqr}$ for the subgroup of this generated by

the first p of $a, b_1, c_1, d_1, e_1, f_1$,
the first q of $a, b_2, c_2, d_2, e_2, f_2$,
and the first r of $a, b_3, c_3, d_3, e_3, f_3$.

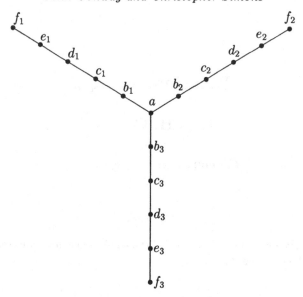

Figure 1: The $\mathbf{M}_{666}$ diagram

The "c" in $c\mathbf{M}_{pqr}$ stands for "Coxeter". To indicate that other relations are to be added we change "c" to some other letter, or omit it to indicate the corresponding subgroups $\mathbf{M}_{pqr}$ of the Bimonster. These subgroups are described in Table 1. In [1] these groups are called $Y_{p-1,q-1,r-1}$, but our reparametrization has many advantages.

The phrase "pqr corrects PQR (in $\mathcal{PQR}$)" means that if we subject the Coxeter group $c\mathbf{M}_{\mathcal{PQR}}$ to the relations that convert $c\mathbf{M}_{pqr}$ to $\mathbf{M}_{pqr}$, then the subgroup $c\mathbf{M}_{PQR}$ is converted to $\mathbf{M}_{PQR}$. If $\mathcal{P}, \mathcal{Q}, \mathcal{R}$ are unspecified, we are to understand their minimal values, namely

$$\begin{aligned}\mathcal{P} &= \max(p, P), \\ \mathcal{Q} &= \max(q, Q), \\ \mathcal{R} &= \max(r, R).\end{aligned}$$

We discuss the first few cases in which $c\mathbf{M}_{pqr}$ differs from $\mathbf{M}_{pqr}$, and indicate the appropriate relations. As is convenient in this subject, we use lower case letters for spherical Dynkin diagrams and upper case letters for Euclidean Dynkin diagrams.

Thus $c\mathbf{M}_{622}$ is the Weyl group of d_8, namely $2^7{:}S_8$, but $\mathbf{M}_{622}$ is its central quotient obtained by adjoining the relation

$$\delta_8 = (ab_1b_2b_3c_1d_1e_1f_1)^7 = 1,$$

where δ_8 is the central inversion of d_8.

Table 1: $\mathbf{M}_{pqr}$ subgroups of $\mathbf{M}_{666}$

pqr	$\mathbf{M}_{pqr}$	# of inequivalent roots
$pq1$	S_{p+q}	$(p+q)(p+q-1)/2$
222	$2^3{:}S_4$	12
322	$2^4{:}S_5$	20
422	$2^5{:}S_6$	30
522	$2^6{:}S_7$	42
622	$2^6{:}S_8$	56
332	$O_6^-(2){:}2 \cong O_5(3){:}2$	36
432	$O_7(2) \times 2$	63
632	$O_8^+(2){:}2$	120
442	$2^7{:}(2 \times O_7(2))$	126
642	$O_9(2) \times 2$	255
662	$O_{10}^-(2){:}2$	528
333	$3^5{:}O_5(3){:}2 \cong 3^5{:}O_6^-(2){:}2$	108
433	$O_7(3) \times 2$	351
633	$O_8^+(3){:}2$	1080
443	$2^2.Fi_{22}$	3510
643	$2 \times Fi_{23}$	31671
663	$3.Fi_{24}$	920808
444	$2^3.{}^2E_6(2)$	3968055
644	$2^2.B$	13571955000
664	$2 \times \mathbf{M}$	$|\mathbf{M}|/2|B| \approx 10^{20}$
666	$(\mathbf{M} \times \mathbf{M}){:}2$	$|\mathbf{M}| \approx 10^{54}$

In general, if a Coxeter group has a central inversion, it is the product of the central inversions of its connected components (which together generate its centre). The central inversion of a connected Coxeter group, when it has one, is the product of its generators raised to the power $\frac{1}{2}h$, where h is the order of this product (the "Coxeter number").

In a similar way, $c\mathbf{M}_{532}$ is the Weyl group of e_8, namely $2G{:}2$, where G is the simple group $O_8^+(2)$, while $\mathbf{M}_{532}$ is its central quotient $G{:}2$, obtained by adjoining $\varepsilon_8 = 1$, where

$$\varepsilon_8 = (ab_1b_2b_3c_1c_2d_1e_1)^{15}$$

is the central inversion of e_8.

We see a new type of relation in the case 632. The Coxeter group $c\mathbf{M}_{632}$ (being associated to the Euclidean diagram E_8) has structure $\mathbb{Z}^8{:}2G{:}2$, while $\mathbf{M}_{632}$ collapses this to the group $G{:}2$ we have just seen. It may be obtained by killing not only ε_8 but also all the *translations* t_v by vectors v of the E_8 root lattice.

Similarly, $c\mathbf{M}_{442}$ is the Weyl group of the Euclidean diagram E_7, namely $\mathbb{Z}^7{:}(2 \times H)$, where $2 \times H$ is the Weyl group of the corresponding spherical diagram e_7. This collapses to

$$2^7{:}(2 \times H)$$

in $\mathbf{M}_{442}$, obtained by killing the translations for which v belongs to *twice* the root lattice of E_7.

Finally, the Weyl group $c\mathbf{M}_{333}$ of the Euclidean diagram E_6 has structure $\mathbb{Z}^6{:}K{:}2$, where $K{:}2$ is the Weyl group of the spherical diagram e_6, and this becomes

$$3^5{:}K{:}2$$

in $\mathbf{M}_{333}$, which we can obtain by killing the translations for which v belongs to *three times* the *weight* lattice of E_6. (The spider relation has precisely this effect.)

2 Some results

Theorem 1 622 *corrects* 532 *(in 632)*.

Proof: We can obtain various spherical Coxeter groups by removing single nodes from a Euclidean Dynkin diagram. If two of these both have central inversions, then the product of these is a translation of the Euclidean group. This is illustrated by the figure for the Euclidean group G_2 (see Fig. 2), defined as

$$G_2 = \langle a,b,c \mid 1 = a^2 = b^2 = c^2 = (ab)^3 = (bc)^6 = (ca)^2 \rangle.$$

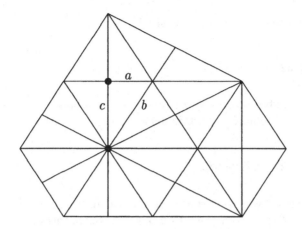

Figure 2: The Euclidean group G_2

Figure 3: The 632 diagram

The subgroups $\langle a, c \rangle$ and $\langle b, c \rangle$ have central inversions $\alpha = ac$ and $\beta = (bc)^3$ which are the half-turns about the two marked vertices, and $\alpha\beta$ is the vertical translation through twice their distance.

In a similar way, by dropping either of the starred nodes in the 632 diagram (see Fig. 3) we obtain the d_8 or e_8 diagrams, having the respective central inversions δ_8 and ε_8. So we have

$$\delta_8\varepsilon_8 = t_v$$

for some vector v in the E_8 root lattice. Therefore to impose the relation $\delta_8 = 1$ is to equate ε_8 with t_v.

Now v happens to be a vector of norm 4 (see [6, 7]), and all such vectors are equivalent under the Weyl group of e_8, and so we also have

$$\varepsilon_8 = t_w$$

for any other norm 4 vector w. We can even choose w so that $v + w$ also has norm 4, so that

$$\varepsilon_8 = t_{v+w} = t_v t_w = (\varepsilon_8)^2,$$

from which we see that $\varepsilon_8 = 1$ as desired. $\qquad\square$

Theorem 2 622 *corrects* 632.

Proof: In the above, since $\varepsilon_8 = t_v$ and $\varepsilon_8 = 1$, we have $t_v = 1$ for any norm 4 vector v. But the norm 4 vectors generate the root lattice. $\qquad\square$

Theorem 3 622 *corrects* 633.

Proof: Since 633 is neither spherical nor Euclidean, we must use a different method, that of explicit enumeration of roots. Since this has been described in detail in [6], we only briefly sketch it here. We observe that (in the so-called "system 1" coordinates of [3]) we have the relation

$$
\begin{bmatrix}
. & . & . & 1 & 1 & 1 \\
. & . & . & 1 & 1 & 1 \\
. & . & . & . & 1 & 2
\end{bmatrix}
=
\begin{bmatrix}
1 & 1 & 1 & . & . & . \\
. & . & . & 1 & 1 & 1 \\
. & . & . & . & 2 & 1
\end{bmatrix}
$$

since the elements represented by the two sides differ by a translation. Using this, we find that the 1080 root elements represented by

$$
\begin{bmatrix}
. & . & . & . & . & . \\
. & . & . & . & . & . \\
. & . & . & . & + & -
\end{bmatrix}
\quad
\begin{bmatrix}
. & . & . & . & + & - \\
. & . & . & . & . & . \\
. & . & . & . & . & .
\end{bmatrix}
\quad
\begin{bmatrix}
. & . & . & . & . & 1 \\
. & . & . & . & . & 1 \\
. & . & . & . & . & 1
\end{bmatrix}
\quad
\begin{bmatrix}
. & . & . & . & 1 & 1 \\
. & . & . & . & 1 & 1 \\
. & . & . & . & 1 & 1
\end{bmatrix}
$$
$$
\ \ \ \ \ \ (6) \qquad\qquad\quad (15) \qquad\qquad\quad (54) \qquad\qquad\quad (135)
$$

$$
\begin{bmatrix}
. & . & . & . & 1 & 2 \\
. & . & . & 1 & 1 & 1 \\
. & . & . & 1 & 1 & 1
\end{bmatrix}
\quad
\begin{bmatrix}
. & . & . & 1 & 1 & 1 \\
. & . & . & 1 & 1 & 1 \\
. & . & . & 0 & 1 & 2
\end{bmatrix}
\quad
\begin{bmatrix}
. & . & . & 1 & 1 & 2 \\
. & . & . & 1 & 1 & 2 \\
. & . & . & 1 & 1 & 2
\end{bmatrix}
\quad
\begin{bmatrix}
. & . & 1 & 1 & 2 & 2 \\
. & . & . & 1 & 2 & 3 \\
. & . & . & 1 & 2 & 3
\end{bmatrix}
$$
$$
\ \ \ \ \ (30) \qquad\qquad\quad (120) \qquad\qquad\quad (540) \qquad\qquad\quad (180)
$$

form a conjugacy class. [The elements are obtained from those shown by permuting all six coordinates of the top row, the last three coordinates of each of the other two rows, or interchanging those two rows. Parentheses show the numbers of elements so obtained.]

Moreover, these roots are transformed the way $O_8^+(3){:}2$ permutes its root elements, and so the group generated is a central extension of the latter group. But the multiplier of the simple subgroup $O_8^+(3)$ is known—it is a 4-group whose elements are annihilated by our relations, which therefore correct 633 to $O_8^+(3){:}2$. $\qquad\square$

3   From 622 to the spider

We check inside $O_8^+(3)$ that the three products

$$
\begin{array}{c}
\cdot\ \cdot\ \cdot\ +\ -\ 0 \\
\cdot\ \cdot\ \cdot\ \cdot\ \cdot\ \cdot \\
\cdot\ \cdot\ \cdot\ \cdot\ \cdot\ \cdot
\end{array}
\cdot
\begin{array}{c}
\cdot\ \cdot\ \cdot\ 0\ 2\ 1 \\
\cdot\ \cdot\ \cdot\ 1\ 1\ 1 \\
\cdot\ \cdot\ \cdot\ 1\ 1\ 1
\end{array}
\cdot
\begin{array}{c}
\cdot\ \cdot\ \cdot\ +\ 0\ - \\
\cdot\ \cdot\ \cdot\ \cdot\ \cdot\ \cdot \\
\cdot\ \cdot\ \cdot\ \cdot\ \cdot\ \cdot
\end{array}
\cdot
\begin{array}{c}
\cdot\ \cdot\ \cdot\ 0\ 1\ 2 \\
\cdot\ \cdot\ \cdot\ 1\ 1\ 1 \\
\cdot\ \cdot\ \cdot\ 1\ 1\ 1
\end{array} ,
$$

$$
\begin{array}{c}
\cdot\ \cdot\ \cdot\ \cdot\ \cdot\ \cdot \\
\cdot\ \cdot\ \cdot\ +\ -\ 0 \\
\cdot\ \cdot\ \cdot\ \cdot\ \cdot\ \cdot
\end{array}
\cdot
\begin{array}{c}
\cdot\ \cdot\ \cdot\ 1\ 1\ 1 \\
\cdot\ \cdot\ \cdot\ 0\ 2\ 1 \\
\cdot\ \cdot\ \cdot\ 1\ 1\ 1
\end{array}
\cdot
\begin{array}{c}
\cdot\ \cdot\ \cdot\ \cdot\ \cdot\ \cdot \\
\cdot\ \cdot\ \cdot\ +\ 0\ - \\
\cdot\ \cdot\ \cdot\ \cdot\ \cdot\ \cdot
\end{array}
\cdot
\begin{array}{c}
\cdot\ \cdot\ \cdot\ 1\ 1\ 1 \\
\cdot\ \cdot\ \cdot\ 0\ 1\ 2 \\
\cdot\ \cdot\ \cdot\ 1\ 1\ 1
\end{array} ,
$$

$$
\begin{array}{c}
\cdot\ \cdot\ \cdot\ \cdot\ \cdot\ \cdot \\
\cdot\ \cdot\ \cdot\ \cdot\ \cdot\ \cdot \\
\cdot\ \cdot\ \cdot\ +\ -\ 0
\end{array}
\cdot
\begin{array}{c}
\cdot\ \cdot\ \cdot\ 1\ 1\ 1 \\
\cdot\ \cdot\ \cdot\ 1\ 1\ 1 \\
\cdot\ \cdot\ \cdot\ 0\ 2\ 1
\end{array}
\cdot
\begin{array}{c}
\cdot\ \cdot\ \cdot\ \cdot\ \cdot\ \cdot \\
\cdot\ \cdot\ \cdot\ \cdot\ \cdot\ \cdot \\
\cdot\ \cdot\ \cdot\ +\ 0\ -
\end{array}
\cdot
\begin{array}{c}
\cdot\ \cdot\ \cdot\ 1\ 1\ 1 \\
\cdot\ \cdot\ \cdot\ 1\ 1\ 1 \\
\cdot\ \cdot\ \cdot\ 0\ 1\ 2
\end{array}
$$

(of four reflections each) are equal. Now the top line is the product of the cM$_{333}$ translations in

$$
\begin{array}{c}
\cdot\ \cdot\ \cdot\ +\ -\ 0 \\
\cdot\ \cdot\ \cdot\ \cdot\ \cdot\ \cdot \\
\cdot\ \cdot\ \cdot\ \cdot\ \cdot\ \cdot
\end{array}
\quad \text{and} \quad
\begin{array}{c}
\cdot\ \cdot\ \cdot\ +\ 0\ - \\
\cdot\ \cdot\ \cdot\ \cdot\ \cdot\ \cdot \\
\cdot\ \cdot\ \cdot\ \cdot\ \cdot\ \cdot
\end{array} ,
$$

and so is the translation in

$$
\begin{array}{c}
\cdot\ \cdot\ \cdot\ 2\ -\ - \\
\cdot\ \cdot\ \cdot\ \cdot\ \cdot\ \cdot \\
\cdot\ \cdot\ \cdot\ \cdot\ \cdot\ \cdot
\end{array} ,
$$

so the quotient of the first two rows is the translation in

$$
\begin{array}{c}
\cdot\ \cdot\ \cdot\ \ 2\ \ -\ - \\
\cdot\ \cdot\ \cdot\ -2\ +\ + \\
\cdot\ \cdot\ \cdot\ \cdot\ \cdot\ \cdot
\end{array} .
$$

But this vector is orthogonal modulo 3 to all the fundamental roots of M$_{333}$, and so it is easy to see that it and its images generate 3 times the weight lattice of E_6. Hence the equality of any two of the above products is a relation equivalent to our spider relation (S).

Since 622 corrects 333, it entails these relations.

4   From the spider to 622

We now work in the other direction, from 333 towards 622.

Theorem 4 333 *half-corrects* 533.

What we mean is that the relations just discussed (which are a way of correcting 333) convert $c\mathbf{M}_{533}$ to the group $2O_8^+(3){:}2$, whereas $\mathbf{M}_{533} \cong O_8^+(3){:}2$ is only half as large.

Proof: This is again proved by root enumeration, and again we only sketch the proof. The relations are used to show that the 1080 elements

```
. . . . .  .        . . . .  +  −        . . . . .  1
. . . . .  .        . . . .  .  .        . . . . .  1
. . . . +  −        . . . . .  .        . . . . .  1
       (6)                (10)                 (45)
```

```
. . . . 1 1         . . . . 1 2          . . . 1 1 1
. . . . 1 1         . . . 1 1 1          . . . 1 1 1
. . . . 1 1         . . . 1 1 1          . . . 0 1 2
     (90)                (20)                (120)
```

```
. . . 1 1 2         . . 1 1 1 1          . . 1 1 1 2
. . . 1 1 2         . . . 1 1 2          . . . 1 1 3
. . . 1 1 2         . . . 0 2 2          . . . 1 2 2
    (270)                (45)                (180)
```

```
. 1 1 1 1 1         . . 1 1 2 2          . 1 1 1 1 2
. . . 1 2 2         . . . 1 2 3          . . . 2 2 2
. . . . 2 3         . . . 1 2 3          . . . . 3 3
     (9)                 (180)                (5)
```

```
. 1 1 1 1 2         . 1 1 1 2 2          . 1 1 1 2 4
. . . 2 2 2         . . . 1 3 3          . . . 3 3 3
. . . 1 1 4         . . . 1 2 4          . . . 1 4 4
     (5)                 (90)                (5)
```

form a conjugacy class in the group they generate, and so as before this must be a central extension of $O_8^+(3){:}2$. [The elements are obtained from those shown by permuting the last five coordinates of the top row, the last three coordinates of each of the other two rows, or interchanging those two rows. Parentheses show the numbers of elements so obtained.] This time the extension that happens is non-trivial, so that $c\mathbf{M}_{533}$ is twice as large as $\mathbf{M}_{533}$.   □

5   Translations

One of the things we must prove is that the E_8 translations of the various subdiagrams 632, 623, ..., 236 are all trivial. One such translation is

$$\boxed{\begin{array}{cccccc} + & - & . & . & . & . \\ . & . & . & . & . & . \\ . & . & . & . & . & . \end{array}} = \boxed{\begin{array}{cccccc} + & - & . & . & . & . \\ . & . & . & . & . & . \\ . & . & . & . & . & . \end{array}} \cdot \boxed{\begin{array}{cccccc} 0 & 2 & 1 & 1 & 1 & 1 \\ . & . & . & 2 & 2 & 2 \\ . & . & . & . & 3 & 3 \end{array}},$$

and we can obtain another for any root vector r of 632, namely

$$\boxed{\boxed{r}} = \boxed{r} \cdot \boxed{\varepsilon_1 - r},$$

where

$$\varepsilon_1 = \boxed{\begin{array}{cccccc} 1 & 1 & 1 & 1 & 1 & 1 \\ . & . & . & 2 & 2 & 2 \\ . & . & . & . & 3 & 3 \end{array}}$$

is the "null vector" of 632. But some root vectors belong to several E_8 subdiagrams—for instance the above example belongs also to 623, whose null vector is

$$\varepsilon_2 = \boxed{\begin{array}{cccccc} 1 & 1 & 1 & 1 & 1 & 1 \\ . & . & . & . & 3 & 3 \\ . & . & . & 2 & 2 & 2 \end{array}}.$$

Fortunately this does not matter:

Theorem 5 *For the above* r, *we have*

$$\boxed{\varepsilon_1 - r} = \boxed{\varepsilon_2 - r},$$

namely

$$\boxed{\begin{array}{cccccc} 0 & 2 & 1 & 1 & 1 & 1 \\ . & . & . & 2 & 2 & 2 \\ . & . & . & . & 3 & 3 \end{array}} = \boxed{\begin{array}{cccccc} 0 & 2 & 1 & 1 & 1 & 1 \\ . & . & . & . & 3 & 3 \\ . & . & . & 2 & 2 & 2 \end{array}}.$$

Proof: Conjugation quickly reduces this to

$$\boxed{\begin{array}{cccccc} . & . & 1 & 1 & 1 & 1 \\ . & . & . & . & 2 & 2 \\ . & . & . & 2 & 1 & 1 \end{array}} = \boxed{\begin{array}{cccccc} . & . & 1 & 1 & 1 & 1 \\ . & . & . & 2 & 1 & 1 \\ . & . & . & . & 2 & 2 \end{array}},$$

which can be established by transforming

$$\boxed{\begin{array}{cccccc} . & . & 1 & 0 & 0 & 0 \\ . & . & . & 1 & 0 & 0 \\ . & . & . & 1 & 0 & 0 \end{array}}$$

by the second and third of the three equivalent products in section 3. □

In a precisely similar way, we can show that if a fundamental root r belongs to two or more of the E_8 diagrams

$$632, 623, 263, 362, 236, 326$$

then the two associated translations are equal. (Up to conjugation there are very few cases.) We can also show that any two of these 16 translations commute, since they can be conjugated into the same copy of E_8.

So the translations

$$\boxed{\boxed{r}}$$

generate some quotient of the free abelian group of rank 16. Since also the corresponding vectors r additively generate a free abelian group of rank 16, this justifies the notation

$$\boxed{\boxed{r_1 + r_2 + r_3 + \ldots}} = \boxed{\boxed{r_1}} \cdot \boxed{\boxed{r_2}} \cdot \boxed{\boxed{r_2}} \cdot \ldots$$

Theorem 6 *The translations are trivial.*

Proof: If we conjugate the translation

$$\begin{array}{|cccccc|} \hline . & . & + & - & 0 & 0 \\ . & . & . & . & . & . \\ . & . & . & . & . & . \\ \hline \end{array}$$

by the last two of our three equivalent expressions from section 3, we obtain

$$\begin{array}{|cccccc|} \hline . & . & 1 & 2 & 3 & 3 \\ . & . & . & 1 & 4 & 4 \\ . & . & . & 3 & 3 & 3 \\ \hline \end{array} = \begin{array}{|cccccc|} \hline . & . & 1 & 2 & 3 & 3 \\ . & . & . & 3 & 3 & 3 \\ . & . & . & 1 & 4 & 4 \\ \hline \end{array},$$

so that their difference

$$\begin{array}{|cccccc|} \hline . & . & . & 0 & 0 & 0 \\ . & . & . & 2 & -1 & -1 \\ . & . & . & -2 & 1 & 1 \\ \hline \end{array} = 0.$$

But a further conjugation connects this to

$$\begin{array}{|cccccc|} \hline . & . & . & . & . & . \\ . & . & -1 & 2 & -1 & 0 \\ . & . & . & -2 & 1 & 1 \\ \hline \end{array} = 0,$$

and so *their* difference, namely

$$\begin{array}{|cccccc|} \hline . & . & . & . & . & . \\ . & . & + & 0 & 0 & - \\ . & . & . & . & . & . \\ \hline \end{array},$$

is also 0. But this is one of the 16 fundamental translations. □

Theorem 7 333 *corrects* 542 *(in 543)*.

Proof: Once we know that the translations are trivial it is very easy to deduce the correctness of 542 by root-enumeration. [No proper central extension can arise, because the index 2 subgroup $O_9(2)$ has trivial multiplier.] □

Corollary 8 333 *corrects* 533 *in* 543.

(Recall that 333 only half-corrected 533 in 533 itself.)
     We now work in 666. In view of the last corollary we have in 666

$$333 \text{ corrects } 532$$

and so since the translations are trivial

$$333 \text{ corrects } 632$$

and so

$$333 \text{ corrects } 622.$$

Now by symmetry, in 666 we have that

$$333 \text{ corrects } 262 \text{ and } 226,$$

and so, since we proved in 666 that 622 implies 333, we obtain our final result:

Theorem 9 *Inside* 666 *the correctness of any one of* 622, 262, 226 *implies that of all three.*

References

[1] J. H. Conway, R. T. Curtis, S. P. Norton, R. A. Parker and R. A. Wilson, ATLAS *of finite groups*, Clarendon Press, Oxford, 1985.

[2] J. H. Conway, S. P. Norton and L. H. Soicher, The Bimonster, the group Y_{555} and the projective plane of order 3, in *Computers in algebra (ed. M. C. Tangora)*, pp. 27–50. Marcel Dekker, 1988.

[3] J. H. Conway and A. D. Pritchard, Hyperbolic reflections for the Bimonster and $3Fi_{24}$, in *Groups, combinatorics and geometry (eds. M. W. Liebeck and J. Saxl)*, pp. 24–45. LMS Lecture Note Series **165**, Cambridge University Press, 1992.

[4] A. A. Ivanov, A geometric characterization of the Monster, in *Groups, combinatorics and geometry (eds. M. W. Liebeck and J. Saxl)*, pp. 46–62. LMS Lecture Note Series **165**, Cambridge University Press, 1992.

[5] S. P. Norton, Constructing the Monster, in *Groups, combinatorics and geometry (eds. M. W. Liebeck and J. Saxl)*, pp. 63–76. LMS Lecture Note Series **165**, Cambridge University Press, 1992.

[6] C. S. Simons, *Hyperbolic reflection groups, completely replicable functions, the Monster and the Bimonster*, Ph. D. thesis, Princeton University, November 1997.

[7] C. S. Simons, Monster roots, in *Monster II and Lie algebras, (eds. J. Ferrar and K. Harada)*, Walter de Gruyter, to appear.

A survey of symmetric generation of sporadic simple groups

Robert T. Curtis

Abstract

Many of the sporadic simple groups possess highly symmetric generating sets which can often be used to construct the groups, and which carry much information about their subgroup structure. We give a survey of results obtained so far.

1 Introduction and motivation

This paper is concerned with groups which are generated by highly symmetric subsets of their elements: that is to say by subsets of elements whose set normalizer within the group they generate acts on them by conjugation in a highly symmetric manner. Rather than investigate the behaviour of various known groups, we turn the procedure around and ask what groups can be generated by a set of elements which possesses certain assigned symmetries. It turns out that this approach enables us to define and construct by hand a large number of interesting groups—including many of the sporadic simple groups.

Accordingly we let $m^{\star n}$ denote $C_m \star C_m \star \ldots \star C_m$, a free product of n copies of the cyclic group of order m. Let $F = T_0 \star T_1 \star \ldots \star T_{n-1}$ be such a group, with $T_i = \langle t_i \rangle \cong C_m$. Certainly permutations of the set of *symmetric generators* $\mathcal{T} = \{t_0, t_1, \ldots, t_{n-1}\}$ induce automorphisms of F. Further automorphisms are given by raising a given t_i to a power of itself coprime to m, while fixing the other symmetric generators. Together these generate the group $\mathcal{M}$ of *monomial automorphisms* of F which is a wreath product $H_r \wr S_n$, where H_r is an abelian group of order $r = \phi(m)$, the number of positive integers less than m and coprime to it. A split extension of the form

$$\mathcal{P} \cong m^{\star n} : \mathcal{N},$$

where $\mathcal{N}$ is a subgroup of $\mathcal{M}$ which acts transitively on the set of cyclic subgroups $\overline{\mathcal{T}} = \{T_0, T_1, \ldots, T_{n-1}\}$, is called a *progenitor*. The group $\mathcal{N}$ is known as the *control subgroup* and its elements are *monomial permutations* or, more informally, *permutations*.

1.1   The Mathieu groups

As is well known, certain subsets of 4 vertices of a regular dodecahedron form the vertex set of a regular tetrahedron. Indeed the 20 vertices can be partitioned (in two ways) into five such tetrahedra, the two partitions being interchanged by reflective symmetries and each set of 5 being acted on by the full group of rotational symmetries isomorphic to the alternating group A_5. Given one such tetrahedron, we define a permutation of the 12 faces of the dodecahedron as follows: each of its four vertices is adjacent to 3 faces, rotate them clockwise to obtain the permutation of cycle shape 3^4 illustrated in Fig. 1. As was described and investigated in detail in [8, 9], the five permutations corresponding to one of the partitions generate a copy of the Mathieu group M_{12}. Thus M_{12} is a homomorphic image of the progenitor

$$3^{*5}{:}A_5,$$

in which the control subgroup $\mathcal{N} \cong A_5$ simply permutes the five symmetric generators. Indeed, we saw further that if $\mathcal{N}$ is extended to the symmetric group S_5 in which the outer elements permute *and invert* the symmetric generators, then Aut $M_{12} \cong M_{12}{:}2$ is a homomorphic image of

$$3^{*5}{:}S_5.$$

A homomorphism which acts faithfully on $\mathcal{N}$ and on the cyclic subgroups T_i, is said to be a *true* homomorphism of the progenitor. We restrict our attention to true homomorphic images and so it causes no confusion to allow $\mathcal{N}$ and T_i to denote both subgroups of the infinite progenitor $\mathcal{P}$, and their isomorphic images. Note that $\mathcal{P} = \langle \mathcal{N}, \mathcal{T} \rangle$ and, since the action by conjugation of $\mathcal{N}$ on $\mathcal{T}$ is known, permutations can be gathered on the left; so each element of $\mathcal{P}$ can be written (essentially uniquely) as πw, where $\pi \in \mathcal{N}$ and w is a word in $\mathcal{T}$. Thus any relator by which we factor $\mathcal{P}$ to obtain a finite group G may be taken to have this form. Now if the five symmetric generators for M_{12} are taken to be $\mathcal{T} = \{t_0, t_1, \ldots, t_4\}$ then it turns out that the element $(t_0^{-1}t_1)^2$ is the permutation of the faces induced by the rotational symmetry of the dodecahedron corresponding to the 3-cycle (2 3 4) of A_5. If we factor our progenitor by this relation we find that

$$\frac{3^{*5}{:}A_5}{(2\ 3\ 4) = (t_0^{-1}t_1)^2} \cong 3 \times M_{12},$$

and that

$$\frac{3^{*5}{:}S_5}{(2\ 3\ 4) = (t_0^{-1}t_1)^2} \cong (3 \times M_{12}){:}2.$$

The normal subgroup of order 3 in either case is generated by, for instance, $[(0\ 1\ 2\ 3\ 4)t_0]^8$.

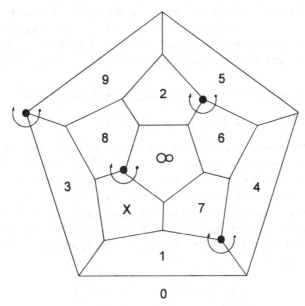

Figure 1: A symmetric generator for M_{12} acting on the 12 faces of a regular dodecahedron

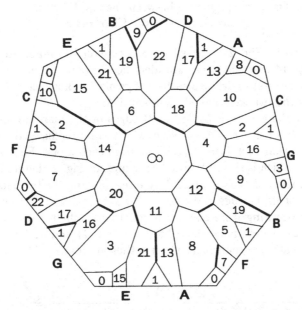

Figure 2: A symmetric generator for M_{24} acting on the 24 faces of the Klein map

To obtain an analogous result for the Mathieu group M_{24} we consider the Klein map, a genus 3 map having 24 heptagonal faces meeting three at each vertex. These may be labelled $\infty, 0, 1, \infty_i, 0_i$ and 1_i, for $i \in \{0, 1, \ldots, 6\}$, where the subscripts are to be read modulo 7 and adjacency of faces is given by:

$$(\infty \; \infty_i), (0 \; 0_i), (1 \; 1_i), (\infty_i \; \infty_{i\pm 1}), (0_i \; 0_{i\pm 2}), (1_i \; 1_{i\pm 4}),$$

$$(0_i \; 1_{i\pm 1}), (1_i \; \infty_{i\pm 2}), (\infty_i \; 0_{i\pm 4}).$$

Its group of rotational symmetries is isomorphic to $L_2(7)$ and a 7-fold symmetry corresponding to $X_i \mapsto X_{i+1}$ for $X \in \{\infty, 0, 1\}$ is exhibited in Fig. 2 whose faces are labelled with the elements of the 24-point projective line, in which the faces $\{\infty, 0, 1\}$ retain their labels. The table below, in which the top row denotes subscripts, shows the correspondence between the labels for the other 21 faces. An element of order 3 normalizing this 7-fold symmetry is given by $X_i \mapsto X_{2i}$ followed by a rotation of the symbols $\{\infty, 0, 1\}$.

i	0	1	2	3	4	5	6
∞_i	11	12	4	18	6	14	20
0_i	22	15	7	3	8	9	10
1_i	2	17	21	5	16	13	19

The correspondence between the two labellings
of the 24 faces of the Klein map.

The Klein map in Fig. 2 is drawn on a heptagon to emphasise the 7-fold symmetry. Its 14 external edges are identified in pairs (as indicated by the capital roman letters) to define a surface of genus 3. Thus face 8, for instance, is split across the two external edges labelled A.

Now under the action of $L_2(7)$ the 84 edges of the map fall into 7 blocks of imprimitivity of size 12, in two ways which are interchanged by reflective symmetries, so that each face has an edge from each block. Each such block, an example of which is shown by the heavy lines in Fig. 2, defines a permutation of the 24 faces of cycle shape 2^{12} by interchanging the two faces incident with any one of its 12 edges. We let $\mathcal{T} = \{t_0, t_1, \ldots, t_6\}$ be the set of seven permutations resulting from one of the partitions, labelled to correspond to the points of the Fano plane whose lines are $\{1, 2, 4\}$ and its translates; they generate the Mathieu group M_{24}, which is thus an image of $2^{\star 7}{:}L_3(2)$. In fact we find that

$$\frac{2^{\star 7}{:}L_3(2)}{[t_0^{t_1}, t_3], (2\ 4)(5\ 6) = (t_0 t_3)^3, [((1\ 2\ 4)(3\ 6\ 5))t_3]^{11}} \cong M_{24}.$$

To show this, first observe that the subgroup generated by four elements of

$\mathcal{T}$ on an oval, such as $\{t_0, t_3, t_6, t_5\}$, is thus an image of

$$\frac{2^{\star 4}:S_4}{[(3\ 6\ 5)t_3]^{11}, (5\ 6) = (t_0 t_3)^3}.$$

The latter group may be shown, either by a mechanical (single) coset enumeration or by a manual double coset enumeration over the control subgroup $\mathcal{N} \cong S_4$, to be isomorphic to $L_2(23)$, and then a mechanical coset enumeration over this subgroup gives the correct index. Of course the labelling has been chosen so that all these permutations on 24 letters lie in the familiar copy of M_{24}.

Further information about how these symmetric generators may be defined combinatorially and algebraically is given in [8, 9], and the six ways in which the binary Golay code can be defined on the faces of the Klein map are described.

1.2 The modular group and triangle groups

As is well known, the modular group $\mathcal{P} = PSL_2(\mathbb{Z})$ is isomorphic to a free product $C_2 \star C_3$. Indeed it can be shown, see for example Jones and Singerman [23, page 296], that

$$\tau \mapsto -\frac{1}{\tau} \text{ and } \tau \mapsto \frac{-1}{1+\tau}$$

are free generators of orders 2 and 3 respectively. This means that any group which is generated by an element of order 2 and an element of order 3 is a homomorphic image of $\mathcal{P}$, which in particular means almost all the finite simple groups, see Liebeck and Shalev, Tamburini and Wilson [24, 28, 29]. To translate this into the language of progenitors let $\mathcal{P} = \langle t, x \mid t^2 = x^3 = 1 \rangle$ and write $t = t_0, t^x = t_1, t^{x^{-1}} = t_2$. Then $\mathcal{T} = \{t_0, t_1, t_2\}$ generates a free product of three copies of C_2, which is normalized by a cyclic automorphism of order 3, thus

$$\mathcal{P} \cong C_2 \star C_3 \cong 2^{\star 3}:C_3.$$

Now $\mathcal{P}$ clearly possesses an outer automorphism which inverts x while commuting with t, and this has the effect of extending it to a progenitor $2^{\star 3}:S_3$. In a homomorphic image of $\mathcal{P}$ this automorphism may survive as either an inner or an outer automorphism, or it may not exist at all. Indeed, $x = (0\ 1\ 2)$ and $t = (1\ 4)(2\ 3)$ generate A_5 in which conjugation by $(1\ 2)(3\ 4) \in \langle x, t \rangle$ realises this automorphism. On the other hand, $x = (1\ 2\ 4)(3\ 6\ 5)$ and $t = (\infty\ 2)(1\ 6)(3\ 5)(0\ 4)$ generate $L_2(7)$ and the automorphism is achieved by conjugating by the element $(\infty\ 0)(2\ 4)(3\ 5)$ of $PGL_2(7)$ but by no inner automorphism.

An image G of $\mathcal{P}$ which does not admit this automorphism must contain a word w in $\mathcal{T}$ which has a different order from its conjugates by odd permutations of S_3; we seek the shortest such w. If $w = w(t_i, t_j)$ involves only two symmetric generators then either it is an involution (if it has odd length) or interchanging i and j inverts it; so we may assume that w involves all three elements of $\mathcal{T}$. But conjugation of w by its leading symmetric generator has the effect of cycling the generators in the word w, and so we may imagine them arranged as an n-gon on which reflections correspond to inversion. For $a, b \in G$ let

$$a \approx b \text{ if, and only if, } a \text{ and } b \text{ have the same order,}$$

and let i stand for t_i. We see that:

$$
\begin{aligned}
102 &\approx 201 \text{ (by inverting)} \\
0102 &\approx 0201 \text{ (by cycling)} \\
12012 &\approx 21021 \text{ (by inverting)} \\
201201 &\approx 102102 \text{ (by inverting)} \\
101202 &\approx 202101 \text{ (by inverting)}
\end{aligned}
$$

so each of these has the same order as its conjugate under $(1\ 2)$. The shortest word w which may satisfy $w \not\approx w^{(1\ 2)}$ is thus 010102. Accordingly, we seek G in which $u = 010102 = [t, x]^2[t, x^{-1}]$ and $v = 020201 = [t, x^{-1}]^2[t, x]$ have different orders. We restrict ourselves to images of the $\triangle$-group $(2, 3, 7)$ in which $xt = (0\ 1\ 2)t_0$ has order 7. If we put $u^3 = 1$ then v assumes order 6 and the image is a non-split extension of shape $2^3 \cdot L_3(2)$. Putting $u^4 = 1$ forces v also to have order 4 and so the image, which is isomorphic to $L_2(41)$, does not possess the property we are seeking. Nor does $L_2(29)$ which is the image when u and v are both required to have order 5. However, if u has order 5 and v has order 6 then we obtain the smallest Janko group J_1, which thus has the following presentation.

$$\langle x, t \mid x^3 = t^2 = (xt)^7 = ([t, x]^2[t, x^{-1}])^5 = ([t, x^{-1}]^2[t, x])^6 = 1 \rangle \cong J_1.$$

Images of $(2, 3, 7)$ are known as *Hurwitz groups* and they have been extensively studied, in particular it is known (except in the case of the Monster) which of the 26 sporadic simple groups can be so generated (see Conder, Wilson and Woldar [4] and Jones [22, SS4], and the extensive bibliography to Jones [22]). It is interesting, however, to see how easily J_1 emerges from these considerations.

Of course we may interchange the roles of x and t and see that

$$\mathcal{P} \cong \langle x, t \mid x^2 = t^3 = 1 \rangle \cong 3^{*2}{:}C_2.$$

Figure 3: A Coxeter diagram with tail

But the full group $\mathcal{M}$ of monomial automorphisms of 3^{*2} is isomorphic to the dihedral group D_8 given by

$$\mathcal{M} = \left\langle \begin{pmatrix} \cdot & 1 \\ 1 & \cdot \end{pmatrix}, \begin{pmatrix} -1 & \cdot \\ \cdot & 1 \end{pmatrix} \right\rangle,$$

where the first matrix corresponds to our element x which interchanges the two symmetric generators, and the second inverts the first generator and fixes the second. It is of interest to ask which finite images of the modular group possess these additional automorphisms. An example which does is afforded by

$$\frac{3^{*2}{:}D_8}{\left[\begin{pmatrix} \cdot & 1 \\ 1 & \cdot \end{pmatrix} t_1 \right]^9} \cong PGL_2(19).$$

1.3   Coxeter diagrams and Y-diagrams

A *Coxeter diagram* of a presentation is a graph in which the vertices correspond to involutory generators and an edge is labelled with the order of the product of its two endpoints. Commuting vertices are not joined, and an edge is left unlabelled if the corresponding product has order 3. Of course a presentation displayed in this manner may have further relations added, and Soicher (see [27]) and others have obtained Coxeter presentations for many interesting groups. Should such a diagram have a "tail" of length at least two, as in Fig. 3, then we see that the generator corresponding to the terminal vertex a_r commutes with the subgroup generated by the subgraph $\mathcal{G}_0$. If we take $\mathcal{N} = \langle \mathcal{G}_0, a_{r-1} \rangle$ as our control subgroup then we see that $G = \langle \mathcal{G}_0, a_{r-1}, a_r \rangle$ is a homomorphic image of $2^{*n}{:}\mathcal{N}$, where n is the index of $\langle \mathcal{G}_0 \rangle$ in $\mathcal{N}$. As an example consider Soicher's presentation for the automorphism group of the Hall–Janko group, see the ATLAS [6, page 42]:

As shown in Fig. 4, we have $\mathcal{N} = \langle a, b, c, d, e \rangle \cong U_3(3){:}2$ and $\langle \mathcal{G}_0 \rangle = \langle a, b, c, d \rangle \cong PGL_2(7)$ with index 36. Thus $HJ{:}2$ is a homomorphic image of $2^{*36}{:}(U_3(3){:}2)$ (see section 2.2).

Now Y-diagrams are Coxeter diagrams in which the graph has the shape of the letter Y with all edges unlabelled, see the ATLAS [6, page 233]. They

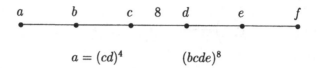

$$a = (cd)^4 \qquad\qquad (bcde)^8$$

Figure 4: A Coxeter diagram for $HJ{:}2$

were investigated by Conway, Norton and others, and Conway's conjecture that Y_{555} together with a single additional relation constituted a presentation of $M\wr2 \cong (M \times M){:}2$, the so-called BiMonster, was finally proved in [5, 18, 25] using further geometric techniques due to Ivanov. Of course such graphs have fine tails and thus lend themselves to the methods of symmetric generation. For instance, together with the necessary additional relations, we see on page 233 of the ATLAS [6] that $Y_{441} \cong O_{10}^-(2){:}2, Y_{431} \cong O_9(2) \times 2$ and $Y_{421} \cong O_8^+(2){:}2$. Thus $O_{10}^-(2){:}2$ is an image of $2^{\star272}{:}(O_9(2) \times 2)$. Note that the action of $O_9(2)$ on 272 letters is imprimitive with blocks of size 2, and the central element of $\mathcal{N}$ transposes each of these pairs.

Campbell, Johnson, Robertson and others have investigated symmetric presentations of groups from a slightly different point of view, see [21, 26, 2].

2 Involutory generators

2.1 The Lemma

In this section we consider true homomorphic images of progenitors of shape $2^{\star n}{:}\mathcal{N}$, where $\mathcal{N}$ is a transitive, and usually primitive, permutation group acting by conjugation on the n involutory generators. Suppose, for now only, that $\mathcal{N}$ is also *perfect*. Then, since $[i, \pi] = ii^\pi$, we have

$$\mathcal{P}' \ge \langle \mathcal{N}, ij \mid i \ne j \rangle \text{ and, therefore, } |\mathcal{P}{:}\mathcal{P}'| = 2.$$

Furthermore $[ij, \pi] = jii^\pi j^\pi = jj^\pi$ if π is chosen to fix i but move j. In particular, if $\mathcal{N}$ is doubly transitive we see that $\mathcal{P}' = \mathcal{P}''$. Consideration of a graph in which i joins j if, and only if, $ij \in \mathcal{P}''$ shows that the same is true provided $\mathcal{N}$ is primitive. So a finite image of $\mathcal{P}$ contains a perfect subgroup to index 1 or 2. Analogous statements hold when $\mathcal{N}$ contains a perfect group to low index, and so the progenitors invariably possess interesting finite images.

We must now decide by what relators to factor $\mathcal{P}$. But if a permutation $\pi \in \mathcal{N}$ can be written as a word in $\{t_{k_1}, t_{k_2}, \ldots, t_{k_r}\}$ then it must commute with all permutations which fix $\{k_1, k_2, \ldots, k_r\}$. The 2-generator form of this is the obvious, but surprisingly effective:

Lemma 2.1 (see Curtis [10])

$$\mathcal{N} \cap \langle t_i, t_j \rangle \leq C_{\mathcal{N}}(\mathcal{N}^{ij}),$$

where $\mathcal{N}^{ij}$ denotes the stabilizer in $\mathcal{N}$ of i and j.

If the centralizer of a 2-point stabilizer in $\mathcal{N}$ is cyclic of order 2 we write $C_{\mathcal{N}}(\mathcal{N}^{ij}) = \langle \pi_{ij} \rangle \cong C_2$, and note that if $\pi_{ij} \in \langle t_i, t_j \rangle$ then $\pi_{ij} t_i$ has odd order in the case when π_{ij} interchanges i and j, and $\pi_{ij} = (t_i t_j)^r$ otherwise.

2.2   The Suzuki chain

As an example using the lemma take $\mathcal{N} \cong G_2(2) \cong U_3(3){:}2$, acting transitively on 36 letters, when $\mathcal{P} \cong 2^{\star 36}{:}(U_3(3){:}2)$ and $\mathcal{N}^i \cong PGL_2(7)$ with orbit lengths 1, 14 and 21. If j is in the 14-orbit we have $\mathcal{N}^{ij} \cong S_4$ and $C_{\mathcal{N}}(\mathcal{N}^{ij}) = \langle \pi_{ij} \rangle \cong C_2$, interchanging i and j. If we factor $\mathcal{P}$ by the relator $(\pi_{ij} t_i)^3$, a manual double coset enumeration as performed in [11, 12, 14] constructs a rank 3 permutation group on $100 = 1+36+63$ letters of order $100 \times 6048 \times 2$, which is of course the automorphism group of the Hall–Janko simple group. Thus

$$\frac{2^{\star 36}{:}(U_3(3){:}2)}{(\pi_{12} t_1)^3} \cong HJ{:}2.$$

A similar procedure, factoring in each case by the analogous relator, yields all groups of the Suzuki chain up to $3 \cdot Suz{:}2$. The Conway group Co_1 is best obtained by a slightly modified approach (see section 3).

2.3   The Higman–Sims group

Now let $\mathcal{N} \cong U_3(5){:}2$, acting transitively on 50 letters, when $\mathcal{P} \cong 2^{\star 50}{:}(U_3(5){:}2)$, and $\mathcal{N}^i \cong S_7$ with orbit lengths 1, 7 and 42. If j is in the 7-orbit, $\mathcal{N}^{ij} \cong S_6$ which has trivial centralizer. However, if k is in the 42-orbit we have $\mathcal{N}^{ik} \cong S_5$ and $C_{\mathcal{N}}(\mathcal{N}^{ik}) = \langle \pi_{ik} \rangle \cong C_2$. Moreover, there is a unique j in the 7-orbits of both i and k and so $\mathcal{N}^{ik} = \mathcal{N}^{ijk}$. If we suppose $\pi_{ik} \in \langle t_i, t_j, t_k \rangle$ we find that the shortest word which does not lead to collapse is $\pi_{ik} = t_i t_k t_i t_j$. A manual double coset enumeration constructs a permutation group on $(176+176)$ letters which is easily seen to be $HS{:}2$, the automorphism group of the Higman–Sims group (see [12]). That is,

$$\frac{2^{\star 50}{:}(U_3(5){:}2)}{\pi_{ik} = ikij} \cong HS{:}2.$$

2.4   The Janko group J_1 and the O'Nan group

Let $\mathcal{N} \cong L_2(11)$, acting doubly transitively on 11 letters, when

$$\mathcal{P} \cong 2^{\star 11}{:}L_2(11),\ \mathcal{N}^i \cong A_5,\ \mathcal{N}^{ij} \cong S_3 \text{ and } C_{\mathcal{N}}(\mathcal{N}^{ij}) = \langle \pi_{ij} \rangle \cong C_2.$$

Now $(\pi_{ij}t_i)^3 = 1$ leads to collapse and so suppose $(\pi_{ij}t_i)^5 = 1$. If $\{i,j,k\}$ is the fixed point set of an involution of $\mathcal{N}$, this relation forces

$$\langle t_i, t_j, t_k \rangle \cong 2 \times A_5,$$

(or a homomorphic image of this). If we now factor by a second relation which sets the central involution here (as a word in t_i, t_j, t_k) equal to the involution of $\mathcal{N}$ fixing i, j and k, a manual double coset enumeration constructs a rank 5 permutation group on 266 letters of order 266×660 which is easily seen to be J_1, the smallest Janko group. It is readily shown, see [11, 15], that the two relations described above are both consequences of the single relation $(\sigma t_i)^5$ where σ is any permutation of cycle shape 2.3.6 and i is in its 3-cycle. Thus, if $\mathcal{N} = \langle (0\ 1\ 2\ 3\ 4\ 5\ 6\ 7\ 8\ 9\ X), (3\ 4)(2\ X)(5\ 9)(6\ 7) \rangle$, then

$$\frac{2^{\star 11}{:}L_2(11)}{(\sigma t_0)^5} \cong J_1,$$

where $\sigma = (3\ 4)(0\ 1\ 8)(2\ 5\ 6\ X\ 9\ 7) \in \mathcal{N}$. It turns out that five of these symmetric generators (one of the 11 special pentads preserved by $L_2(11)$) are sufficient to generate J_1 and we have

$$\frac{2^{\star 5}{:}A_5}{(\pi t_1)^7} \cong J_1,$$

where $\pi = (1\ 2\ 3\ 4\ 5) \in A_5$.

Now let $\mathcal{N}$ be isomorphic to the Mathieu group M_{11} acting transitively on the 12 element set $\Lambda = \{\infty, 0, 1, ..., X\}$. So $\mathcal{P} \cong 2^{\star 12}{:}M_{11}$ and $\mathcal{N}^i \cong L_2(11)$. Let $s_i = (t_\infty t_i)^2$ be an involution for each $i \in \{0, 1, 2, .., X\}$, and define $\sigma = (\infty)(3\ 4)(0\ 1\ 8)(2\ 5\ 6\ X\ 9\ 7) \in \mathcal{N}$. Factoring $\mathcal{P}$ by $(\sigma s_0)^5$ ensures that $\langle s_0, s_1, ..., s_X \rangle$ is isomorphic to J_1 and commutes with t_∞. A further relation $(\sigma^3 t_\infty t_3)^5$ forces $\langle t_\infty, t_0, t_1, t_8 \rangle \cong 4{\cdot}L_3(4){:}2_3$ (see Table 5 of [14]), and comparison with Soicher's Coxeter-style presentation for the O'Nan group (now proved complete) on page 132 of the ATLAS shows that

$$\frac{2^{\star 12}{:}M_{11}}{(t_\infty t_0)^4 = (\sigma^3 t_\infty t_3)^5 = (\sigma(t_\infty t_0)^2)^5 = 1} \cong \text{O'N}{:}2.$$

3 Generators of higher order

Lemma 2.1 (suitably modified) can still be used when the symmetric generators are of higher order, but it is usually more profitable to consider what the subgroups $\langle t_i, t_j \rangle$ can be in a finite image. These groups often have several automorphisms visible in the control subgroup, which restricts the possibilities. A generic example for linear fractional groups has significant sporadic ramifications:

Example 3.1 Let λ be a primitive element in the multiplicative group of the field $\mathbb{Z}_p$, p prime. As is well known, see for example Dickson [16],

$$t_1 \equiv \tau \mapsto \tau + 1, x \equiv \tau \mapsto \lambda\tau \text{ and } y \equiv \tau \mapsto \frac{-1}{\tau}$$

generate the group $PGL_2(p)$. Taking $\langle x, y \rangle \cong D_{2(p-1)}$ as control subgroup $\mathcal{N}$ it is readily shown that

$$\frac{p^{\star 2} : D_{2(p-1)}}{\left[\begin{pmatrix} . & 1 \\ 1 & . \end{pmatrix} t_1 \right]^3} \cong PGL_2(p), \text{ where } D_{2(p-1)} \cong \left\langle \begin{pmatrix} \lambda & . \\ . & \lambda^{-1} \end{pmatrix}, \begin{pmatrix} . & 1 \\ 1 & . \end{pmatrix} \right\rangle.$$

If we take $\mathcal{N} = \langle x^2, y \rangle \cong D_{p-1}$ instead, then the same relator yields $PSL_2(p)$ or $2 \times PSL_2(p)$ in the cases $p \equiv 3 \pmod 4$ and $p \equiv 1 \pmod 4$ respectively.

Sporadic simple groups can be obtained directly in this manner. For example,

$$\frac{7^{\star 2} : (3 \times D_8)}{\left[\begin{pmatrix} . & 3 \\ 3 & . \end{pmatrix} t_1 \right]^7} \cong HJ{:}2,$$

where

$$3 \times D_8 = \left\langle \begin{pmatrix} 2 & . \\ . & 2 \end{pmatrix}, \begin{pmatrix} 1 & . \\ . & -1 \end{pmatrix}, \begin{pmatrix} . & 1 \\ 1 & . \end{pmatrix} \right\rangle,$$

and

$$\frac{7^{\star 2} : D_{12}}{\left[\begin{pmatrix} . & 1 \\ 1 & . \end{pmatrix} t_1 \right]^5, \left[\begin{pmatrix} . & 3 \\ 5 & . \end{pmatrix} t_1 t_2^{-1} t_1 \right]^3} \cong J_1,$$

where

$$D_{12} = \left\langle \begin{pmatrix} 5 & . \\ . & 3 \end{pmatrix}, \begin{pmatrix} . & 1 \\ 1 & . \end{pmatrix} \right\rangle.$$

Note that in the latter case two relators are necessary for a perfect image since $\mathcal{P}/\mathcal{P}' \cong V_4$.

3.1 Monomial modular representations

As our notation indicates, seeing how a group can act as control subgroup in a progenitor amounts to constructing a monomial modular representation for it over a suitable finite field. The two matrices

$$
x = \begin{bmatrix} . & 1 & . & . \\ 1 & . & . & . \\ . & . & 1 & . \\ . & . & . & -1 \end{bmatrix}, y = \begin{bmatrix} -1 & . & . & . \\ . & . & -1 & . \\ . & . & . & -1 \\ . & -1 & . & . \end{bmatrix},
$$

written over $\mathbb{Z}_3$, yield a 3-modular monomial representation of the group $2^{\cdot}S_4^+$ (see [13]) and so define a unique progenitor $\mathcal{P} \cong 3^{*4}{:}2^{\cdot}S_4^+$. Factoring this by the relator $(xt_1)^5$, which we may write as $[(1\ 2)(4\ \bar{4})t_1]^5$, yields the smallest Mathieu group M_{11}.

3.2 The Held group

Sporadic groups will generally follow when we start with an 'unusual' control subgroup, and so it is natural to consider $\mathcal{N} \cong 3^{\cdot}A_7$, the triple cover of the alternating group on 7 letters. In order to get a *faithful* monomial modular representation for $\mathcal{N}$ we induce a non-trivial linear representation of a subgroup isomorphic to $3 \times L_2(7)$, of index 15, up to $\mathcal{N}$. Clearly the field in which we work must possess cube roots of unity, and so we choose $\mathbb{Z}_7$. This defines a progenitor

$$\mathcal{P} \cong 7^{*15}{:}3^{\cdot}A_7,$$

but we prefer to 'double up' to

$$\mathcal{P} \cong 7^{*(15+15)}{:}3^{\cdot}S_7,$$

in which a 'central' element of order 3 squares one set of fifteen 7-elements, while fourth-powering the other. Now, when S_7 acts on $30 = (15 + 15)$ letters, the $L_2(7)$ stabilizing a point has orbits 1, 14, 7 and 8, and the subgroup isomorphic to S_4 fixing a further point in the 7-orbit commutes with a unique involution in S_7, π say, which interchanges these two fixed points. The subgroup generated by the corresponding symmetric generators r_1, s_1 possesses the D_6 of automorphisms described in Example 3.1 with $p = 7$, and factoring by $(\pi r_1)^3$ yields the image $L_2(7)$. Factoring the whole progenitor by this same relator (and noting that π has three preimages in $\mathcal{N}$) gives

$$\frac{7^{*(15+15)}{:}3^{\cdot}S_7}{(\pi r_1)^3} \cong He,$$

the Held simple group. The outer automorphism may be realised by adjoining an involution which commutes with $\mathcal{N}$ and inverts all 30 symmetric generators. For further details the reader is referred to [13].

3.3   The Harada–Norton group

Now take $\mathcal{N} \cong 2\dot{}HS$, the double cover of the Higman–Sims group. Then $\mathcal{N}$ has a subgroup U to index 176 of shape $(2 \times U_3(5))\dot{}2$, which contains outer elements normalizing the copy of $U_3(5)$ and squaring to the central involution. Thus $H/H' \cong C_4$. If we induce a faithful linear representation of this group up to $\mathcal{N}$ we obtain a 176-dimensional monomial modular representation of $\mathcal{N}$ over any field which possesses fourth roots of unity, such as $\mathbb{Z}_5$. As in the previous case we double up to obtain a progenitor

$$\mathcal{P} \cong 5^{*(176+176)}{:}2\dot{}HS{:}2.$$

As we saw in the construction of $HS{:}2$, the point stabilizer in the action on $352 = (176 + 176)$ letters has orbits 1, 175, 50 and 126. The subgroup of this stabilizer fixing a further point in the 50 orbit is isomorphic to S_7; it centralizes a unique involution in $HS{:}2$ which interchanges the two fixed points. If π denotes one of its two preimages in $\mathcal{N}$ and r_1, s_1 the two fixed symmetric generators of order 5, then as in Example 3.1 with $p = 5$ the relator $(\pi r_1)^3$ implies $\langle r_1, s_1 \rangle \cong A_5$. Factoring the whole progenitor by this relator gives

$$\frac{5^{*(176+176)}{:}2\dot{}HS.2}{(\pi r_1)^3} \cong HN,$$

the Harada–Norton simple group. If we extend our control subgroup to a group of shape $4\dot{}HS{:}2$ (a central product of a cyclic group C_4 with $2\dot{}HS$, extended by an involution normalizing the $2\dot{}HS$ and inverting the central elements of order 4), by adjoining an element of order 4 which squares one set of 176 symmetric generators and cubes the other, we obtain the group $\mathrm{Aut}(HN) \cong HN{:}2$. The action of HN on the cosets of $2\dot{}HS.2$ is described by Ivanov, Linton, Lux, Saxl and Soicher [19] as are many of the graphs relevant to this work.

3.4   Non-cyclic symmetric generators

Of course there is no need to restrict our monomial representations to being over prime fields. Indeed, when we require cube roots of unity we can take GF_4, the galois field of order 4, rather than $\mathbb{Z}_7$. The cyclic groups T_i are then replaced by copies of the Klein fourgroup labelled $V_i = \langle t_{i1}, t_{i2}, t_{i3} \rangle$ and we write a progenitor as $(2^2)^{*n}{:}\mathcal{N}$. As an example consider

$$\mathcal{P} \cong (2^2)^{*6}{:}3\dot{}A_6,$$

where a central element of order 3 fixes each of the six fourgroups while cycling its non-trivial elements. The control subgroup $\mathcal{N} \cong 3\dot{}A_6$ contains a unique class of elements of order 4, and the element $\pi \sim (1\ 2\ 3\ 4)(5\ 6)$ is a uniquely defined element in this class. Bray has demonstrated that if we factor $\mathcal{P}$ by $(\pi t_{11})^6$ we obtain the Hall–Janko group. On the other hand if $\sigma \sim (1\,2\,3\,4\,5)$ is an element of order 5 in $\mathcal{N}$, then factoring $\mathcal{P}$ by $(\sigma t_{11})^6$ yields $G_2(4)$.

It is now apparent that groups whose multipliers have order divisible by 3 are excellent candidates to be control subgroups over the fields GF_4 and $\mathbb{Z}_7$. The triple cover $3{\cdot}Suz$ of the Suzuki group contains a class of subgroups isomorphic to $3 \times G_2(4)$ and so we obtain a progenitor

$$\mathcal{P} \cong (2^2)^{\star 1782}{:}3{\cdot}Suz.$$

Now the outer automorphism of $3{\cdot}Suz$ inverts central elements of order 3; elements in the outer half of the group $3{\cdot}Suz{:}2$ can thus act as permutations of the 1782 Klein fourgroups followed by the *field automorphism* σ of GF_4. So an involution in the outer half, together with the central element of order 3, complete a Klein fourgroup V_i, which it fixes, to a copy of S_4. With this understanding we extend the progenitor to

$$\mathcal{P} \cong (2^2)^{\star 1782}{:}3{\cdot}Suz{:}2,$$

which can readily be seen to contain a perfect subgroup to index 2. Now the centralizer in $\mathcal{N}$ of one of the fourgroups is isomorphic to $G_2(4)$, with orbits $1 + 416 + 1365$. Centralizing a further fourgroup in the 416-orbit is a subgroup isomorphic to HJ whose centralizer in $\mathcal{N}$ is a copy of S_3. Elements of order 3 in this centralizer cycle the involutions in each of the two fixed fourgroups, while its involutions (which lie *outside* $\mathcal{N}'$) interchange them and apply the field automorphism σ. We thus seek an image of

$$(2^2)^{\star 2}{:}S_3 \text{ where } S_3 \cong \langle x = \begin{pmatrix} \omega & \cdot \\ \cdot & \omega \end{pmatrix}, y = \begin{pmatrix} \cdot & 1 \\ 1 & \cdot \end{pmatrix} \sigma \rangle.$$

Factoring this by the relator $(yt_{11})^3$ is easily seen to give the image A_5 (with for example $x = (3\,4\,5), y = (1\,2)(4\,5), t_{11} = (1\,3)(4\,5)$). If the progenitor $\mathcal{P}$ is factored by a corresponding relator we obtain the largest Conway group Co_1. This construction is being investigated by Mohamed Sayed.

3.5 The Fischer groups

The Fischer groups Fi_{22}, Fi_{23} and Fi_{24} are generated by a class of involutions called *transpositions* which have the property that the product of any two of them has order 1, 2 or 3. Indeed, in each case the action by conjugation on the transpositions is rank 3 with the two non-trivial orbits consisting of those elements which generate a copy of V_4 with the fixed transposition, and those which generate a copy of S_3. Clearly elements in the former orbit commute with the fixed transposition, and so cannot generate the whole group. However, the $2n$ elements in the latter orbit (which are *paired* under conjugation by the fixed transposition) do generate. We could take as progenitor $2^{\star 2n}{:}\mathcal{N}$, where $\mathcal{N}$ is the centralizer of a transposition, but prefer to take $\mathcal{P} \cong 3^{\star n}{:}\mathcal{N}$, where each symmetric generator is a product of the two involutions in a pair. This leads us to the homomorphisms

$$3^{\star 1408}{:}2 \dot{} U_6(2) \;\; \mapsto \;\; Fi_{22},$$
$$3^{\star 14080}{:}2 \dot{} Fi_{22} \;\; \mapsto \;\; Fi_{23},$$
$$3^{\star 137632}{:}Fi_{23} \;\; \mapsto \;\; Fi'_{24},$$

which are being investigated by John Bray. Although the numbers involved are large, the low rank facilitates construction. In the third case the outer automorphism is achieved by adjoining a transposition which commutes with $\mathcal{N}$ and inverts all the symmetric generators.

4   Larger groups

The claims made so far in this paper can be verified either by the manual double coset enumeration techniques described in [11, 12, 14], or by mechanical Todd–Coxeter single coset enumeration using for instance the implementation due to Havas in the MAGMA package, see [3]. The larger sporadic groups are out of range of such methods and more sophisticated geometric techniques are required.

4.1   The Fischer–Griess Monster

As stated at the bottom of page 230 of the ATLAS the BiMonster is in fact generated by 26 involutions corresponding to the points and lines of the projective plane of order 3, yielding a Coxeter diagram in which the 13 points and 13 lines are joined by incidence. That is to say, $M \wr 2$ is an image of

$$\frac{2^{\star(13+13)}{:}(L_3(3){:}2)}{(p_1 p_2)^2, (pl)^2 \text{ for } p \notin l, (pl)^3 \text{ for } p \in l}.$$

(Note that, since our control subgroup interchanges points and lines, the relator $(l_1 l_2)^2$ is implied.) Consider now $\mathcal{N} \cong 3 \dot{} Fi'_{24}$, the triple cover of the derived group of the largest Fischer group; let $f = |Fi'_{24}|$, and $h = |He|$, the order of the Held group. $\mathcal{N}$ possesses a subgroup $H \cong 3 \times He{:}2$ with $H/H' \cong C_6$. Take a faithful linear representation of this cyclic group over the field $\mathbb{Z}_7$ and induce it up to $\mathcal{N}$ to obtain a 7-modular monomial representation of $\mathcal{N}$ of degree $f/2h$. As in previous examples we 'double up' to obtain a progenitor

$$\mathcal{P} \cong 7^{\star(\frac{f}{2h}+\frac{f}{2h})}{:}3 \dot{} Fi_{24},$$

in which a 'central' 3-element squares one set of $f/2h$ symmetric generators while fourth powering the other set. Now the centralizer in $\mathcal{N}$ of a symmetric generator t_1, which is isomorphic to He, possesses a subgroup isomorphic to $S_4(4){:}2$ which centralizes another symmetric generator, s_1 say, *in the other set*. The centralizer in $\mathcal{N}$ of this subgroup is a copy of S_3 generated by the centre of $\mathcal{N}'$ together with three involutions lying outside $\mathcal{N}'$, which thus interchange $\langle t_1 \rangle$ and $\langle s_1 \rangle$. If π is one of these involutions the relator $(\pi t_1)^3$ ensures that $\langle t_1, s_1 \rangle \cong PSL_2(7)$, as in Example 3.1 with $p = 7$. We *conjecture* that

$$\frac{7^{\star(\frac{f}{2h}+\frac{f}{2h})}{:}3^{\cdot}Fi_{24}}{(\pi t_1)^3} \cong M.$$

Again we can consider $\mathcal{N} \cong 2^{\cdot}B$, the double cover of the Baby Monster; let $b = |B|$ and $r = |HN|$, the order of the Harada–Norton group. By restricting the 4371-character of B to $HN{:}2$ we see that the unique outer class of involutions in $HN{:}2$ fuses to class $2C$ in B. Since pre-images of elements in this class have order 4 we see that a subgroup of shape $HN{:}2$ in B lifts to a subgroup $H \cong (2 \times HN)^{\cdot}2$ in $2^{\cdot}B$, where $H/H' \cong C_4$. Take a faithful linear representation of this cyclic group over $\mathbb{Z}_5$ and induce it up to $\mathcal{N}$ to obtain a 5-modular monomial representation of $\mathcal{N}$ of degree $b/2r$. This defines the progenitor

$$\mathcal{P} \cong 5^{\star\frac{b}{2r}}{:}2^{\cdot}B,$$

in which the central involution of the control subgroup inverts all the symmetric generators. Now $\mathcal{N}^1 = C_{\mathcal{N}}(t_1) \cong HN$, and there exists a further symmetric generator t_2 such that $\mathcal{N}^{12} = C_{\mathcal{N}}(\langle t_1, t_2 \rangle) \cong U_3(8){:}3$. Furthermore $C_{\mathcal{N}}(\mathcal{N}^{12}) \cong V_4$ generated by the central element of $\mathcal{N}$, which inverts t_1 and t_2, and an involution π which interchanges them. Similarly, there exists a symmetric generator t_3, say, such that $\mathcal{N}^{13} \cong A_{12}$, and $C_{\mathcal{N}}(\mathcal{N}^{13}) \cong V_4$ generated by the central element of $\mathcal{N}$ again and an involution σ which interchanges them. As in Example 3.1 with $p = 5$, factoring by $(\pi t_1)^3$ ensures that $\langle t_1, t_2 \rangle \cong A_5$, and factoring by $(\sigma t_1)^3$ ensures that $\langle t_1, t_3 \rangle \cong A_5$. We *conjecture* that

$$\frac{5^{\star\frac{b}{2r}}{:}2^{\cdot}B}{(\pi t_1)^3, (\sigma t_1)^3} \cong M,$$

although either of these additional relators may be redundant. Finally with $\mathcal{N} \cong 2^{\cdot}B$, let $t = |Th|$, the order of the Thompson group. Inducing a non-trivial linear representation over $\mathbb{Z}_3$ of the subgroup $2 \times Th$ up to $\mathcal{N}$ we obtain a 3-modular monomial representation of $\mathcal{N}$ of degree b/t. This defines the progenitor

$$\mathcal{P} \cong 3^{\star\frac{b}{t}}{:}2^{\cdot}B,$$

which also has M as a homomorphic image.

5   Conclusion

In this paper we have described a method of constructing new groups by using a familiar group as control group over a suitable field. If the control subgroup has an unusual permutation action (such as $L_2(11)$ on 11 letters) or an exceptional multiplier (such as $3 \cdot A_7$) then we are frequently led to a sporadic simple group, and so in a sense the construction 'explains' the existence of the sporadic groups.

Although we have not described the process here, the double coset construction referred to can readily be performed by hand for the smaller sporadic groups. Elements of the new group can then be represented as πw, where $\pi \in \mathcal{N}$ and w is a word in the symmetric generators, and it is possible to work with elements of the group so represented (see our work with Hasan [15]).

Finally I should like to acknowledge the considerable contribution made by Bray, Hammas and Jabbar [1, 17, 20] to this programme.

References

[1] J. N. Bray, *Symmetric presentations of finite groups*, M. Phil. (Sc., Qual.) thesis, Birmingham, 1995.

[2] C. M. Campbell, G. Havas, S. A. Linton and E. F. Robertson, Symmetric presentations and orthogonal groups, *these proceedings*, pp. 1–10.

[3] J. Cannon and C. Playoust, *An introduction to MAGMA*, School of Mathematics and Statistics, University of Sydney, 1993.

[4] M. D. E. Conder, R. A. Wilson and A. J. Woldar, The symmetric genus of sporadic groups, *Proc. Amer. Math. Soc.* **116** (1992), 653–663.

[5] J. H. Conway, Y_{555} and all that, *Groups, combinatorics and geometry (eds. M. W. Liebeck and J. Saxl)*, pp. 22–23. Cambridge University Press, 1992.

[6] J. H. Conway, R. T. Curtis, S. P. Norton, R. A. Parker and R. A. Wilson, *An* ATLAS *of Finite Groups*, Oxford University Press, 1985.

[7] H. S. M. Coxeter and W. O. J. Moser, *Generators and relations for discrete groups*, 4th edition, Springer-Verlag, 1984.

[8] R. T. Curtis, Natural constructions of the Mathieu groups, *Math. Proc. Cambridge Philos. Soc.* **106** (1989), 423–29.

[9] R. T. Curtis, Geometric interpretations of the 'natural' generators of the Mathieu groups, *Math. Proc. Cambridge Philos. Soc.* **107** (1990), 19–26.

[10] R. T. Curtis, Symmetric presentations I: Introduction, with particular reference to the Mathieu groups M_{12} and M_{24}, *Groups, combinatorics and geometry (eds. M. W. Liebeck and J. Saxl)*, pp 380–396. Cambridge University Press, 1992.

[11] R. T. Curtis, Symmetric presentations II: The Janko group J_1, *J. London Math. Soc. (2)* **47** (1993), 294–308.

[12] R. T. Curtis, Symmetric generation of the Higman–Sims group, *J. Algebra* **171** (1995), 567–586.

[13] R. T. Curtis, Monomial modular representations and construction of the Held group, *J. Algebra* **184** (1996), 1205–1227.

[14] R. T. Curtis, A. Hammas and J. N. Bray, A systematic approach to symmetric presentation I: involutory generators, *Math. Proc. Cambridge Phil. Soc.* **119** (1996), 23–34.

[15] R. T. Curtis and Z. Hasan, Symmetric representation of elements of the Janko group J_1, *J. Symbolic Comput.* **22** (1996), 201–214.

[16] L. E. Dickson, *Linear groups with an exposition of the Galois field theory*, Teubner, Leipzig, 1901; reprinted, Dover, 1958.

[17] A. Hammas, *Symmetric presentations of some finite groups*, Ph. D. thesis, Birmingham, 1991.

[18] A. A. Ivanov, Geometric characterization of the Monster, *Groups, combinatorics and geometry (eds. M. W. Liebeck and J. Saxl)*, pp. 46–62. Cambridge University Press, 1992.

[19] A. A. Ivanov, S. A. Linton, K. Lux, J. Saxl and L. H. Soicher, Distance-transitive representations of the sporadic groups, *Comm. Algebra* **23** (1995), 3379–3427.

[20] A. Jabbar, *Symmetric presentations of subgroups of the Conway groups and related topics*, Ph. D. thesis, Birmingham, 1992.

[21] D. L. Johnson, *Presentations of groups*, London Math. Soc. Lecture Notes 22, Cambridge University Press, 1976.

[22] G. A. Jones, Characters and surfaces: a survey, *these proceedings*, pp. 90–118.

[23] G. A. Jones and D. Singerman, *Complex functions—an algebraic and geometric viewpoint*, Cambridge University Press, 1987.

[24] Martin W. Liebeck and Aner Shalev, Probabilistic methods in simple groups, *these proceedings*, pp. 163–173.

[25] S. P. Norton, Constructing the Monster, in *Groups, combinatorics and geometry (eds. M. W. Liebeck and J. Saxl)*, pp. 63–76. LMS Lecture Note Series **165**, Cambridge University Press, 1992.

[26] E. F. Robertson and C. M. Campbell, Symmetric presentations, in *Proceedings of the 1987 Singapore conference on finite groups (eds. K. N. Cheng and Y. K. Leung)*, pp. 497–506. Walter de Gruyter, 1989.

[27] L. H. Soicher, *Presentations of some finite groups*, Ph. D. thesis, Cambridge, 1985.

[28] M. C. Tamburini and J. S. Wilson, On the (2,3)-generation of some classical groups, II, *J. Algebra* **176** (1995), 667–680.

[29] J. S. Wilson, Economical generating sets for finite simple groups, in *Groups of Lie type and their geometries (eds. W. M. Kantor and L. Di Martino)*, Cambridge University Press, 1995.

Harish-Chandra theory, q-Schur algebras, and decomposition matrices for finite classical groups

Gerhard Hiss

Abstract

We give a survey on the present state of knowledge in the representation theory of finite groups of Lie type in the non-defining characteristic case. A systematic use of Harish-Chandra induction leads to the definition of q-Schur algebras. Their representation theory yields explicit information on the decomposition matrices for classical groups.

1 Introduction

This paper is an extended version of my talk given at the conference. It is intended as a brief survey on recent developments in the representation theory of finite groups of Lie type in non-defining characteristic.

Let $G = G(q)$ denote a finite group of Lie type. Examples are classical groups such as $\mathrm{GL}_n(q)$ or $\mathrm{Sp}_n(q)$, but also exceptional groups such as $E_8(q)$. Let k be an algebraically closed field of characteristic $\ell \geq 0$ and assume that ℓ does not divide q. We are interested in the category kG-**mod** of finitely generated left kG-modules.

If $\ell = 0$ (or, more generally, $\ell \nmid |G|$, the order of G), every finite dimensional kG-module is completely reducible. Hence it suffices to find the simple kG-modules. In 1955, Green determined the ordinary character table for the general linear groups $\mathrm{GL}_n(q)$ [26], thus solving the problem for these groups. Beginning with the seminal paper [7] of Deligne and Lusztig in 1976, Lusztig arrived at a complete classification of the simple kG-modules for all finite groups of Lie type G (see Lusztig [36]).

The representation theory for these groups in characteristics $\ell \neq 0$, $\ell \nmid q$, started in 1982 with the fundamental paper [16] by Fong and Srinivasan, in which these authors determined the ℓ-blocks of the general linear groups and the unitary groups. This work was continued in [18] with the description of the ℓ-blocks in the other classical groups. Since then, their work has been put into a conceptual framework and the ℓ-blocks of all finite groups of Lie type arising from algebraic groups with connected centre, have been found up to a few exceptions (see the work of Broué, Cabanes, Enguehard, Malle and Michel [4, 3, 5]).

The computation of decomposition matrices was initiated by work of Fong and Srinivasan [17] and by Dipper [9]. Much work has been done on the decomposition matrices for classical groups in the cyclic defect case [19], and, without restrictions on the defect, for the general linear groups. Substantial progress was made in the latter case with the introduction of the q-Schur algebra by Dipper and James [13], leading to the determination of the decomposition matrices for $\mathrm{GL}_n(q)$ for all $n \leq 10$ by James [34].

In this paper we indicate how the results of Dipper and James can be extended to other classical groups. This is achieved under additional conditions on ℓ. The methods include the introduction of q-Schur algebras associated to Weyl groups of types B and D.

2   Harish-Chandra theory

In this section we present some of those aspects of Harish-Chandra theory which can be formulated and proved entirely in the framework of groups with split BN-pairs.

Let G be a finite group with a split BN-pair of characteristic p, and let k be a field of characteristic $\ell \geq 0$, $\ell \neq p$. Let L be a Levi subgroup of G and P a parabolic subgroup with Levi complement L, i.e., $P = UL$, where U is the largest normal p-subgroup of P, and $U \cap L = \{1\}$. There are two functors

$$R_{L,P}^G : kL\text{-}\mathbf{mod} \to kG\text{-}\mathbf{mod}, \quad M \mapsto \mathrm{Ind}_P^G(M),$$

and

$${}^*R_{L,P}^G : kG\text{-}\mathbf{mod} \to kL\text{-}\mathbf{mod}, \quad M \mapsto \mathrm{Inv}_U(M).$$

Here, Ind_P^G is the usual induction functor from P to G, and M is considered as a kP-module via inflation. Also, $\mathrm{Inv}_U(M)$ denotes the U-invariants of the kG-module M, considered as a kL-module. The first one of these functors is called *Harish-Chandra induction*, the second one *Harish-Chandra restriction* or *truncation*. They constitute a pair of mutually adjoint functors, are transitive, and satisfy a Mackey formula (see Dipper and Fleischmann [11]).

The next theorem is of fundamental importance in Harish-Chandra theory.

Theorem 2.1 (Deligne, Dipper/Du, Howlett/Lehrer) *Let P and Q be two parabolic subgroups of G such that L is a Levi complement in P and in Q. Then the two functors $R_{L,P}^G$ and $R_{L,Q}^G$ are naturally isomorphic. The same is then true for their adjoint functors ${}^*R_{L,P}^G$ and ${}^*R_{L,Q}^G$.*

This theorem was first proved in the case of $\ell = 0$ by Deligne (see Lusztig and Spaltenstein [37]), and in the case of $\ell > 0$ by Dipper and Du [10] and, independently, by Howlett and Lehrer [33]. Because of this result we may, and will, omit the subscript P from the notation for Harish-Chandra induction and restriction. One consequence of Theorem 2.1 is the fact that $R_L^G(X) \cong R_{L'}^G(X')$ if the pairs (L, X) and (L', X') are conjugate in G.

Definition 2.2 (Harish-Chandra, Dipper/Fleischmann) A kG-module M is called *cuspidal*, if ${}^*R_L^G(M) = \{0\}$ for all Levi subgroups L properly contained in G.

The concept of cuspidal modules was introduced by Harish-Chandra in [30, Section 3] in the case of $\ell = 0$, and by Dipper and Fleischmann in general [11]. The following theorem gives structural information about modules Harish-Chandra induced from simple cuspidal modules.

Theorem 2.3 (Geck/Hiss/Malle, Geck/Hiss) *Let L be a Levi subgroup of G. Suppose that X is a simple cuspidal kL-module, and that Y is an indecomposable direct summand of $R_L^G(X)$. Then*

(a) *Y has a unique simple quotient and a unique simple submodule; these are isomorphic to each other.*

Moreover, if L' is another Levi subgroup, X' a cuspidal simple kL'-module and Y' an indecomposable direct summand of $R_{L'}^G(X')$, we have

(b) *Y is isomorphic to Y' if and only if the simple quotient of Y is isomorphic to the simple quotient of Y'.*

(c) *If Y and Y' are isomorphic, then the pairs (L, X) and (L', X') are conjugate in G.*

We give a few references for the proof of the above theorem. The first part is proved in [24, Theorem 2.9] building upon earlier work in [25]. The last two statements follow from [25, Theorem 2.4] and [31, Theorem 5.8].

Note the following important corollary, which gives a classification of the simple kG-modules.

Corollary 2.4 (Geck/Hiss/Malle [25, Theorem 2.4]) *There is a bijection between the set of isomorphism classes of simple kG-modules and the set of triples (L, X, ψ), where L is a Levi subgroup of G, X is a simple cuspidal kL-module and ψ a simple $\mathrm{End}_{kG}(R_L^G(X))$-module (where the pair (L, X) has to be taken modulo conjugation in G and ψ modulo isomorphism).*

Let us observe how this corollary is derived from the statements of Theorem 2.3. Start with a simple kG-module M. Let L be a Levi subgroup, minimal with the property that $^*R_L^G(M) \neq \{0\}$. Take a simple kL-submodule X of $^*R_L^G(M)$. Then X is cuspidal by minimality of L and the transitivity of Harish-Chandra restriction. It follows from adjointness that M is a quotient of $R_L^G(X)$. By Theorem 2.3, M determines an indecomposable direct summand Y of $R_L^G(X)$, up to isomorphism. By Fitting's Lemma, Y corresponds to a principal indecomposable module of $\mathrm{End}_{kG}(R_L^G(X))$, which in turn determines a simple module for this algebra. By Part (c) of the theorem, Y (and hence M) determine the pair (L, X) up to conjugation in G.

The corollary allows one to partition the set of isomorphism classes of simple kG-modules into Harish-Chandra series. Two simple kG-modules M and M' belong to the same Harish-Chandra series, if and only if they correspond to pairs (L, X, ψ) and (L', X', ψ') such that (L, X) and (L', X') are conjugate in G.

We next turn to the question of finding the structure of the endomorphism algebras $\mathrm{End}_{kG}(R_L^G(X))$, where X is a simple cuspidal kL-module. We put $H(L, X) := \mathrm{End}_{kG}(R_L^G(X))$. Let $W(L, X)$ be the image of the inertia subgroup of X in $(N_G(L) \cap N)L/L$. Then, by a result of Dipper and Fleischmann [11], the dimension of $H(L, X)$ equals the order of $W(L, X)$.

Theorem 2.5 (Howlett/Lehrer, Geck/Hiss/Malle) (a) *The inertia group $W(L, X)$ has a semidirect product decomposition*

$$W(L, X) = R(L, X)C(L, X) \tag{1}$$

with normal subgroup $R(L, X)$ and complement $C(L, X)$. Moreover, $R(L, X)$ is a Coxeter group.

(b) *$H(L, X)$ is a $C(L, X)$ graded k-algebra, i.e., there are subspaces H_x of $H(L, X)$, $x \in C(L, X)$, such that*

$$H(L, X) = \bigoplus_{x \in C(L, X)} H_x,$$

H_1 is a subalgebra of $H(L, X)$ and $H_x H_y = H_{xy}$ for all $x, y \in C(L, X)$.

(c) *H_1 is an Iwahori–Hecke algebra associated to the Coxeter group $R(L, X)$, with multiplication possibly twisted by a 2-cocycle.*

Special cases of this theorem for $\mathrm{char}(k) = 0$ were proved by Iwahori, Harish-Chandra, Lusztig and Kilmoyer. The general case was settled by Howlett and Lehrer in characteristic 0 [32] and, building on their work, by Geck, Hiss and Malle [25, Section 3] in positive characteristic.

If $R_L^G(X)$ has an indecomposable direct summand with multiplicity 1, or if X can be extended to its inertia subgroup in $(N_G(L) \cap N)L$, then the 2-cocycle appearing in Theorem 2.5(c) is trivial (see [25, Corollary 3.13]).

We recall the definition of an Iwahori–Hecke algebra associated to a Coxeter system. Let S be a finite set and for $s, s' \in S$, let $m(s, s') \in \mathbb{Z}$, such that $m(s, s') = m(s', s)$, $m(s, s) = 1$ and $m(s, s') \geq 2$ for $s \neq s'$. The Coxeter group associated to the matrix of the $m(s, s')$ is the group with the presentation

$$W = \langle s \in S \mid \overbrace{ss's\cdots}^{m(s,s')} = \overbrace{s'ss'\cdots}^{m(s,s')}, s \neq s' \in S, s^2 = 1, s \in S\rangle.$$

Now let R be a commutative ring with 1, and let $q_s, s \in S$, be invertible elements of R such that $q_s = q_{s'}$, whenever s and s' are conjugate in W. Let us denote the vector of the q_s by $\mathbf{q}$. The Iwahori–Hecke algebra associated to these data is the associative R-algebra presented by

$$H_{R,\mathbf{q}}(W) := \left\langle T_s, s \in S \mid \begin{array}{c} \overbrace{T_sT_{s'}T_s\cdots}^{m(s,s')} = \overbrace{T_{s'}T_sT_{s'}\cdots}^{m(s,s')}, s \neq s' \in S, \\ T_s^2 = q_sT_1 + (q_s - 1)T_s, s \in S \end{array}\right\rangle,$$

where T_1 denotes the identity of the algebra. The elements q_s, $s \in S$, of R are called the *parameters* of $H_{R,\mathbf{q}}(W)$.

We can now formulate three fundamental problems in the representation theory of finite groups with spit BN-pairs in non-defining characteristic.

(A) Determine the pairs (L, X) of Levi subgroups L of G having a simple cuspidal kL-module X (up to conjugation).

(B) For all pairs (L, X) as in (A), determine the structure of $H(L, X) = \mathrm{End}_{kG}(R_L^G(X))$, that is, determine the inertia subgroup $W(L, X)$, the semidirect product decomposition (1), and the values of the parameters of the Iwahori–Hecke algebra associated to the Coxeter group $R(L, X)$.

(C) For all pairs (L, X) as in (A), find the indecomposable direct summands of $R_L^G(X)$ or, equivalently, the principal indecomposable modules of $H(L, X)$.

3   Cuspidal modules of classical groups

Let G be a finite group of Lie type and let k be a splitting field for G of characteristic 0. In this case all three problems at the end of the preceding section have been completely solved. By far the hardest one was to find the simple cuspidal kG-modules. This problem has been settled by Lusztig in a long series of papers. An excellent source for an account of Lusztig's results is the book by Carter [6]. The endomorphism algebras of Harish-Chandra induced cuspidal simple modules are all known. It turns out that there is no 2-cocycle twisting the multiplication in Theorem 2.5(c) (see Geck [21, Corollary 2]). Moreover, the Iwahori–Hecke algebra H_1 is isomorphic to the group algebra $kR(L, X)$ by the Tits deformation theorem. This observation solves (C).

To describe the knowledge in the case of char$(k) = \ell > 0$, we introduce a splitting ℓ-modular system $(K, \mathcal{O}, k)$ for G. That is, $\mathcal{O}$ is a complete, discrete valuation ring with field of fractions K of characteristic 0, and residue class field k of characteristic $\ell > 0$. Furthermore, K and k are splitting fields for G and all of its subgroups. We restrict our attention to the classical groups. By a classical group we shall understand here one of the following groups:

(a) The general linear group $\mathrm{GL}_n(q)$, $n \geq 2$.

(b) The general unitary group $\mathrm{GU}_n(q)$, $n \geq 3$.

(c) The symplectic group $\mathrm{Sp}_n(q)$, $n \geq 4$, n even.

(d) The orthogonal group $\mathrm{SO}_n(q)$, $n \geq 7$ odd, q odd.

(e) The orthogonal group $\mathrm{SO}_n^+(q)$, $n \geq 8$, n even.

(f) The orthogonal group $\mathrm{SO}_n^-(q)$, $n \geq 8$, n even.

Suppose first that $G = \mathrm{GL}_n(q)$ for some prime power q, and that, as always, $\ell \nmid q$. By work of Dipper and James, all cuspidal simple kG-modules are known.

Theorem 3.1 (Dipper/James) *Let* $G = \mathrm{GL}_n(q)$. *A simple KG-module is cuspidal, if and only if its character is* $(-1)^{n+1}R_T^G(\vartheta)$, *where T is the Coxeter torus of G, ϑ is an irreducible character of T in general position (and R_T^G denotes Deligne–Lusztig induction).*

Every simple cuspidal kG-module is obtained from some cuspidal KG-module by reduction modulo ℓ. Two such cuspidal KG-modules with characters $(-1)^{n+1}R_T^G(\vartheta)$ *and* $(-1)^{n+1}R_T^G(\vartheta')$ *reduce to isomorphic kG-modules, if and only if ϑ and ϑ' have the same ℓ'-part.*

For a proof of this theorem see the papers by Dipper and James [9, 12].

The knowledge for the other classical groups is far less complete. To say more, we have to impose further conditions on the characteristic ℓ.

Definition 3.2 Let G be one of the classical groups of the list given above. Suppose first that G is neither a general linear nor a unitary group. A prime ℓ is called *linear* for G, if ℓ is odd, and if the order of q modulo ℓ is odd. Moreover, if $G = \mathrm{GL}_n(q)$, all primes are called *linear*, and, in case $G = \mathrm{GU}_n(q)$, ℓ is *linear* if the order of $-q$ modulo ℓ is even.

Theorem 3.3 (Geck/Hiss/Malle, Gruber/Hiss) *Let $G \neq \mathrm{GL}_n(q)$ be a classical group, and let ℓ be a linear prime for G. Then there is a bijection between the isomorphism types of the cuspidal simple KG-modules and the isomorphism types of the cuspidal simple kG-modules, given by reduction modulo ℓ.*

In the case of unipotent characters, the above theorem is due to Geck, Hiss and Malle [25] and Gruber [27]. A publication of the proof in the general case is in preparation and will appear elsewhere [29].

We close this section with a result which indicates the existence of many cuspidal simple kG-modules, which do not arise as reductions modulo ℓ of KG-modules.

Theorem 3.4 (Geck, Geck/Hiss/Malle, Gruber) *Let $G = \mathrm{GU}_n(q)$. The unipotent kG-modules have a natural parametrization by partitions of n.*

If ℓ is an odd prime dividing $q+1$, then a unipotent kG-module is cuspidal if and only if it corresponds to a partition μ whose consecutive parts differ by at most 1. (This is the same as saying that the partition conjugate to μ is 2-regular, i.e., has no equal parts).

The first part of the theorem on the parametrization follows from a theorem of Geck on the decomposition matrices of $\mathrm{GU}_n(q)$ [20]. The second part was proved in [25, Theorem 4.12] in case $\ell > n$ and by Gruber in general [28].

4   Decomposition numbers

To describe the matrices of decomposition numbers of a classical group we introduce some more notation. Let G be a finite group of Lie type of characteristic p, and let ℓ be a prime different from p. For a semisimple element s in the (finite) group G^* dual to G, let $\mathcal{E}(G, s)$ denote the rational series of characters corresponding to the G^*-conjugacy class of s (see Digne and Michel [8, p. 136]). Moreover, if the order of s is prime to ℓ, let

$$\mathcal{E}_\ell(G, s) := \bigcup_{t \in C_{G^*}(s)_\ell} \mathcal{E}(G, st).$$

Theorem 4.1 (Broué/Michel [4]) *$\mathcal{E}_\ell(G, s)$ is a union of ℓ-blocks of G.*

To find the decomposition matrix of G, it therefore suffices to determine the decomposition numbers of the characters in the sets $\mathcal{E}_\ell(G, s)$, where s runs over a system of representatives of the G^*-conjugacy classes of elements of order prime to ℓ.

Now let G be a classical group as in section 3, and suppose that ℓ is odd, if G is neither a linear nor a unitary group.

Theorem 4.2 (Geck/Hiss) *The characters of $\mathcal{E}(G, s)$ form a basic set for $\mathcal{E}_\ell(G, s)$. This means that every irreducible Brauer character of $\mathcal{E}_\ell(G, s)$ can be expressed uniquely as a $\mathbb{Z}$-linear combination of the restrictions of the characters of $\mathcal{E}(G, s)$ to the ℓ-regular classes of G.*

This is a special case of a result of Geck and Hiss [23, Theorem 5.1]. Moreover, if the decomposition numbers for the characters in $\mathcal{E}(G, s)$ are known, one can determine all decomposition numbers of $\mathcal{E}_\ell(G, s)$. They can be deduced from a knowledge of the Lusztig map R_L^G for various (non-split) Levi subgroups L of G.

Thus the problem of finding decomposition matrices for G is reduced to finding the square matrix of decomposition numbers for the characters in $\mathcal{E}(G, s)$. These square matrices have been given a nice interpretation in terms of q-Schur algebras by Dipper and James [13].

One way of defining the q-Schur algebra is as follows. Let R denote a commutative ring with 1, and let q be an invertible element of R. For a non-negative integer n let $\mathcal{H} := \mathcal{H}_{R,q}(S_n)$ denote the Iwahori–Hecke algebra over R with all parameters equal to q, associated to the symmetric group S_n on n letters. Let $\mathcal{P}(n)$ denote the set of partitions of n. For a partition λ of n, let S_λ denote the Young subgroup of S_n determined by λ. Finally, define $y_\lambda \in \mathcal{H}$ by

$$y_\lambda := \sum_{w \in S_\lambda} (-q)^{-\ell(w)} T_w. \tag{2}$$

Here, T_w is the basis element of $\mathcal{H}$ corresponding to w, and $\ell(w)$ denotes the length of $w \in S_n$.

Definition 4.3 (Dipper/James [13]) The R-algebra

$$\mathcal{S}_{R,q}(n) := \mathrm{End}_{\mathcal{H}} \left(\bigoplus_{\lambda \in \mathcal{P}(n)} y_\lambda \mathcal{H} \right)$$

with $\mathcal{H} = \mathcal{H}_{R,q}(S_n)$ is called q-*Schur algebra.*

Now let, as in section 3, $(K, \mathcal{O}, k)$ be a splitting ℓ-modular system for G. Decomposition matrices of $\mathcal{O}$-orders will always be with respect to this modular system. Suppose first that $G = \mathrm{GL}_n(q)$ and consider the set $\mathcal{E}(G, 1)$ of unipotent characters of G. They have a natural labelling by partitions of n, resulting from the fact that the unipotent modules of G are the composition factors of M, the permutation module of G over K on the cosets of a Borel subgroup. We have $\mathrm{End}_{KG}(M) = \mathcal{H}_{K,q}(S_n)$. The latter K-algebra is isomorphic to the group algebra KS_n by the Tits deformation theorem. We thus obtain a natural bijection between the unipotent characters of G and the simple KS_n-modules.

Since $\mathcal{H}_{K,q}(S_n)$ is semisimple, it is Morita equivalent to the q-Schur algebra $\mathcal{S}_{K,q}(n)$.

Theorem 4.4 (Dipper/James, [13]) *Let R be one of the fields K or k. The unipotent $R\mathrm{GL}_n(q)$-modules and the simple modules of $\mathcal{S}_{R,q}(n)$ can be parameterized by partitions of n such that the decomposition matrix of $\mathcal{S}_{\mathcal{O},q}(n)$ is the same (with respect to these parameterizations) as the decomposition matrix of $\mathcal{E}(\mathrm{GL}_n(q), 1)$.*

More generally, consider $\mathcal{E}_\ell(G, s)$ for some semi-simple ℓ'-element s of $G = \mathrm{GL}_n(q)$. Let $C = C_G(s)$. Then $C \cong \mathrm{GL}_{n_1}(q^{d_1}) \times \mathrm{GL}_{n_2}(q^{d_2}) \times \cdots$.

Theorem 4.5 (Dipper/James, [13]) *The decomposition matrix of $\mathcal{E}(G, s)$ is equal to the decomposition matrix of $\mathcal{S}_{\mathcal{O},q^{d_1}}(n_1) \otimes \mathcal{S}_{\mathcal{O},q^{d_2}}(n_2) \otimes \cdots$.*

Let v be an indeterminate, put $\mathcal{A} := \mathbb{Z}[v, v^{-1}]$ and $u := v^2$. Consider the generic Iwahori–Hecke algebra $\mathcal{H}_{\mathcal{A},u}(S_n)$ and the corresponding q-Schur algebra $\mathcal{S}_{\mathcal{A},u}(n)$. Let $U(\mathfrak{gl}_n)$ denote the quantized universal enveloping algebra over $\mathcal{A}$ corresponding to the Lie algebra $\mathfrak{gl}_n$. Beilinson, Lusztig and MacPherson have constructed a surjective homomorphism of $\mathcal{A}$-algebras from $U(\mathfrak{gl}_n)$ to an associative $\mathcal{A}$-algebra, free over $\mathcal{A}$ and of finite rank [1]. This quotient of $U(\mathfrak{gl}_n)$ turns out to be Morita equivalent to $\mathcal{S}_{\mathcal{A},u}(n)$ (see Du [15]). Thus, in a certain sense the representation theory of the q-Schur algebra is governed by the representation theory of the quantum group $U(\mathfrak{gl}_n)$.

By specializing the indeterminate u to an invertible element x of K, we obtain a ring homomorphism $\mathcal{H}_{\mathcal{A},u}(S_n) \to \mathcal{H}_{K,x}(S_n)$. Any simple $\mathcal{H}_{\mathbb{Q}(v),u}(S_n)$-module can be realized over the ring $\mathcal{A}$, and thus gives rise to an $\mathcal{H}_{K,x}(S_n)$-module. We thus obtain a decomposition map from the Grothendieck group of $\mathcal{H}_{\mathbb{Q}(v),u}(S_n)$-modules to the Grothendieck group of $\mathcal{H}_{K,x}(S_n)$-modules (see Geck, [22, Kapitel 4] for a general discussion of such decomposition maps). A decomposition map can, of course, also be defined for the q-Schur algebra $\mathcal{S}_{\mathcal{A},u}(n)$.

Suppose now that ξ is an eth root of unity in K for some positive integer e. Specializing u to ξ, we obtain a decomposition map from the Grothendieck group of $\mathcal{H}_{\mathbb{Q}(v),u}(S_n)$-modules to the Grothendieck group of $\mathcal{H}_{K,\xi}(S_n)$-modules. Let M_e denote the corresponding decomposition matrix, also called the Φ_e-modular decomposition matrix of $\mathcal{H}_{\mathbb{Q}(v),u}(S_n)$. Similarly, one can define a Φ_e-modular decomposition matrix N_e of $\mathcal{S}_{\mathbb{Q}(v),u}(n)$.

The following conjecture of James is relevant to the problem of finding the decomposition numbers of the general linear groups. For a positive integer q and a prime ℓ not dividing q, let $e(q, \ell)$ denote the smallest positive integer r such that ℓ divides $1 + q + q^2 + \cdots + q^{r-1}$. Now suppose that q is a power of a prime, ℓ a prime not dividing q, and $e := e(q, \ell)$. Then the conjecture of James can be stated as follows (see [34, Section 4]). Suppose that $n < e\ell$. Then the decomposition matrix of the unipotent characters of $\mathrm{GL}_n(q)$, in other words, the decomposition matrix of $\mathcal{S}_{\mathcal{O},q}(n)$, is equal to N_e, the Φ_e-modular decomposition matrix of $\mathcal{S}_{\mathbb{Q}(v),u}(n)$. Geck has shown [22, Satz 4.2.8] that the analogous statement is true for the decomposition matrix of the Iwahori–Hecke algebra $\mathcal{H}_{\mathcal{O},q}(S_n)$, provided ℓ is large enough (without giving an explicit bound).

The Φ_e-modular decomposition matrix M_e of the Iwahori–Hecke algebra $\mathcal{H}_{\mathbb{Q}(v),u}(S_n)$ is contained as a submatrix in the decomposition matrix N_e of $\mathcal{S}_{\mathbb{Q}(v),u}(n)$. In fact, M_e consists of those columns of N_e which correspond to e-regular partitions. There is a very beautiful conjecture of Lascoux, Leclerc and Thibon on the Φ_e-modular decomposition matrix of $\mathcal{H}_{\mathbb{Q}(v),u}(S_n)$ in terms of the canonical basis of an integrable highest weight representation of the quantized universal enveloping algebra $U(\widehat{\mathfrak{sl}}_e)$ corresponding to the affine Lie algebra $\widehat{\mathfrak{sl}}_e$ (for more details see [35]).

We next turn to the decomposition numbers of the other classical groups. To describe these for linear primes, we recall the definition of q-Schur algebras of type B (see Geck and Hiss [24]). Let W_n denote the Weyl group of type B_n, generated by fundamental reflections corresponding to the following Dynkin diagram.

Let S_n be the subgroup of W_n generated by $s_1, \ldots, s_{n-1}$. Let R be a commutative ring with 1 and let Q and q be two invertible elements of R. Denote by $\mathcal{H}$ the Iwahori–Hecke algebra $\mathcal{H} := \mathcal{H}_{R,Q,q}(W_n)$, where the parameter Q corresponds to the fundamental reflection t.

Definition 4.6 (Geck/Hiss [24]) The R-algebra

$$\mathcal{S}_{R,Q,q}(B_n) := \mathrm{End}_{\mathcal{H}}\left(\bigoplus_{\lambda \in \mathcal{P}(n)} y_\lambda \mathcal{H}\right)$$

with $\mathcal{H} = \mathcal{H}_{R,Q,q}(W_n)$, is called q-*Schur algebra of type B*. Here, the y_λ are defined as in (2), considered as elements of $\mathcal{H}_{R,Q,q}(W_n)$ lying in the parabolic subalgebra spanned by the T_w for $w \in S_n$.

One may also define q-Schur algebras associated to Weyl groups of type D. Consider the following Dynkin diagram of type D_n.

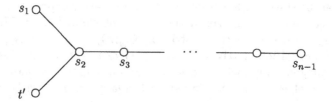

The corresponding Weyl group will be denoted by W_n'. We may, and shall, view it as the subgroup of W_n generated by $t' := ts_1t, s_1, s_2, \ldots, s_{n-1}$. Then the subgroup S_n of W_n generated by $s_1, \ldots, s_{n-1}$ is contained in W_n'. For an invertible element $q \in R$ let $\mathcal{H} := \mathcal{H}_{R,q}(W_n')$ denote the Iwahori–Hecke algebra of type D_n.

Definition 4.7 (Gruber/Hiss [29]) The R-algebra

$$\mathcal{S}_{R,q}(D_n) := \operatorname{End}_{\mathcal{H}} \left(\bigoplus_{\lambda \in \mathcal{P}(n)} y_\lambda \mathcal{H} \right)$$

with $\mathcal{H} := \mathcal{H}_{R,q}(W_n')$ is called *q-Schur algebra of type D.*

Theorem 4.8 (Geck/Hiss, Gruber/Hiss) *Let $G \neq \operatorname{GL}_n(q)$ be one of the classical groups introduced in section 3. If G is unitary, let $\delta = 2$, otherwise let $\delta = 1$. Let ℓ be a linear prime for G. Then the decomposition matrix M of $\mathcal{E}(G, 1)$ is a block diagonal matrix*

$$M = \begin{pmatrix} M_1 & 0 & \cdots & & 0 \\ 0 & M_2 & & & \\ & & \ddots & & \vdots \\ \vdots & & & M_{r-1} & 0 \\ 0 & & \cdots & 0 & M_r \end{pmatrix}.$$

The M_is correspond to the Harish-Chandra series of ordinary characters in $\mathcal{E}(G, 1)$.

Except in case (e), each M_i is the decomposition matrix of an algebra $\mathcal{S}_{\mathcal{O}, Q_i, q^\delta}(B_m)$, where Q_i is some power of q and $m \leq n$.

In case (e), one of the M_is, say M_1, is the decomposition matrix of $\mathcal{S}_{\mathcal{O}, q}(D_{n/2})$, and $M_2, \ldots, M_r$ are decomposition matrices of q-Schur algebras of type B as above.

This theorem was proved by Geck and Hiss in [24, Theorems 4.7, 4.11] for groups of types (b)–(d). A proof for the remaining cases will be given in a forthcoming publication by Gruber and Hiss [29].

The result of Theorem 4.8 can be generalized to arbitrary Lusztig series $\mathcal{E}(G, s)$ for semi-simple ℓ'-elements $s \in G^*$. The decomposition matrices occuring here are again given by those of certain $\mathcal{O}$-orders. These are tensor products with factors of the form $\mathcal{S}_{\mathcal{O},q^d}(m)$, $\mathcal{S}_{\mathcal{O},q^d}(D_m)$ and $\mathcal{S}_{\mathcal{O},Q',q^d}(B_m)$ (Q' a power of q) for various $m \leq n$ and $d \geq 0$. There are also factors which are defined like the q-Schur algebras in Definitions 4.3, 4.6 and 4.7, but with the Iwahori–Hecke algebras $\mathcal{H}$ replaced by more general algebras.

The following theorem reduces the determination of decomposition matrices of classical groups for linear primes to the problem of finding decomposition numbers for the q-Schur algebras associated to symmetric groups.

Theorem 4.9 (Gruber [27]) *Let q be a root of unity in $\mathcal{O}$ of odd order and let Q be a power of q. Then $\mathcal{S}_{\mathcal{O},Q,q}(B_n)$ is Morita equivalent to $\oplus_{i=0}^{n}\mathcal{S}_{\mathcal{O},q}(i) \otimes \mathcal{S}_{\mathcal{O},q}(n - i)$. In particular, the decomposition matrix M of $\mathcal{S}_{\mathcal{O},Q,q}(B_n)$ is a block diagonal matrix*

$$
M = \begin{pmatrix}
N_0 \otimes N_n & 0 & \cdots & & & 0 \\
0 & N_1 \otimes N_{n-1} & & & & \\
& & \ddots & & & \vdots \\
\vdots & & & & & \\
& & & N_{n-1} \otimes N_1 & 0 \\
0 & & \cdots & & 0 & N_n \otimes N_0
\end{pmatrix},
$$

where N_i is the decomposition matrix of the unipotent modules of $\mathrm{GL}_i(q)$.

Most of the decomposition numbers of the q-Schur algebra $\mathcal{S}_{\mathcal{O},q}(D_{n/2})$ can also be computed from those of general linear groups, but the result is more technical to state. For details see Gruber and Hiss [29]. We remark that the decomposition matrices for the q-Schur algebras $\mathcal{S}_{\mathcal{O},q}(n)$ have been computed by James up to $n = 10$ (see [34]). From the above results we obtain the decomposition matrices for linear primes for all classical groups up to $\mathrm{GL}_{10}(q)$, $\mathrm{GU}_{21}(q)$, $\mathrm{Sp}_{20}(q)$, $\mathrm{SO}_{21}(q)$ and $\mathrm{SO}_{20}^{\pm}(q)$.

5 On a conjecture of Broué

We finally indicate that our results give a remarkable property of the Jordan decomposition of characters. Let $\mathbf{G}$ denote a connected simple algebraic group, defined over the finite field $\mathbb{F}_q$. Let F denote the Frobenius morphism corresponding to this $\mathbb{F}_q$-structure of $\mathbf{G}$ and write $G := \mathbf{G}^F$ for the finite group of F-fixed points of $\mathbf{G}$. As always, ℓ is a prime not dividing q, and $(K, \mathcal{O}, k)$ is a splitting ℓ-modular system for G. Let $s \in \mathbf{G}^*$ denote a semisimple, F-stable ℓ'-element such that $C_{\mathbf{G}^*}(s)$ is a Levi subgroup of $\mathbf{G}^*$. Let $\mathbf{C}$ be an F-stable Levi subgroup of $\mathbf{G}$, dual to $C_{\mathbf{G}^*}(s)$, and write $C := \mathbf{C}^F$.

Broué has conjectured (see [2, Section 3]) that the two algebras $e_s^G \mathcal{O}G$ and $e_1^C \mathcal{O}C$ are Morita equivalent. Here, e_s^G and e_1^C denote the central idempotents of $\mathcal{O}G$, respectively $\mathcal{O}C$, corresponding to $\mathcal{E}_\ell(G, s)$ respectively $\mathcal{E}_\ell(C, 1)$.

If Broué's conjecture were true, then $\mathcal{E}_\ell(G, s)$ and $\mathcal{E}_\ell(C, 1)$ would have the same decomposition matrices. If G is one of the classical groups introduced in section 3 and ℓ a linear prime for G, this statement can be made precise and proved in the following way.

There is a bijection (a special case of the Jordan decomposition of characters)

$$\mathcal{L}_{z,\ell} : \mathcal{E}_\ell(C, 1) \to \mathcal{E}_\ell(G, z)$$

induced by the Lusztig map R_C^G (see Digne and Michel [8], Theorem 13.25(ii)). Let $\mathrm{CF}(\mathcal{E}_\ell(G, z))$ denote the K-vector space of class functions on G spanned by the elements of $\mathcal{E}_\ell(G, z)$, and let $\mathrm{CF}_{\ell'}(\mathcal{E}_\ell(G, z))$ denote the subspace of functions vanishing on ℓ-singular conjugacy classes. Brauer characters are extended to functions on all of G by letting their values be 0 on ℓ-singular elements. Hence Brauer characters are elements of $\mathrm{CF}_{\ell'}(\mathcal{E}_\ell(G, z))$. The map $\mathcal{L}_{z,\ell}$ is extended to $\mathrm{CF}(\mathcal{E}_\ell(C, 1))$ by linearity. Note that it sends $\mathrm{CF}_{\ell'}(\mathcal{E}_\ell(C, 1))$ to $\mathrm{CF}_{\ell'}(\mathcal{E}_\ell(G, z))$ by Digne and Michel [8, Proposition 12.6].

Theorem 5.1 (Gruber/Hiss) *Let G be one of the classical groups of section 3 and let ℓ be a linear prime for G. Then $\mathcal{L}_{z,\ell}$ sends the set of irreducible Brauer characters of $\mathcal{E}_\ell(C, 1)$) to the set of irreducible Brauer characters of $\mathcal{E}_\ell(G, z)$. With respect to the bijections between the irreducible characters induced by $\mathcal{L}_{z,\ell}$, the decomposition numbers of $\mathcal{E}_\ell(G, z)$ and of $\mathcal{E}_\ell(C, 1)$ are the same.*

The Lusztig map R_C^G commutes with reduction modulo ℓ, and $\mathcal{E}(G, z)$ and $\mathcal{E}(C, 1)$ are basic sets for $\mathcal{E}_\ell(G, z)$ respectively $\mathcal{E}_\ell(C, 1)$. Thus it suffices to prove that $\mathcal{E}(G, z)$ and $\mathcal{E}(C, 1)$ have the same decomposition matrices with respect to suitable orderings of the irreducible Brauer characters. This latter fact is then established by showing that the two (q-Schur) algebras describing the decomposition matrices of $\mathcal{E}(G, s)$ and $\mathcal{E}(C, 1)$ respectively, are isomorphic.

ACKNOWLEDGEMENTS

I thank Jochen Gruber for many helpful conversations on the subject of this paper.

References

[1] A. A. Beilinson, G. Lusztig and R. MacPherson, A geometric setting for the quantum deformation of GL_n, *Duke Math. J.* **61** (1990), 655–677.

[2] M. Broué, Isométries de caractères et équivalences de Morita ou dérivées, *Inst. Hautes Études Sci. Publ. Math.* **71** (1990), 45–63.

[3] M. Broué, G. Malle and J. Michel, Generic blocks of finite reductive groups, in *Représentations unipotentes génériques et blocs des groupes réductifs finis, Astérisque* **212** (1993), 7–92.

[4] M. Broué and J. Michel, Blocs et séries de Lusztig dans un groupe réductif fini, *J. reine angew. Math.* **395** (1989), 56–67.

[5] M. Cabanes and M. Enguehard, Local methods for blocks of reductive groups over a finite field, in *Finite reductive groups: related structures and representations (ed. M. Cabanes)*, pp. 141–163. Birkhäuser, 1997.

[6] R. W. Carter, *Finite groups of Lie type: Conjugacy classes and complex characters*, Wiley, 1985.

[7] P. Deligne and G. Lusztig, Representations of reductive groups over finite fields, *Ann. of Math.* **103** (1976), 103–161.

[8] F. Digne and J. Michel, *Representations of finite groups of Lie type*, London Math. Soc. Students Texts 21, Cambridge University Press, 1991.

[9] R. Dipper, On the decomposition numbers of the finite general linear groups, *Trans. Amer. Math. Soc.* **290** (1985), 315–343; II, *ibid.* **292** (1985), 123–133.

[10] R. Dipper and J. Du, Harish-Chandra vertices, *J. reine angew. Math.* **437** (1993), 101–130.

[11] R. Dipper and P. Fleischmann, Modular Harish-Chandra theory I, *Math. Z.* **211** (1992), 49–71.

[12] R. Dipper and G. D. James, Identification of the irreducible modular representations of $GL_n(q)$, *J. Algebra* **104** (1986), 266–288.

[13] R. Dipper and G. D. James, The q-Schur algebra, *Proc. London Math. Soc.* **59** (1989), 23–50.

[14] R. Dipper and G. D. James, Representations of Hecke algebras of type B_n, *J. Algebra* **146** (1992), 454–481.

[15] J. Du, A note on quantized Weyl reciprocity at roots of unity, *Algebra Colloq.* **4** (1995), 363–372.

[16] P. Fong and B. Srinivasan, The blocks of finite general linear and unitary groups, *Invent. Math.* **69** (1982), 109–153.

[17] P. Fong and B. Srinivasan, Brauer trees in $GL(n, q)$, *Math. Z.* **187** (1984), 81–88.

[18] P. Fong and B. Srinivasan, The blocks of finite classical groups, *J. reine angew. Math.* **396** (1989), 122–191.

[19] P. Fong and B. Srinivasan, Brauer trees in classical groups, *J. Algebra* **131** (1990), 179–225.

[20] M. Geck, On the decomposition numbers of the finite unitary groups in non-defining characteristic, *Math. Z.* **207** (1991), 83–89.

[21] M. Geck, A note on Harish-Chandra induction, *Manuscripta Math.* **80** (1993), 393–401.

[22] M. Geck, *Beiträge zur Darstellungstheorie von Iwahori–Hecke-Algebren*, Habilitationsschrift, Aachen, 1994. Aachener Beiträge zur Mathematik, Band 11, Aachen, 1995.

[23] M. Geck and G. Hiss, Basic sets of Brauer characters of finite groups of Lie type, *J. reine angew. Math.* **418** (1991), 173–188.

[24] M. Geck and G. Hiss, Modular representations of finite groups of Lie type in non-defining characteristic, in *Finite reductive groups: related structures and representations (ed. M. Cabanes)*, pp. 195–249. Birkhäuser, 1997.

[25] M. Geck, G. Hiss and G. Malle, Towards a classification of the irreducible representations in non-defining characteristic of a finite group of Lie type, *Math. Z.* **221** (1996), 353–386.

[26] J. A. Green, The characters of the finite general linear groups, *Trans. Amer. Math. Soc.* **80** (1955), 402–447.

[27] J. Gruber, *Cuspidale Untergruppen und Zerlegungszahlen klassischer Gruppen*, Dissertation, Heidelberg, 1995.

[28] J. Gruber, Green vertex theory, Green correspondence and Harish-Chandra induction, *J. Algebra* **186** (1996), 476–521.

[29] J. Gruber and G. Hiss, Decomposition numbers of finite classical groups for linear primes, *J. reine angew. Math.*, to appear.

[30] Harish-Chandra, Eisenstein series over finite fields, in *Functional analysis and related fields*, pp. 76–88. Springer, New York, 1970.

[31] G. Hiss, Harish-Chandra series of Brauer characters in a finite group with a split BN-pair, *J. London Math. Soc.* **48** (1993), 219–228.

[32] R. B. Howlett and G. I. Lehrer, Induced cuspidal representations and generalized Hecke rings. *Invent. Math.* **58** (1980), 37–64.

[33] R. B. Howlett and G. I. Lehrer, On Harish-Chandra induction for modules of Levi subgroups. *J. Algebra* **165** (1994), 172–183.

[34] G. D. James, The decomposition matrices of $GL_n(q)$ for $n \leq 10$, *Proc. London Math. Soc.* **60** (1990), 225–265.

[35] A. Lascoux, B. Leclerc and J.-Y. Thibon, Une conjecture pour le calcul des matrices de décomposition des algèbres de Hecke du type A aux racines de l'unité, *C. R. Acad. Sci. Paris Sér. I* **321** (1995), 511–516.

[36] G. Lusztig, *Characters of reductive groups over a finite field*, Ann. Math. Studies **107**, Princeton University Press, 1984.

[37] G. Lusztig and N. Spaltenstein, Induced unipotent classes, *J. London Math. Soc.* **19** (1979), 41–52.

The Meataxe as a tool in computational group theory

Derek F. Holt

Abstract

The Meataxe is a practical algorithm, first introduced by Richard Parker, for testing finite dimensional modules over finite fields for irreducibility, and for finding explicit submodules in the reducible case. This and associated algorithms are described briefly, together with more recent improvements. The possibility of extending these methods to fields of characteristic zero, such as the rational numbers, is also discussed.

1 Chopping up modules

The problem of explicitly finding the irreducible constituents of a finite dimensional KG-module, where K is a field and G is a finite group, is without doubt the most basic problem in computational group-representation theory. It corresponds roughly to finding the orbits of a finite permutation group, except that it is considerably more difficult.

Most of the research on this problem to date has been restricted to the case where $K = GF(q)$ is finite, and we shall assume this to be true in the first two sections of this paper. The characteristic zero case will be discussed in section 3. We shall denote the degree of the representation by d, throughout.

The theoretical complexity of the problem was proved to be polynomial in $d \log(q)$ by Rónyai in [12], but the algorithm described there does not appear to be practical as it stands, and has complexity at least as bad as $O(d^6 \log(q))$. For current applications, it is essential to find methods that are practical for d equal to at least several thousand and, to achieve this, we must aim for complexity $O(d^3 \log(q))$. In practice, this is equal to the complexity of multiplying two matrices, inverting a matrix, or performing a Gaussian reduction.

The Meataxe algorithm, due to Richard Parker and described in [8], still appears to be the only method with efficient implementations that comes anywhere close to achieving this aim. The main application for which it was conceived was constructing the irreducible representations of some of the

74

larger (usually sporadic) finite simple groups over small fields, and it proved to be very well-suited for this purpose. For example, it played an essential rôle in Norton's construction and existence proof of the simple group J_4 (see [7]). It was used frequently and more routinely as one of the principal tools for calculating the Brauer characters and Brauer trees of the sporadic finite simple groups (see [6] and [2]) in cases where it was necessary or expedient to find explicit irreducible representations of these groups over finite fields.

In addition to Parker's implementation in FORTRAN, there has been an efficient implementation in C by Ringe and Lux (see [11]) that has the additional facility of being able to compute the complete submodule lattice of the given module. Each of these implementations is strongly geared towards potentially large degree representations over very small fields, with particular emphasis on the field of order 2, where only a single bit is used for each matrix entry, and the number of matrices that are held in main memory at any one time is kept as small as possible (usually, at most 2 or 3).

It is clear from the above that the range of application of these specific methods is rather restricted. In fact, analysis shows that their efficiency depends not only on q being small, but on the splitting fields of the irreducible constituents of the module being small. The theoretical complexity is linear in the order of these splitting fields, and therefore exponential in d.

One other major application of the module chopping process occurs in the calculation of the composition factors of a finite group. If the group H has a chief factor K/L which is an elementary abelian p-group for some prime p, then to find the composition factors of H that lie between L and K, we need to regard K/L as an H/K-module over $GF(p)$, and then find its irreducible constituents. The original Meataxe did not always perform satisfactorily in this situation; for example, when H/K is a soluble group, then relatively large splitting fields can easily arise, even when the dimension of K/L is quite small.

Fortunately, in [4] Holt and Rees describe and analyse a relatively simple extension of the basic algorithm, which renders it efficient and practical in a much larger range of situations. Some further ideas by Charles Leedham-Green appear to make it work well in all situations.

As with the Parker Meataxe, these new techniques can also be used to provide efficient procedures for calculating endomorphism algebras of modules, testing modules for isomorphism and, more generally, for calculating $\mathrm{Hom}_{KG}(V, W)$, for modules V and W. They are extremely fast for irreducibles (or when at least one of V and W is irreducible), but are quite practical in full generality.

A brief description of all of these methods will be given in the next section. They have been implemented within the MAGMA computational algebra system (see [1]) by Allan Steel, with the help of Holt and Leedham-Green.

As well as being usable explicitly on KG-modules, they are used implicitly by some MAGMA functions which find composition factors and, in particular, by the solvable group functions. Although there has been no systematic comparison using a large range of examples, it appears as though the general endomorphism algebra method may be superior in many situations, and generally more reliable, than that proposed by Schneider in [13].

These algorithms have also been implemented by Holt and Rees within the GAP system (see [14]) as part of the external SMASH package (see [5]). Furthermore, the extensions proposed by Holt and Rees have been incorporated within the standalone Meataxe package written by Ringe and Lux (see [11]), which is also accessible as an external GAP package. This implementation and the one within MAGMA are particularly well adapted for working with representations of very large degree, since they store as many matrix entries as possible within a single machine word, and use lookup tables to facilitate fast matrix multiplication, by processing several entries at a time. Finally, it should be mentioned that the Lux–Ringe package is accessible from the QUOTPIC interactive graphics package for studying finite quotients of finitely presented groups (see [3]). It is used there for finding the composition factors within elementary abelian sections.

2 A description of the Meataxe methods

Parker's standard Meataxe uses the following test, known as Norton's irreducibility test. Let M be a right KG-module of dimension d defined by explicit $d \times d$-matrices for the generators of G, let A denote the K-algebra generated by these matrices, and let M^T denote the KG-module defined by the transposes of these matrices. (So M^T, also regarded as a right module, is the contragredient module of M, where the defining matrices correspond to the inverses of the generators of G.) An element ξ of A is chosen, and the nullspaces N of ξ, and N' of its transpose ξ^T are computed. Then provided that

(a) N is nonzero,

(b) every nonzero vector v in N generates the whole of M as a KG-module, and

(c) at least one vector w of N' generates the whole of M^T as a KG-module,

M is proved irreducible. Otherwise M is reducible, and a proper submodule has been found explicitly in (b) or (c).

Note that only the basic matrix operations, multiplication, transposing, and Gaussian elimination (for calculating nullspaces, and performing part (c))

are involved here. Part (b) of the test is clearly the one which could be time-consuming if ξ were not well chosen. In Parker's Meataxe, elements are chosen until ξ is found with a nontrival nullspace of low dimension, preferably 1-dimensional. In [8], a list of choices of ξ that work well for 2-generator groups over small fields, and involve only one extra matrix multiplication for each new choice, is given. A general problem with this approach however is that, if the field has order q, then the probability of the nullspace being nontrivial is about $1/q$ and so, for large q, we might have to make a large number of choices. Another, more serious, problem arises when the dimension, e say, of the endomorphism ring of M is large because, for example, when M is irreducible we cannot find a ξ with nontrivial nullspace of dimension smaller than e.

In the Holt–Rees extension, rather than generate ξ itself randomly, we select a random element θ of the K-algebra A, calculate the characteristic polynomial $c(x)$ of θ, and then factorize it. Then we set $\xi = p(\theta)$, where $p(x)$ is an irreducible factor of $c(x)$. In this way, ξ will always have nontrivial nullspace N. In the situation where N is irreducible as a $K\langle\theta\rangle$-module (which is the case, for example, whenever $p(x)$ is a non-repeating factor of $c(x)$), it is sufficient to carry out part (b) of the test for a single nonzero vector v in the nullspace of ξ. In other cases, examination of a single vector will not give a conclusive test for irreducibility, but might prove reducibility. If not, another factor $p(x)$ of $c(x)$, or another element θ is selected.

Note that the only additional calculations compared with Parker's algorithm are the calculation and factorization of the characteristic polynomial. Both of these can be accomplished in time $O(d^3 \log(q))$, where d is the dimension of M, which is the same complexity as that of the other parts of the process, such as calculating nullspaces.

As one might expect from the above description, this is a *Las Vegas* algorithm, which means that there is no certainty that it will ever stop. However, if it does stop, then it will always return the correct answer. It is proved in [4] that, with high probability, the algorithm will complete in time $O(d^3 \log(q))$ in almost all situations (where the probability rapidly approaches 1 as the constant in the time estimate increases).

The exceptional situation in which this analysis fails is where the module is reducible but uniserial, with all of its composition factors being isomorphic and having a large endomorphism algebra. A specific example of this type for which the method really does come to grief is described in [4]. Fortunately, a fix has been suggested for this situation by Charles Leedham-Green, and it has been successfully implemented in MAGMA, so there now appears to be a satisfactory solution to the chopping problem that works in all situations.

The idea is that when it becomes clear that the main algorithm is not working (because the dimensions of all of the nullspaces found are too large), we give up and calculate the full endomorphism algebra E of M. (There are tricks for doing this quickly based on the methods for irreducible modules described in [4].) Then, since we are in the exceptional uniserial reducible situation, E will certainly contain nonzero singular matrices ρ, and the nullspace of any such matrix will be a submodule of M. The only problem is to find such a ρ. If E is not commutative (which is the case in the difficult example mentioned above), then we can find ρ as $\alpha\beta - \beta\alpha$, where α and β are randomly chosen elements of E. Otherwise, a random element α of E is likely to have order divisible by p, and a suitable power α^n of α will have order a power of p and then $\rho := \alpha^n - I_d$ will be nonzero and singular. The power n might be very large but, fortunately, raising elements to power n requires only about $\log(n)$ multiplications. The complexity of this fix is probably slightly worse than $O(d^3 \log(q))$ (I suspect it might be $O(d^4 \log(q))$ in the worst case), but fortunately the bad situation does not arise very often (or indeed at all) in the context of current applications.

3   Reducing modules in characteristic zero

In this section we consider and speculate on to what extent, if any, the Meataxe techniques discussed so far for finite fields can be made to apply to representations over the field $\mathbb{Q}$ of rational numbers, or to algebraic extensions thereof. There has been very little work on this problem to date, and even less attempt at implementations, and so we shall have to be content for the time being with pinpointing the likely difficulties, and speculating on means of overcoming them.

We are not yet at the stage of being able to propose a push-button type general purpose algorithm, and so we should be prepared to assume that the user can apply experience and discretion (possibly based on knowledge of the particular group involved), in the choice of which techniques to employ in a particular situation.

Plesken and Souvignier consider this problem in [10], and describe methods that should work for dimensions up to about 200. They justify these claims by finding representatives of all irreducible $\mathbb{Q}G$-representations of the group $S_4(3)$, which involves chopping up modules of degrees up to 240. (Here, they make use of the known character table of the group.) Roughly speaking, their method is to compute all or part of the endomorphism algebra of the module M to be reduced, and to find submodules as nullspaces of endomorphisms. In addition, Richard Parker is working on a characteristic zero meataxe using p-adic methods (see [9]).

In contrast, the author, working with Allan Steel in MAGMA, has been trying to extend the Meataxe techniques themselves to the rational case. In fact, the same two major problems are involved in both of these approaches.

The first is the theoretical problem that, unlike in the finite case, the endomorphism algebra of an irreducible representation in characteristic zero can be a noncommutative division ring; that is, the Schur index of the representation can be greater than 1. This makes it very difficult to distinguish between such an irreducible representation, and the sum of isomorphic irreducible representations of smaller dimensions, even if one succeeds in computing a basis for the endomorphism algebra E. Of course, in the second case, E will contain nonzero singular elements, but they are not easy to find.

The second problem is practical, and is that the integers arising in the matrices with which one is working tend to grow very rapidly with each step in the computation, and each individual reduction. Unless one finds means of preventing this happening, the basic matrix operations involved rapidly grind to a halt. One possibility is to use the LLL-algorithm, but more refined and goal-directed methods are needed to find a satisfactory solution. For example, Plesken and Souvignier suggest keeping an invariant bilinear form for each representation, because this allows one to manipulate the basis by various reduction routines to keep small coefficients for the matrix entries in the representation.

The direct use of the Meataxe techniques described in section 2 on rational representations is not very easy or practical for degrees above about 50 or 100 (depending on the sparsity of the matrices) since, for example, the calculation of the characteristic polynomial of a matrix becomes much too slow.

One can, however, consider the representation M modulo p, where p is a prime number not dividing any of the denominators arising in the representation, reduce it over $GF(p)$, and then attempt to rewrite the reduction over $\mathbb{Q}$ (by rewriting $p-1$ as -1, $(p+1)/2$ as $1/2$, and so on). It can then be checked easily whether the result is a genuine reduction of the $\mathbb{Q}G$-module M. This simple technique has been implemented in MAGMA by Allan Steel. It is of course limited, but it does work surprisingly often, especially for large sparse representations, such as permutation representations, and it is very fast, so it should certainly not be discarded completely. In fact, by intelligent application and correct choice of the prime (which can be very large if necessary), one can usually reduce M to homogeneous constituents (i.e. direct sums of isomorphic irreducibles), and certainly to direct sums of irreducibles of the same degree. Of course, if the reduction of a representation over $GF(p)$ is irreducible, then it is irreducible over $\mathbb{Q}$, but the converse is not true. For irreducible rational representations with Schur index greater than 1, when p is coprime to the group order, the reduction over $GF(p)$ will be a direct sum of modules of the same dimension for all primes p, and so we are up against the theoretical difficulty described above.

In this situation, we propose the following approach. Let the algebra A generated by the matrices of M be defined as in section 2. Since all representations in characteristic zero are completely reducible, the part of the Meataxe involving the dual module, where we transpose the matrices, can be omitted completely. Furthermore, we can use sums and intersections of nullspaces of elements of A in place of the nullspaces themselves. So, we find a number of nontrivial nullspaces N_i of elements $\xi_i \in A$ (for example, choose $\xi_i = g_i - 1$, where g_i is the matrix of an element of G itself). By replacing N_i by translates under G if necessary, we seek to construct a subspace $N := \cap N_i$ that is nontrivial, but has as small a dimension as possible. We can also use sums of some of the N_i. (For example, if we have N_1 of dimension 4, and N_2 of dimension 6, and M itself has degree 24, then by taking appropriate translates, sums and intersections, we can always find an N of dimension 2.)

We now calculate the endomorphism algebra E of M, using the fact that all of its elements must fix the subspace N. If N is not too large, then this calculation is very quick. If, at this stage, $\dim(E) \geq \dim(N)$, then we can determine reducibility of M as follows. If E acts freely on N then, as in the extension of the Meataxe described in section 2, it is enough to calculate w^G for a single nonzero vector w in N; either it is a proper submodule of M, or it is equal to M and M is irreducible. If E does not act fixed-point-freely, then we can find a nonzero singular matrix in E, and its nullspace will be a submodule of M.

By combining the techniques of reducing to $GF(p)$ and lifting back to $\mathbb{Q}$, and working with nullspaces as just described, the author succeeded, with some diffciulty, in constructing the irreducible rational representations of $S_4(3)$ (the same example as in [10]). The difficulties encountered resulted purely from integer entry explosion. There was no inherent difficulty in finding nullspaces with the required properties, but the related sum and intersection calculations tended to result in huge integers.

References

[1] W. Bosma and J. Cannon, **MAGMA** *Handbook*, University of Sydney, 1993.

[2] G. Hiß and K. Lux, *Brauer trees of sporadic groups*, Oxford University Press, 1989.

[3] Derek F. Holt and Sarah Rees, A graphics system for displaying finite quotients of finitely presented groups, in *Groups and computation: workshop on groups and computation, October 7–10, 1991 (eds. L. Finkelstein and W. M. Kantor)*, pp. 113–126. DIMACS Series, Vol. 11, American Mathematical Society, 1993.

[4] Derek F. Holt and Sarah Rees, Testing modules for irreducibility, *J. Austral. Math. Soc. Ser. A* **57** (1994), 1–16.

[5] Derek F. Holt, C. R. Leedham-Green, E. A. O'Brien and Sarah Rees, Computing matrix group decompositions with respect to a normal subgroup, *J. Algebra* **184** (1996), 818–838.

[6] C. Jansen, K. Lux, R. Parker and R. Wilson, *An Atlas of Brauer characters*, London Mathematical Society Monographs, New Series 11, Clarendon Press, Oxford, 1995.

[7] Simon Norton, The construction of J_4, in *The Santa Cruz conference on finite groups (eds. B. Cooperstein and G. Mason)*, pp. 271–277. Proceedings of Symposia in Pure Mathematics, Vol. 37, American Mathematical Society, 1980.

[8] R. Parker, The computer calculation of modular characters—the MeatAxe, in *Computational group theory (ed. M. D. Atkinson)*, pp. 267–274. Academic Press, London, 1984.

[9] R. Parker, An integral meataxe, *these proceedings*, pp. 215–228.

[10] W. Plesken and B. Souvignier, Constructing rational representations of finite groups, in *Proceedings of the Magma conference, London, 1993*, to appear.

[11] Michael Ringe, *The C MeatAxe*, Lehrstuhl D für Mathematik, RWTH Aachen, 1992.

[12] Lajos Rónyai, Computing the structure of finite algebras, *J. Symbolic Comput.* **9** (1990), 355–373.

[13] Gerhard Schneider, Computing with endomorphism rings of modular representations, *J. Symbolic Comput.* **9** (1990), 607–636.

[14] M. Schönert *et al.*, GAP—*Groups, Algorithms, and Programming*, Lehrstuhl D für Mathematik, RWTH Aachen, 1992.

Branching rules for modular projective representations of the symmetric groups

John Humphreys

Abstract

In this survey we shall discuss known results concerning the decomposition numbers for the p-modular projective representations of the symmetric groups (for p odd). Several open questions are mentioned.

Introduction

It is well-known (see [4]), that the problem of determining the decomposition numbers for p-modular projective representations of the symmetric groups (for odd p) is the same problem as determining the decomposition numbers for the faithful representations of a double cover of the symmetric groups. There are, in general, two non-isomorphic such double covers, one of which is denoted by $\widetilde{S}(n)$ and is generated by elements $t_1, \ldots, t_{n-1}$ together with a central involution z satisfying the relations

$$
\begin{aligned}
(t_i)^2 &= 1 \text{ (for } 1 \le i \le n-1), \\
(t_i t_{i+1})^3 &= 1 \text{ (for } 1 \le i \le n-2), \\
(t_i t_j)^2 &= z, \text{ (for } |i-j| \ge 2).
\end{aligned}
$$

A result of G. D. James [6] shows that it is possible to index the irreducible p-modular Brauer characters and the ordinary characters in such a way that the decomposition matrix is upper uni-triangular. This is not the case for projective representations, and we first use the modular ATLAS [7] to look at some of the situations that arise. The situation is more complicated here than in the linear case since no analogue has yet been found of the construction of the irreducible modules given by Specht modules. It is a well-known result of Schur (see Theorem 8.6 of [4]) that the irreducible complex projective representations of the symmetric group $S(n)$ may be labelled by partitions of n into distinct parts. However, this method of labelling is not bijective. If λ has an even number of even parts then there is a unique irreducible associated with λ denoted by $\langle \lambda \rangle$. When λ has an odd number of even parts there are two irreducibles associated with λ denoted $\langle\lambda\rangle$ and $\langle\lambda\rangle^a$. The characters of these two representations agree everywhere except on elements which project into permutations of cycle type λ and have the value zero for all other elements which project into odd permutations.

This survey is divided into three sections. In section 1, we look at examples to try to formulate a sensible idea of what the decomposition matrix might look like. In section 2, the most general known result is discussed, and finally, in section 3, the branching rules of the title are considered.

1   Experimental evidence

In this section, we use the modular ATLAS [7] to consider how near the decomposition matrix is to being an upper-triangular matrix. We shall discuss the decomposition matrix associated with each p-block. These are easily explained using [5]. First of all a *p-bar* associated with the Young diagram of a partition λ of n into distinct parts arises in two ways. The first of these occurs when one of the rows of the Young diagram has more than p nodes. In this case the *p-bar* is *deleted* by removing p nodes from this row provided that the resulting diagram has distinct parts. The second type of p-bar occurs when the number of nodes in two distinct rows adds up to p. In this case, the p-bar is removed by deleting these two rows. The *p-core* is obtained by repeating this process as often as possible. For example, when $n = 12$ and $p = 3$ the 3-core of each diagram is empty except for the diagram corresponding to the partition $(7\ 4\ 1)$, which cannot have a 3-bar removed without violating the requirement for distinct parts. According to the results of [5], there are three blocks in this case, that containing $\langle n\rangle$ together with all characters with empty 3-core (in general, we shall call the block containing $\langle n\rangle$ the *main* block), as well as two further blocks, one containing the character corresponding to $\langle 7\ 4\ 1\rangle$, the other containing the character corresponding to $\langle 7\ 4\ 1\rangle^a$.

The first interesting case occurs when $n = 6$ when the decomposition matrix for the only block of projective representations modulo 3 is

	ϕ_1	ϕ_2
$\langle 6\rangle$	1	
$\langle 6\rangle^a$	1	
$\langle 5\ 1\rangle$	1	1
$\langle 4\ 2\rangle$	2	1
$\langle 3\ 2\ 1\rangle$	1	
$\langle 3\ 2\ 1\rangle^a$	1	

where, as usual, a blank entry in the matrix represents the number zero and

ϕ_1, ϕ_2 denote the irreducible modular projective Brauer characters of degrees 4 and 12 respectively.

Note here that the irreducible ϕ_1 arises from the restriction to p-regular classes of two ordinary irreducibles ($\langle 6 \rangle$ and $\langle 6 \rangle^a$), because these differ on a conjugacy class which contains elements of order divisible by p but which is α-regular. It follows that the decomposition matrix can have small "cliffs" of height 2.

Next consider the case of 3-modular representations when $n = 9$. In this case the decomposition matrix for the unique block modulo 3 is

	ϕ_1	ϕ_2	ϕ_3	ϕ_4	ϕ_5	ϕ_6
$\langle 9 \rangle$	1	1				
$\langle 8\ 1 \rangle$	1		1			
$\langle 8\ 1 \rangle^a$		1		1		
$\langle 7\ 2 \rangle$	1		1		1	
$\langle 7\ 2 \rangle^a$		1		1		1
$\langle 6\ 3 \rangle$	1	1			1	1
$\langle 6\ 3 \rangle^a$	1	1			1	1
$\langle 6\ 2\ 1 \rangle$	2	2			1	1
$\langle 5\ 4 \rangle$	1				1	
$\langle 5\ 4 \rangle^a$		1				1
$\langle 5\ 3\ 1 \rangle$	2	2	1	1	1	1
$\langle 4\ 3\ 2 \rangle$			1	1		

where $\phi_1, \ldots, \phi_6$ are the Brauer characters of the modular irreducible projective representations of $S(9)$ of degrees 8, 8, 48, 48, 104 and 104 respectively.

The feature to note here is that the irreducible characteristic zero projective representation corresponding to $\langle 9 \rangle$ has two irreducible modular constituents so that the decomposition matrix has a small "plateau" of width 2.

Next, consider the case $n = 12$ again when $p = 3$. As we have seen, there are three blocks in this case. In writing down the decomposition matrix for the main block, we do not use lexicographic order of partitions as in the previous examples, but rather first consider Schur-regular partitions, ordered lexicographically, then Schur-irregular partitions, also ordered lexicographically (a partition $\lambda = (\lambda_1, \ldots, \lambda_\ell)$ is *Schur-regular* if $\lambda_i - \lambda_{i+1} \geq 3$ for $1 \leq i \leq \ell - 1$ and $\lambda_i - \lambda_{i+1} > 3$ whenever $\lambda_i \equiv 0 \pmod 3$, see [3]). The decomposition matrix for the main block is then as in Table 1, where $\phi_1, \ldots, \phi_5$ are the Brauer characters of the modular irreducible characters in the main block of a double cover of $S(12)$ of degrees 32, 288, 640, 1008, 2880 and 7776 respectively. The feature to note here is the entry of 2 on the "diagonal".

As can be seen from these examples, the decomposition matrix of each block of projective characters of $S(n)$ is only approximately upper triangular. It has been conjectured (see [2]) that apart from "cliffs" of height 2 and "plateaux" of width 2, the decomposition matrix is of upper-triangular shape. A more precise conjecture about the way to label irreducible modular projective characters is formulated in [2], and this is also established in the case $p = 3$. The case $p = 5$ is established in [1], where it is also explained why this more precise conjecture fails for $p = 7$ and $p = 11$. The labelling of the modular projective characters remains a very difficult problem.

Table 1: Decomposition of the main block for $n = 12$ and $p = 3$

	ϕ_1	ϕ_2	ϕ_3	ϕ_4	ϕ_5
$\langle 12 \rangle$	1				
$\langle 12 \rangle^a$	1				
$\langle 11\ 1 \rangle$	1	1			
$\langle 10\ 2 \rangle$	1	1	1		
$\langle 9\ 3 \rangle$	2		2	2	
$\langle 8\ 4 \rangle$	1	1	2		1
$\langle 9\ 2\ 1 \rangle$	2		1	1	
$\langle 9\ 2\ 1 \rangle^a$	2		1	1	
$\langle 7\ 5 \rangle$	2			2	1
$\langle 7\ 3\ 2 \rangle$	3	1	1	2	1
$\langle 7\ 3\ 2 \rangle^a$	3	1	1	2	1
$\langle 6\ 5\ 1 \rangle$	1	1		1	1
$\langle 6\ 5\ 1 \rangle^a$	1	1		1	1
$\langle 6\ 4\ 2 \rangle$	4	1	2	3	1
$\langle 6\ 4\ 2 \rangle^a$	4	1	2	3	1
$\langle 6\ 3\ 2\ 1 \rangle$	2		2	2	
$\langle 5\ 4\ 3 \rangle$	1		1	1	
$\langle 5\ 4\ 3 \rangle^a$	1		1	1	
$\langle 5\ 4\ 2\ 1 \rangle$	1		2	1	

2   Wales' results

In this section, we discuss the most general result yet proved about the reduction modulo p of characteristic zero projective characters. This is a result of Wales [10], which may be stated as follows.

Theorem 2.1 *Let p be an odd prime. There is an irreducible p-modular character $\phi(\langle n \rangle)$ associated with the projective character $\langle n \rangle$ as follows:*

(i) if n is even, or if n is odd and p does not divide n, then $\phi(\langle n \rangle)$ is equal to the restriction to p-regular classes of $\langle n \rangle$;

(ii) if n is odd and p divides n then the restriction to p-regular classes of $\langle n \rangle$ *is* $\phi(\langle n \rangle) + \phi(\langle n \rangle)^a$ *where the restriction of the character* $\phi(\langle n \rangle) \downarrow \widetilde{S(n-1)}$ *to p-regular classes is equal to* $\langle n - 1 \rangle$.

There is a similar result for the character $\langle n - 1, 1 \rangle$, but no such result is known for any other projective representation in general.

Theorem 2.2 *The restriction of* $\langle n - 1, 1 \rangle$ *to p-regular conjugacy classes is an irreducible p-modular representation if p does not divide n and p does not divide* $n - 1$.

If p divides n, the restriction of $\langle n - 1, 1 \rangle$ *to p-regular conjugacy classes is equal to* $\phi(\langle n - 1, 1 \rangle) + \phi(\langle n \rangle)$. *Finally if p divides* $n - 1$ *the restriction of* $\langle n - 1, 1 \rangle$ *to p-regular conjugacy classes is equal to* $\phi(\langle n - 1, 1 \rangle) + \phi(\langle n \rangle)$ *if n is odd and is* $\phi(\langle n - 1, 1 \rangle) + \phi(\langle n - 1, 1 \rangle)^a + \phi(\langle n \rangle) + \phi(\langle n \rangle)^a$ *if n is even.*

Remark: In a future publication, we hope to present an alternative proof of these facts which is more straightforward than the original proof.

3   Branching rules

In this section, we discuss the branching rules for projective modular representations. We first discuss representations of the symmetric groups. Thus, the James indexing of modular irreducible representations of the symmetric groups (according to which the decomposition matrix is upper-unitriangular) is used to label the irreducibles. The motivation for considering branching rules for projective representations is the Jantzen–Seitz conjecture [8] for representations of the symmetric group which gives necessary and sufficient conditions for the modular representation associated with a partition λ to be irreducible on restriction to the subgroup $S(n - 1)$ of $S(n)$. This was established in one direction in [8], and proved in the reverse direction independently in [3] and [9]. There is as yet no analogue of this result for modular projective representations. However, the following partial result may be obtained from the result of Wales discussed in the last section.

Theorem 3.1 *The p-modular irreducible* $\phi(\langle n \rangle)$ *is irreducible on restriction to* $\widetilde{S(n-1)}$ *if n is even and p divides* $n - 1$, *or if n is odd and p divides n.*

The irreducible $\phi(\langle n - 1, 1 \rangle)$ *is also irreducible on restriction to* $\widetilde{S(n-1)}$ *if either p divides* $n - 1$, *or n is odd and p divides n.*

Proof: We first consider the character $\langle n \rangle$. This has degree $2^{(n-2)/2}$ if n is even and has degree $2^{(n-1)/2}$ if n is odd. As we have seen in Theorem 2.1, $\langle n \rangle$ is irreducible modulo p if n is even or if n is odd and p does not divide n. If n is odd and p divides n, then $\phi(\langle n \rangle)$ is of degree $2^{(n-3)/2}$. To consider branching rules, first suppose that n is even so that $\langle n \rangle$ is irreducible modulo p. Thus $n - 1$ is odd, and by the ordinary branching rules,

$$\langle n \rangle \downarrow \widetilde{S(n-1)} = \langle n - 1 \rangle.$$

Since the restriction of $\langle n \rangle \downarrow \widetilde{S(n-1)}$ to p-regular classes is equal to the character obtained by restricting to $\widetilde{S(n-1)}$ the restriction of $\langle n \rangle$ to p-regular classes, we see that $\langle n - 1 \rangle$ is also irreducible modulo p unless p divides $n - 1$, in which case $\langle n - 1 \rangle$ is reducible. Now consider the case when n is odd so that

$$\langle n \rangle \downarrow \widetilde{S(n-1)} = \langle n - 1 \rangle + \langle n - 1 \rangle^a.$$

The p-modular character is therefore reducible unless p divides n, in which case $\phi(\langle n \rangle) \downarrow \widetilde{S(n-1)} = \langle n - 1 \rangle$ and is irreducible because $n - 1$ is even.

Now consider $\langle n - 1, 1 \rangle$ which has degree $2^{(n-2)/2}(n - 2)$ if n is even and has degree $2^{(n-3)/2}(n - 2)$ when n is odd. Consider first the case when n is even so that $\phi(\langle n - 1, 1 \rangle)$ has degree $2^{(n-2)/2}(n - 2)$ if p divides neither n nor $n - 1$; the degree is $2^{(n-2)/2}(n - 3)$ if p divides n and is $2^{(n-4)/2}(n - 4)$ if p divides $n - 1$. Since $n - 1$ is odd, $\phi(\langle n - 2, 1 \rangle)$ has degree $2^{(n-4)/2}(n - 3)$, $2^{(n-4)/2}(n - 4)$, or $2^{(n-4)/2}(n - 5)$ respectively in the three situations when p divides neither $n - 1$ nor $n - 2$, p divides $n - 1$, or p divides $n - 2$. On degree grounds, we see that $\phi(\langle n - 1, 1 \rangle)$ could only possibly be irreducible on restriction to $\widetilde{S(n-1)}$ when n is even and p divides $n - 1$, so we now consider this case in more detail. By Theorem 2.2, the reduction modulo p of $\langle n - 1, 1 \rangle$ is

$$\phi(\langle n - 1, 1 \rangle) + \phi(\langle n - 1, 1 \rangle)^a + \phi(\langle n \rangle) + \phi(\langle n \rangle)^a.$$

The ordinary branching rule shows that

$$\langle n - 1, 1 \rangle \downarrow \widetilde{S(n-1)} = \langle n - 1 \rangle + \langle n - 2, 1 \rangle + \langle n - 2, 1 \rangle^a,$$

so when this restriction is reduced modulo p we obtain (since p divides $n - 1$)

$$\phi(\langle n - 1 \rangle) + \phi(\langle n - 1 \rangle)^a + \phi(\langle n - 1 \rangle) + \phi(\langle n - 2, 1 \rangle)$$

$$+ \phi(\langle n - 1 \rangle)^a + \phi(\langle n - 2, 1 \rangle)^a.$$

Since $n - 1$ is odd and p divides $n - 1$,

$$\phi(\langle n \rangle) \downarrow \widetilde{S(n-1)} = \phi(\langle n \rangle)^a \downarrow \widetilde{S(n-1)} = \phi(\langle n - 1 \rangle) + \phi(\langle n - 1 \rangle)^a,$$

so we see that $\phi(\langle n - 1, 1 \rangle) \downarrow \widetilde{S(n-1)}$ is equal to $\phi(\langle n - 2, 1 \rangle)$, as required.

Finally, consider the case when n is odd. Then the degree of $\phi(\langle n-1,1\rangle)$ is $2^{(n-3)/2}(n-2)$, $2^{(n-3)/2}(n-3)$, or $2^{(n-3)/2}(n-4)$ in the three cases when p divides neither n nor $n-1$, p divides n, or p divides $n-1$ respectively. Since $n-1$ is even the degree of $\phi(n-2,1))$ is $2^{(n-3)/2}(n-3)$, $2^{(n-3)/2}(n-4)$, or $2^{(n-3)/2}(n-5)$ respectively in the cases p divides neither $n-1$ nor $n-2$, p divides $n-1$, or p divides $n-2$. Thus on degree grounds, we see that $\phi(\langle n-1,1\rangle)$ can only possibly be irreducible on restriction to $\widetilde{S(n-1)}$ either if p divides n and at the same time p divides neither $n-1$ nor $n-2$, or if p divides $n-1$. However, if p divides n, since p is odd it is greater than or equal to 3 and so cannot divide either $n-1$ or $n-2$. Thus the condition that p divides n implies that p cannot divide either $n-1$ or $n-2$. When n is odd, we therefore only need to consider the cases that p divides n or that p divides $n-1$. In either of these two cases the p-modular constituents of $\langle n-1,1\rangle$ are $\phi(\langle n\rangle)$ and $\phi(\langle n-1,1\rangle)$ whereas

$$\langle n-1,1\rangle \downarrow \widetilde{S(n-1)} = \langle n-1\rangle + \langle n-2,1\rangle.$$

Since $\langle n\rangle \downarrow \widetilde{S(n-1)} = \langle n-1\rangle$, it follows that the restriction of $\phi(\langle n-1,1\rangle)$ must contain $\phi(\langle n-2,1\rangle)$ and since we have just shown that these have equal degrees, they must be equal. $\square$

References

[1] G. E. Andrews, C. Bessenrodt and J. B. Olsson, Partition identities and labels for some modular characters, *Trans. Amer. Math. Soc.* **344** (1994), 597–613.

[2] C. Bessenrodt, A. Morris and J. B. Olsson, Decomposition matrices for spin characters at characteristic 3, *J. Algebra* **164** (1994), 146–172.

[3] B. Ford, Irreducible restrictions of representations of the symmetric groups, *Bull. London Math. Soc.* **27** (1995), 453–459.

[4] P. N. Hoffman and J. F. Humphreys, *Projective representations of the symmetric groups*, Oxford University Press, 1992.

[5] J. F. Humphreys, Blocks of projective representations of the symmetric groups, *J. London Math. Soc. (2)* **33** (1986), 441–452.

[6] G. D. James, The irreducible representations of the symmetric groups, *Bull. London Math. Soc.* **8** (1976), 229–232.

[7] C. Jansen, K. Lux, R. Parker and R. Wilson, *An* ATLAS *of Brauer characters*, Oxford University Press, 1995.

[8] J. C. Jantzen and G. M. Seitz, On the representation theory of the symmetric groups, *Proc. London Math. Soc. (3)* **65** (1992), 475–504.

[9] A. S. Kleshchev, On restrictions of irreducible modular representations of semisimple algebraic groups and symmetric groups to some natural subgroups, I, *Proc. London Math. Soc. (3)* **69** (1994), 515–540.

[10] D. B. Wales, Some projective representations of S_n, *J. Algebra* **61** (1979), 37–57.

Characters and surfaces: a survey

Gareth A. Jones

Abstract

This is a survey of some recent applications of the character theory of finite groups to the theory of surfaces. The emphasis is on compact Riemann surfaces, together with their associated automorphism groups, coverings, homology groups, combinatorial structures, fields of definition, and length spectra.

1   Introduction

With the publication of the ATLAS *of Finite Groups* [16], and subsequently of its companion, the ATLAS *of Brauer Characters* [44], there is a now considerable wealth of information available about the character theory of finite simple groups and related groups. This means that the great apparatus of representation theory developed by Frobenius, Schur, Brauer and others can be applied more effectively than ever to construct examples, to solve specific problems and to test general conjectures. Most of these applications have been within finite group theory itself, but mathematicians in other areas are also beginning to make use of these methods. My aim here is to give a survey of some of the ways in which character theory can contribute to the study of surfaces, with particular emphasis on compact Riemann surfaces. I have not attempted to be comprehensive: the applications I have chosen are simply those which I have recently found useful or interesting. Nevertheless, I hope that these brief comments, together with the references, will provide the interested reader with at least a sketch-map for further exploration.

2   Counting solutions of equations

One of the most effective character-theoretic techniques is the enumeration of the solutions of an equation in a finite group. Traditionally, this is a purely algebraic activity, as in the investigation by Frobenius [22] of the number of solutions of $x^n = a$, but classical results of Hurwitz [38] imply that these methods can also be applied to the relationship between finite groups and surfaces. Before considering these applications, let me state a well-known character-theoretic formula.

Let $K_1, \ldots, K_r$ be conjugacy classes (not necessarily distinct) in a finite group G, and let $n(K_1, \ldots, K_r)$ denote the number of r-tuples $(g_1, \ldots, g_r) \in K_1 \times \cdots \times K_r$ such that $g_1 \ldots g_r = 1$. Then we have

$$n(K_1, \ldots, K_r) = \frac{|K_1| \ldots |K_r|}{|G|} \sum_{\chi} \frac{\chi(g_1) \ldots \chi(g_r)}{\chi(1)^{r-2}}, \tag{1}$$

where $g_i \in K_i$ for each i, and where χ ranges over the irreducible complex characters of G. Serre gives a proof in [87, Theorem 7.2.1], and a formula equivalent to the most useful case $r = 3$ is proved in [27, §4.2] and [43, §28]. The basic idea of the proof is to work in the centre of the group-algebra $\mathbb{C}G$, using the orthogonality relations to compute $n(K_1, \ldots, K_r)$ as the coefficient of 1 in the product of the class-sums corresponding to the classes K_i.

Heuristically, one can regard $|K_1| \ldots |K_r|/|G|$ as a first approximation to the number of solutions of $g_1 \ldots g_r = 1$, obtained by assuming that each element of G occurs equally often as a value of $g_1 \ldots g_r$ for $(g_1, \ldots, g_r) \in K_1 \times \cdots \times K_r$. Of course, groups rarely behave as straightforwardly as this, and the summation factor in (1) compensates for the fact that the products $g_1 \ldots g_r$ may not be uniformly distributed. This sum is often dominated by the contribution (equal to 1) from the principal character of G, in which case the first approximation is not far from the truth. The evaluation of this sum is often simplified by the tendency for many characters χ to take the value 0 on at least one of the classes K_i.

Formula (1) is useful in determining whether a given finite group G is an epimorphic image of the group

$$\Delta = \Delta(k_1, \ldots, k_1) = \langle X_1, \ldots, X_r \mid X_1^{k_1} = \cdots = X_r^{k_r} = X_1 \ldots X_r = 1 \rangle.$$

First one computes $n(K_1, \ldots, K_r)$ for each choice of classes K_i of elements of order dividing k_i in G; the sum of these numbers is the number of solutions in G of the defining relations of Δ, that is, the number of homomorphisms $\theta : \Delta \to G$; by applying the same formula to appropriate subgroups of G, one estimates how many solutions generate proper subgroups, and hence one can determine whether any such θ is an epimorphism. (A more refined version of this procedure, due to Philip Hall, is described in section 6.)

3   Finite groups and surfaces

Many of the basic results in this area were proved by Hurwitz in [38]; for modern treatments, see [51, 98, 107]. Let S be a compact, connected, oriented surface without boundary, and let G be a finite group of orientation-preserving homeomorphisms from S to itself. Then G preserves a complex structure on S, so we can assume that S is a Riemann surface and G is contained in the group $\text{Aut}\, S$ of conformal automorphisms of S. There are just finitely many non-regular orbits $\Omega_1, \ldots, \Omega_r$ of G on S, the fibres of the branch-points of the natural projection $S \to T := S/G$. If T has genus h then G has generators $a_1, b_1, \ldots, a_h, b_h, x_1, \ldots, x_r$ satisfying

$$x_1^{k_1} = \cdots = x_r^{k_r} = \prod_{i=1}^{h}[a_i, b_i].\prod_{i=1}^{r} x_i = 1\,,$$

where each x_i generates the stabilizer in G of some point in Ω_i. The abstract group

$$\Gamma = \langle A_1, B_1, \ldots, A_h, B_h, X_1, \ldots, X_r \mid X_i^{k_i} = \prod_{i=1}^{h}[A_i, B_i].\prod_{i=1}^{r} X_i = 1 \rangle$$

is said to have *signature* $(h; k_1, \ldots, k_r)$. The epimorphism $\Gamma \to G$ given by $A_i, B_i, X_i \mapsto a_i, b_i, x_i$ has kernel Λ, a normal subgroup of Γ with $\Gamma/\Lambda \cong G$. The only elements of finite order in Γ are the conjugates of the powers of the generators X_i, so Λ is torsion-free; in fact, Λ is isomorphic to the fundamental group $\pi_1(S)$ of S, so it has signature $(g; -)$ where g is the genus of S.

Conversely, every torsion-free normal subgroup Λ of finite index in Γ has signature $(g; -)$ for some g, and it arises in the above way from a finite group action. If we define $M(\Gamma) = 2h - 2 + \sum_i(1 - k_i^{-1})$, then Γ acts as a discontinuous group of automorphisms of a simply-connected Riemann surface

$$X = \begin{cases} \mathcal{U} & \text{(the upper half-plane)} & \text{if } M(\Gamma) > 0, \\ \mathbb{C} & \text{(the complex plane)} & \text{if } M(\Gamma) = 0, \\ \Sigma & \text{(the Riemann sphere } \mathbb{C} \cup \{\infty\}) & \text{if } M(\Gamma) < 0, \end{cases}$$

each generator X_i inducing a rotation through $2\pi/k_i$ about some point in X. The quotient-surface $S = X/\Lambda$ is a compact Riemann surface of genus g, the quotient-group $G = \Gamma/\Lambda$ acts as a group of automorphisms of S with point-stabilizers $\langle X_i\Lambda \rangle$ of orders k_i, and $S/G = X/\Gamma$ has genus h. The *Riemann–Hurwitz formula* (a more general version of which is proved in section 8) asserts that $M(\Lambda) = |G|M(\Gamma)$, or equivalently

$$g = 1 + |G|\left(h - 1 + \frac{1}{2}\sum_{i=1}^{r}(1 - k_i^{-1})\right). \tag{2}$$

We have seen that faithful actions of G on compact Riemann surfaces S correspond to epimorphisms $\Gamma \to G$ with torsion-free kernels Λ. There is particular interest in the case where S/G is the Riemann sphere Σ. Then

$h = 0$, so $\Gamma = \Delta(k_1, \ldots, k_r)$ and the character-theoretic method of section 2 can be applied to determine whether a given finite group G is an epimorphic image of Γ. To study torsion-free kernels Λ, one restricts the classes K_i in G to consist of elements of order exactly k_i. These groups $\Delta = \Delta(k_1, \ldots, k_1)$ are sometimes called *polygonal groups* (and those with $r = 3$ *triangle groups*) since their generators X_i are rotations around the vertices of an r-sided polygon in X with internal angles π/k_i.

The notion of the signature of a group was developed mainly by Macbeath and his former students [37, 64, 92, 100], in rather more generality than here, to provide an algebraic approach to NEC groups (non-Euclidean crystallographic groups) and their actions on surfaces. The signature can be regarded as a combinatorial description of the quotient-orbifold, or alternatively as an encoded presentation for the group which reveals much of the geometry of its action. Comprehensive accounts of group actions on surfaces (including anticonformal actions) can be found in [4, 107], while Tucker gives a concise summary in [98].

4 Hurwitz groups

These are the groups attaining Hurwitz's upper bound $|G| \leq 84(g-1)$ for the automorphism group G of a compact Riemann surface S of genus $g > 1$. Hurwitz used (2) to show that this is equivalent to G being a non-trivial finite epimorphic image of $\Delta(2, 3, 7)$ (see [51, Theorem 5.11.2], for example). Since $2, 3$ and 7 are mutually coprime, a Hurwitz group must be perfect, with a nonabelian simple Hurwitz group as an epimorphic image, so much of the effort has gone into determining the simple Hurwitz groups. Malle [69] has shown that $G_2(q)$ is a Hurwitz group for $q > 4$, as is the Ree group ${}^2G_2(3^e)$ for odd $e > 1$ (see also [45, 82] for the Ree groups). Among the sporadic simple groups, $J_1, J_2, J_4, Fi_{22}, Fi'_{24}, Co_3, He, Ru, HN, Ly$ and Th are Hurwitz groups, the status of the Monster M is unknown, while the other fourteen are not Hurwitz groups [15, 101, 103, 104].

In all these cases, one can use (1) to count solutions of $x_1 x_2 x_3 = 1$ in G where x_1, x_2 and x_3 have orders 2, 3 and 7. Some solutions may generate proper subgroups of G, and these are often of type $L_2(q)$. Macbeath [63] has shown that $L_2(q)$ is a Hurwitz group precisely for $q = 7$, for primes $q \equiv \pm 1 \bmod 7$, and when $q = p^3$ for primes $p \equiv \pm 2$ or $\pm 3 \bmod 7$. His proof uses a more direct method than (1), applicable only to low-dimensional linear groups; Cohen [10] has used a similar method to show that among the groups $L_3(q)$, only $L_3(2)$ ($\cong L_2(7)$) is a Hurwitz group. Conder [12] has used Higman's method of coset diagrams to show that the alternating group A_n is a Hurwitz group for all $n \geq 168$ and for all but 64 integers $n < 168$; he gives a useful survey of Hurwitz groups, correcting some errors in the literature, in [14].

5   Symmetric genus

The *strong symmetric genus* $\sigma^0(G)$ of a finite group G is the least genus of any compact Riemann surface S such that $G \le \operatorname{Aut} S$. (By allowing elements of G to act anticonformally, one obtains the *symmetric genus* $\sigma(G) \le \sigma^0(G)$; if G has no subgroup of index 2 then it must act conformally and so $\sigma(G) = \sigma^0(G)$.) In most cases, including all finite simple groups, the Riemann–Hurwitz formula implies that $\sigma^0(G)$ is attained with S/G having genus $h = 0$, that is, with G represented as an image of some polygonal group Δ, so the method of section 2 can be applied to determine which of these groups Δ map onto G and hence to compute $\sigma^0(G)$. For example, every Hurwitz group G has (strong) symmetric genus $1 + |G|/84$, and it is shown in [48] that the Suzuki groups $^2B_2(q)$, being images of $\Delta(2,4,5)$, have symmetric genus $1 + |G|/40$. Conder, Wilson and Woldar have now dealt with all of the sporadic simple groups except the Monster; a good account of their results is given in [15], but see also [101, 102] for $G = B$ and Fi_{23}, which have symmetric genus $1 + |G|/48$ as images of $\Delta(2,3,8)$. Using different methods, Harvey [36] and Maclachlan [67] respectively have dealt with cyclic and non-cyclic abelian groups, Tucker [98] and Conder [13] with the alternating and symmetric groups, and Glover and Sjerve [31] with $L_2(q)$. Tucker gives a good introduction to the general theory of the symmetric genus and related topics in [98].

6   Hall's theory

One often needs to know in how many ways a finite group G is an image of a group Γ; more precisely, one requires the number $|\operatorname{Epi}(\Gamma, G)|$ of epimorphisms $\Gamma \to G$, or equivalently the number $n_\Gamma(G)$ of normal subgroups Λ of Γ with $\Gamma/\Lambda \cong G$. For instance, Macbeath [63] has shown that if $\Gamma = \Delta(2,3,7)$ and $G = L_2(q)$ then $n_\Gamma(G) = 3$ when $q = p \equiv \pm1 \bmod 7$, the three kernels (and the three associated surfaces) corresponding to the three classes of elements x_3 of order 7 in G; in the other cases where $G = L_2(q)$ is a Hurwitz group we have $n_\Gamma(G) = 1$, all such elements x_3 being conjugate in $\operatorname{Aut} G$.

One can often evaluate $n_\Gamma(G)$ by combining character theory with Hall's group-theoretic generalisation of the Möbius inversion formula [33]. If Γ is a finitely generated group, and G a finite group, then epimorphisms $\theta, \theta' : \Gamma \to G$ have the same kernel if and only if $\theta' = \theta \circ \alpha$ for some $\alpha \in \operatorname{Aut} G$, so $n_\Gamma(G)$ is the (finite) number of orbits of $\operatorname{Aut} G$ acting by composition on $\operatorname{Epi}(\Gamma, G)$.

This action is semi-regular, so

$$n_\Gamma(G) = \frac{|\text{Epi}\,(\Gamma, G)|}{|\text{Aut}\,G|}.$$

Now $\text{Hom}\,(\Gamma, G)$ is the disjoint union of the sets $\text{Epi}\,(\Gamma, H)$ where H ranges over the subgroups of G, so

$$|\text{Hom}\,(\Gamma, G)| = \sum_{H \leq G} |\text{Epi}\,(\Gamma, H)|.$$

Hall inverted this equation to obtain

$$|\text{Epi}\,(\Gamma, G)| = \sum_{H \leq G} \mu_G(H)|\text{Hom}\,(\Gamma, H)|,$$

where μ_G is the Möbius function on the lattice of subgroups of G, defined recursively by

$$\sum_{H \geq K} \mu_G(H) = \delta_{K,G} = \begin{cases} 1 & \text{if } K = G \\ 0 & \text{if } K < G, \end{cases}$$

so we have

$$n_\Gamma(G) = \frac{1}{|\text{Aut}\,G|} \sum_{H \leq G} \mu_G(H)|\text{Hom}\,(\Gamma, H)|. \tag{3}$$

Given sufficient knowledge of G, one can compute $|\text{Aut}\,G|$ and $\mu_G(H)$, thus reducing the problem to that of determining $|\text{Hom}\,(\Gamma, H)|$ for subgroups H of G. As we have seen, character-theoretic methods such as formula (1) will do this for certain groups Γ, and further examples will be given in section 7.

As an illustration, if $\Gamma = \Delta(2, 4, 5)$ and G is the Suzuki group $^2B_2(2^e)$ for odd $e > 1$, then

$$n_\Gamma(G) = \frac{1}{e} \sum_{d|e} \mu\left(\frac{e}{d}\right)(2^d - \delta(d)2^{(d+1)/2}),$$

where μ is the Möbius function on $\mathbf{N}$, and $\delta(d) = 1$ or -1 as $d \equiv \pm 1$ or ± 3 mod 8; this formula, proved in [48], gives the number of isomorphism classes of Riemann surfaces of genus $\sigma^0(G) = 1 + |G|/40$ with automorphism group G. A similar formula for the Ree groups $^2G_2(3^e)$, with $\Gamma = \Delta(2, 3, 7)$, is given in [45].

7   Surface groups

Besides the polygonal groups Γ, character theory also allows one to count homomorphisms $\Gamma \to G$ when Γ is a *surface group*. If $\Gamma = \pi_1(S)$ where S

is orientable and of genus $g > 0$, then Γ has signature $(g; -)$, so homomorphisms $\Gamma \to G$ correspond to solutions in G of the equation $\prod_{i=1}^{g}[a_i, b_i] = 1$. Frobenius [21] (for $g = 1$) and Mednykh [74] (for $g > 1$) have shown that

$$|\text{Hom}\,(\Gamma, G)| = |G|^{2g-1} \sum_{\chi} \chi(1)^{2-2g}, \qquad (4)$$

where χ ranges over the irreducible complex characters of G. (As with (1), one can regard $|G|^{2g-1}$ as a first approximation, based on the assumption that the values of $\prod[a_i, b_i]$ are uniformly distributed in G.) This formula, together with Hall's theory in section 6, has been used in [46] to determine $n_\Gamma(G)$ for various finite groups G, thus giving the number of equivalence classes of regular coverings of S with covering group G. (These are the coverings $R \to S = R/G$ induced by an action of G on a surface R.) Typical examples are

$$n_\Gamma(C_n) = \frac{1}{\phi(n)} \sum_{m|n} \mu\left(\frac{n}{m}\right) m^{2g}$$

where μ and ϕ are the Möbius and Euler functions on $\mathbf{N}$, and

$$n_\Gamma(D_p) = \frac{(2^{2g} - 1)(p^{2g-2} - 1)}{(p - 1)}$$

where D_p is the dihedral group of order $2p$ for an odd prime p.

There is a formula analogous to (4) for a non-orientable surface S of genus $g > 0$, where $\Gamma = \pi_1(S)$ has generators $R_1, \ldots, R_g$ with a single defining relation $R_1^2 \ldots R_g^2 = 1$. Frobenius and Schur [23] showed that

$$|\text{Hom}\,(\Gamma, G)| = |G|^{g-1} \sum_{\chi} c_\chi^g \chi(1)^{2-g}, \qquad (5)$$

where χ ranges over the irreducible complex characters of G, and c_χ is the Frobenius–Schur indicator of χ,

$$c_\chi = \frac{1}{|G|} \sum_{x \in G} \chi(x^2) = \begin{cases} 1, & \text{if } \chi \text{ is the character of a real representation,} \\ -1, & \text{if } \chi \text{ is real but its representation is not real,} \\ 0, & \text{if } \chi \text{ is not real.} \end{cases}$$

As in the orientable case, one can combine (5) with Hall's theory to find the number $n_\Gamma(G)$ of equivalence classes of regular coverings of S with a given finite covering-group G. Examples given in [46] include

$$n_\Gamma(C_n) = \frac{1}{\phi(n)} \sum_{m|n} \mu\left(\frac{n}{m}\right) \eta_m m^{g-1}$$

and

$$n_\Gamma(D_n) = \frac{1}{\phi(n)} \sum_{m|n} \mu\left(\frac{n}{m}\right) m^{g-2}(m + \eta_m(2^g - 2)),$$

where $\eta_m = 1$ or 2 as m is odd or even.

Mednykh [76] has combined the ideas underlying (1) and (4) to count homomorphisms $\Gamma \to G$ where Γ has signature $(g; k_1, \ldots, k_r)$. Hall's theory then allows one to enumerate the regular branched coverings of an orientable surface, with a given covering group G, with various restrictions on the branching-pattern; by replacing (4) with (5), one can also deal with regular branched coverings of non-orientable surfaces. These results, given in [46], are extended in [47] to enumerate normal subgroups of NEC groups without reflections, or equivalently regular coverings of 2-dimensional orbifolds without boundary.

8 Monodromy groups

Finite groups can occur in connection with surfaces as monodromy groups of finite-sheeted coverings. Let $\psi : S \to T$ be an n-sheeted covering, let $P = \{p_1, \ldots, p_r\} \subset T$ be the set of branch-points of ψ, let $T_0 = T \setminus P$ and let $S_0 = \psi^{-1}(T_0)$. Then ψ restricts to an n-sheeted unbranched covering $\psi_0 : S_0 \to T_0$, which induces an isomorphism from $\pi_1(S_0)$ to a subgroup M of index n in $\Pi := \pi_1(T_0)$. The *monodromy group* G of ψ is the transitive permutation group of degree n induced by Π on the cosets of M. One can identify these cosets with the sheets of the covering, or more precisely, with the fibre above the chosen base-point, so that each homotopy class $[\gamma] \in \Pi$ permutes the sheets by following the lifted path $\psi_0^{-1}(\gamma)$ from start to finish.

Now Π has a presentation

$$\Pi = \langle A_1, B_1, \ldots, A_h, B_h, X_1, \ldots, X_r \mid [A_1, B_1] \ldots [A_h, B_h]X_1 \ldots X_r = 1 \rangle,$$

where h is the genus of T, so G is generated by the corresponding permutations a_i, b_i, x_i, satisfying

$$[a_1, b_1] \ldots [a_h, b_h]x_1 \ldots x_r = 1. \tag{6}$$

Each generator X_i is the homotopy class of a loop around p_i, and the disjoint cycles of x_i correspond to the points in $\psi^{-1}(p_i)$, the length of each cycle giving the number of sheets meeting at such a point.

Conversely, every transitive finite permutation group (G, Ω) arises in this way. Let $|\Omega| = n$ and let G be generated by permutations a_i, b_i, x_i satisfying (6). If we choose a set $P = \{p_1, \ldots, p_r\} \subset T$ and let $\Pi = \pi_1(T_0)$ where $T_0 = T \setminus P$, then the mapping $A_i, B_i, X_i \mapsto a_i, b_i, x_i$ extends to a transitive permutation representation $\theta : \Pi \to G$ of Π on Ω. The stabilizer $M = \theta^{-1}(G_\alpha)$

of any $\alpha \in \Omega$ is a subgroup of index n in Π, so by general covering space theory [72, Chapter 5], it corresponds to an n-sheeted unbranched covering $\psi_0 : S_0 \to T_0$; this can be extended to an n-sheeted covering $\psi : S \to T$, branched at P, by compactifying S_0 with finitely many additional points, one point over p_i for each disjoint cycle of x_i. Different choices of $\alpha \in \Omega$ give conjugate stabilizers and hence equivalent coverings, all with monodromy group G.

Each cycle of length l_{ij} of x_i corresponds to a critical point of order $l_{ij} - 1$ on S, so the order of branching of ψ above p_i is $\beta(x_i) = \sum_j (l_{ij} - 1) = n - \nu(x_i)$, where $\nu(x_i)$ is the number of cycles of x_i on Ω, and hence the total order of branching of ψ is $B = \sum_i \beta(x_i)$; since S is an n-sheeted branched covering of T, it has Euler characteristic $\chi(S) = n\chi(T) - B = n(2 - 2h) - B$, so its genus is given by the Riemann–Hurwitz formula

$$g = 1 - \frac{\chi(S)}{2} = 1 + n(h-1) + \frac{1}{2}\sum_{i=1}^{r}\beta(x_i)\,. \tag{7}$$

To summarize, n-sheeted branched coverings ψ correspond to solutions of (6) which generate transitive subgroups G of the symmetric group S_n, the cycle-structures of the permutations x_i giving the pattern of branching. Two such $(2h + r)$-tuples correspond to equivalent coverings if and only if they are conjugate in S_n, a fact which has been exploited by Mednykh [74–78] to enumerate equivalence classes of coverings with given branching-patterns in terms of the irreducible characters of the symmetric group.

The covering transformations of ψ are the self-homeomorphisms α of S such that $\alpha \circ \psi = \psi$; these form the automorphism group Aut ψ of ψ, which can be identified with the centralizer of G in S^Ω, and so Aut $\psi \cong N_\Pi(M)/M$. Regular coverings ψ correspond to regular permutation groups (G, Ω), and hence to normal subgroups M of Π, with Aut $\psi \cong \Pi/M \cong G$; in this case $n = |G|$ and $\nu(x_i) = |G|/k_i$ where k_i is the order of x_i, so that (7) implies formula (2) of section 3.

If T is the Riemann sphere Σ then by using ψ to lift the complex structure from Σ to S, one can assume that S is a Riemann surface and ψ is meromorphic. The *monodromy genus* $\gamma(G)$ of a finite group G is defined to be the least genus of any connected covering of Σ with monodromy group isomorphic to G. This is obtained by putting $h = 0$, and minimising the value of g in (7) over all faithful transitive permutation representations of G, and over all sets of generators $x_1, \ldots, x_r$ for G satisfying $\prod x_i = 1$. First one calculates $\beta(x)$ for each conjugacy class of elements $x \in G$ and for each transitive representation (in fact, it is sufficient to consider just the primitive representations); then one uses the character-theoretic methods of section 2, and in some cases Hall's theory, to determine the possible generating sets, comparing their corresponding values of g. Kuiken [58] has shown that $G = C_n, D_n, A_n$ and S_n

all have $\gamma(G) = 0$, attained by their natural representations. On the other hand, Guralnick and Thompson [32] have conjectured that only finitely many simple groups of Lie type can be composition factors of monodromy groups of a given genus γ; Liebeck and Saxl [60] have shown that the order of the underlying field is bounded in terms of γ, and there is some evidence that the Lie rank is also bounded; for more on this problem, see [30]. In [49] it is shown that the Suzuki group $^2B_2(q)$ has monodromy genus $(q^2 + 16)/40$, attained by its doubly transitive representation of degree $q^2 + 1$ as an image of $\Delta(2,4,5)$; the corresponding value given in [45] for the Ree group $^2G_2(q)$ is $(q^3 - 21q + 36)/84$, in its doubly transitive representation of degree $q^3 + 1$, as an image of $\Delta(2,3,7)$.

9 Maps and hypermaps

Triangle groups and their finite images play a major role in the theories of orientable maps and hypermaps, so formula (1) has proved to be an effective tool in studying these objects. Let $\mathcal{M}$ be a map on S, that is, an embedding of a finite graph $\mathcal{G}$ in S without crossings, so that the faces (connected components of $S \setminus \mathcal{G}$) are homeomorphic to discs. We define two permutations x_1 and x_2 of the set Ω of directed edges of $\mathcal{M}$: x_1 reverses the direction of each $\alpha \in \Omega$, while x_2 uses the orientation of S to rotate each α to the next directed edge pointing to the same vertex as α. Thus the cycles of x_1 and x_2 correspond to the edges and vertices of $\mathcal{M}$, and one easily sees that the cycles of $x_3 := (x_1x_2)^{-1}$ correspond to its faces. Our topological assumptions imply that $\mathcal{G}$ is connected, so x_1 and x_2 generate a transitive subgroup $G = \langle x_1, x_2 \rangle$ of S^Ω. In general, G does not consist of automorphisms of $\mathcal{M}$ or $\mathcal{G}$, since x_1 and x_2 do not preserve incidence. Indeed, the automorphism group $\operatorname{Aut} \mathcal{M}$ of $\mathcal{M}$ (preserving orientation) can be identified with the centralizer of G in S^Ω. We say that $\mathcal{M}$ is *regular* (or *rotary*) if $\operatorname{Aut} \mathcal{M}$ acts transitively on Ω, or equivalently if G is a regular permutation group, in which case $G \cong \operatorname{Aut} \mathcal{M}$.

Clearly $x_1^2 = 1$, and the orders of x_2 and x_3 are the least common multiples m and n of the vertex-valencies and the face-valencies of $\mathcal{M}$, so there is an epimorphism $\theta : \Delta = \Delta(2,m,n) \to G, X_i \mapsto x_i$ with torsion-free kernel Λ; extending the terminology of [19] for regular maps, we say that $\mathcal{M}$ has *type* $\{n,m\}$. Conversely, any transitive finite permutation representation $\theta : \Delta \to G \leq S^\Omega$ with torsion-free kernel corresponds to a map $\mathcal{M}$ of type $\{n,m\}$ in the above way. To construct $\mathcal{M}$, we use the *universal map* $\tilde{\mathcal{M}}$ of type $\{n,m\}$; this is a polygonal tessellation of $X = \Sigma, \mathbb{C}$ or $\mathcal{U}$ as $2^{-1} + m^{-1} + n^{-1} > 1, = 1$ or < 1, and it is invariant under Δ, its vertices, edge-centres and face-centres forming the three non-regular orbits of Δ on X. We then take $\mathcal{M} = \tilde{\mathcal{M}}/M$ where $M = \theta^{-1}(G_\alpha)$ is the stabilizer in Δ of some $\alpha \in \Omega$, so that the kernel Λ of θ is the core of M in Δ. (Any fixed-points of X_1 on Ω will correspond to 'free edges', incident with one vertex and carrying one directed edge, as in Fig. 3.) We call G the *monodromy group* of $\mathcal{M}$, since it is permutation-isomorphic to the monodromy group of the branched covering $S \cong X/M \to X/\Delta \cong \Sigma$. The genus g of $\mathcal{M}$ is given by

$$2 - 2g = \chi(S) = \nu(x_1) + \nu(x_2) + \nu(x_3) - |\Omega|.$$

If we start with any topological or combinatorial map $\mathcal{M}$ and form the permutation representation $\Delta \to G$, the above process yields a map $\tilde{\mathcal{M}}/M$ isomorphic to $\mathcal{M}$ lying on a Riemann surface $S = X/M$. This can be regarded as a canonical conformal model for $\mathcal{M}$: for instance, its edges are all geodesics of S.

These general ideas, developed more fully in [50, 68], show that isomorphism classes of maps of type $\{n, m\}$ correspond to conjugacy classes of subgroups M of finite index in $\Delta(2, m, n)$ with torsion-free core, and that regular maps correspond to normal subgroups of Δ. Thus a finite group G occurs as the monodromy group of such a map, or equivalently as the automorphism group of such a regular map, if and only if it is an image of Δ with torsion-free kernel. As before, this allows the character-theoretic methods of section 2 (with $r = 3$) to be applied effectively. For instance the results in section 4 imply that Hurwitz groups such as $^2G_2(q)$ arise in this way in connection with maps of type $\{7, 3\}$ (and their duals, of type $\{3, 7\}$), and likewise the Suzuki groups for maps of type $\{4, 5\}$; the value of $n_\Delta(G)$ provided by Hall's theory gives the number of regular maps of the given type associated with each group G.

As examples of non-regular maps, the planar (genus 0) map in Fig. 1 has monodromy group $G = L_2(7)$ in its natural representation of degree 8, with $\Delta = \Delta(2, 3, 7)$, while Figs. 2 and 3 have the Mathieu group M_{12} as monodromy group, acting with degree 12, with $\Delta = \Delta(2, 3, 11)$.

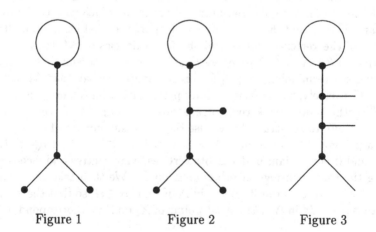

Figure 1   Figure 2   Figure 3

In the first case, $n_\Delta(L_2(7)) = 1$ and $L_2(7)$ has a single transitive representation of degree 8, so the map is unique; however, if we use the *regular* representation of $L_2(7)$ we obtain Klein's map $\mathcal{K}$, a regular map of genus 3 and type $\{7, 3\}$ consisting of 24 heptagons, with automorphism group $L_2(7)$ [57, §§ 11, 14]. In the second case $n_\Delta(M_{12}) = 2$: the methods of section 2 and section 6 show that Aut M_{12} has four orbits on the non-trivial homomorphisms $\theta : \Delta \to M_{12}$, and (in the ATLAS notation) the choices of conjugacy classes $(2A, 3B)$ and $(2B, 3A)$ for (x_1, x_2) give maximal and non-maximal proper subgroups $L_2(11)$ as images $\theta(\Delta) = \langle x_1, x_2 \rangle$, while $(2A, 3A)$ and $(2B, 3B)$, corresponding to Figs. 2 and 3, give M_{12}. Since M_{12} has two inequivalent representations of degree 12, transposed by an outer automorphism fixing x_1 and inverting x_2, each of the two kernels $\Lambda \lhd \Delta$ yields a chiral pair of maps, consisting of Figs. 2 or 3 and their mirror-images.

Another application of character theory is to the enumeration of maps. A map can be identified with a pair $x_1, x_2 \in S^\Omega$ satisfying $x_1^2 = 1$ and generating a transitive group; since isomorphism of maps corresponds to conjugacy of such pairs in S^Ω, one can use the character theory of the symmetric group to count maps (rooted or unrooted) with specified properties, as shown by Jackson and Visentin in [41, 42] for instance. By regarding a map as a covering $S \to X/\Delta \cong \Sigma$, branched over three points, one can view this as part of the general theory of enumeration of coverings considered in sections 7–8.

There is an unexpected application of this character-theoretic enumeration of maps to the computation of the orbifold characteristics of moduli spaces. Any pairing of the sides of a $2n$-gon gives rise, by identification of paired sides, to an orientable map with n edges and a single $2n$-gonal face. Let $\varepsilon_g(n)$ denote the number of pairings for which this map has genus g; for instance $\varepsilon_0(n)$ is the nth Catalan number $C(n) = \binom{2n}{n}/(n + 1)$, and $\sum_g \varepsilon_g(n) = (2n - 1)!! = (2n - 1)(2n - 3) \ldots 3.1$, the total number of such pairings. The preceding theory, applied to the dual map, shows that $\varepsilon_g(n)$ is the number of fixed-point-free involutions $x_1 \in S_{2n}$ such that $\nu(x_1 x_2) = n + 1 - 2g$, where x_2 is a fixed $2n$-cycle in S_{2n}. In [35], Harer and Zagier performed a lengthy computation of the numbers $\varepsilon_g(n)$, involving integration over spaces of matrices, to show that the moduli space $\mathcal{M}_g^1$ of Riemann surfaces of genus g with one base-point has orbifold characteristic $\chi(\mathcal{M}_g^1) = \zeta(1 - 2g)$; this is also the Euler characteristic of the mapping class group Γ_g^1 of isotopy classes of orientation- and base-point-preserving self-homeomorphisms of such a surface. Jackson [40] and Zagier [106] have subsequently used the character theory of S_{2n} to determine $\varepsilon_g(n)$ from formula (1), thus greatly simplifying the proof.

Hamilton originated the use of permutations to study maps in [34], where he used them to construct Hamiltonian circuits in the icosahedron; he noted that the regular polyhedra can all be described by permutations in this way. Coxeter and Moser give a detailed study of regular maps in [19, Chapter 8], though they use automorphism groups rather than monodromy groups, which are more effective for non-regular maps. There is a generalisation of this theory to triangle groups $\Delta(l, m, n)$ with l not necessarily equal to 2; the corresponding combinatorial objects are *hypermaps*, introduced by Cori in [17] as surface-embeddings of hypergraphs, where an edge may be incident with an arbitrary number of vertices and faces. (A typical example of a hypergraph is a projective plane, with vertices and edges given by the points and lines.) The general theory of hypermaps is very similar to that of maps, which are special cases of hypermaps; Cori and Machì give a comprehensive survey in [18], and connections with triangle groups are discussed in [52].

10   Dessins d'enfants

A Riemann surface S is compact if and only if it is isomorphic to the Riemann surface of an algebraic curve $f(x, y) = 0$ for some polynomial $f \in \mathbb{C}[x, y]$. We say that S is defined over a subfield K of $\{\mathbb{C}$ if we can take $f \in K[x, y]$. Belyĭ's Theorem [2] states that S is defined over the field $\overline{\mathbb{Q}}$ of algebraic numbers if and only if there is a *Belyĭ function* on S, that is, a meromorphic function $\beta : S \to \Sigma$ which is unbranched outside $\{0, 1, \infty\}$. (In fact, Belyĭ proved that this condition is necessary; its sufficiency follows from Weil's criterion [99], as explained by Wolfart in [105].) This condition is equivalent to S having the form X/M where M is a subgroup of finite index in a triangle group Δ acting on a simply-connected Riemann surface X, in which case β corresponds to the projection $X/M \to X/\Delta \cong \Sigma$. It follows that the Riemann surfaces defined over $\overline{\mathbb{Q}}$ are precisely those obtained from maps or hypermaps as in section 9.

Example 1 The *Fermat curve* $x^n + y^n = 1$ defines a Riemann surface S of genus $(n-1)(n-2)/2$ defined over $\overline{\mathbb{Q}}$ (in fact, over $\mathbb{Q}$), which has a Belyĭ function $\beta : (x, y) \mapsto x^n$ of degree n^2. We can take $S = X/\Delta_n'$ where Δ_n' is the derived group of the triangle group $\Delta_n := \Delta(n, n, n)$, and X is $\mathcal{U}, \mathbb{C}$ or Σ as $n \geq 4, n = 3$ or $n \leq 2$. This surface S carries a regular hypermap of type (n, n, n), called the *Fermat hypermap* $\mathcal{F}_n$, with monodromy group and automorphism group $\Delta_n/\Delta_n' \cong C_n \times C_n$; the vertices and edges of the embedded hypergraph are the rows and columns of an $n \times n$ chess-board, and the automorphisms are obtained by multiplying x and y by powers of $\zeta_n = \exp(2\pi i/n)$ [52].

Example 2 A *Lefschetz surface* S has equation $y^p = x^m(x - 1)$, where p is a prime greater than 3 and $1 \leq m < p - 1$; it has genus $(p-1)/2$ and has an automorphism $a : y \mapsto y\zeta_p$ of order p. It is visibly defined over $\mathbb{Q}$, and it has a Belyĭ function $(x, y) \mapsto x$, a regular branched covering $S \to \Sigma$ of degree p induced by the group $G = \langle a \rangle$. This corresponds to a normal subgroup M of index p in Δ_p, so it defines a regular hypermap of type (p, p, p) on S, of the form $\mathcal{F}_p/C_p$ since M contains Δ_p' as a normal subgroup of index p.

Example 3 *Klein's quartic curve*, given by $u^3v + v^3w + w^3u = 0$ in homogeneous projective coordinates, defines a Riemann surface $S = \mathcal{U}/K$ of genus 3, where K is the unique normal subgroup of $\Delta = \Delta(2,3,7)$ with $\Delta/K \cong L_2(7)$ [6, §§267, 303; 57; 80, pp. 130–133]. The projection $S \to \mathcal{U}/\Delta \cong \Sigma$ is a Belyĭ function of degree 168, corresponding to Klein's map $\mathcal{K}$ on S, but a simpler function is obtained from the fact that $K < \Delta_7 < \Delta$: putting $u/w = -x/y^2$ and $v/w = -x/y^3$ we get $y^7 = x^2(x-1)$, so by Example 2 there is a Belyĭ function $(x,y) \mapsto x$ of degree 7, corresponding to a regular hypermap of type $(7,7,7)$ on S.

These combinatorial structures associated with Riemann surfaces over $\overline{\mathbb{Q}}$ are called *dessins d'enfants* by Grothendieck [29], no doubt motivated by pictures like Figs. 1, 2 and 3. He first pointed out their connections with algebraic geometry and Galois theory, and a number of interesting examples were developed by Shabat and Voevodsky [88]. Much of the current interest in these objects lies in the insight they provide into the Galois group of $\overline{\mathbb{Q}}$ over $\mathbb{Q}$: this uncountable profinite group (the projective limit of the Galois groups of the algebraic number fields) has a faithful action on *dessins*, induced by its natural action on Belyĭ pairs (S, β). Examples 1–3, defined over $\mathbb{Q}$, are fixed by $\mathrm{Gal}\,\overline{\mathbb{Q}}$, but the *dessins* in Figs. 2 and 3 correspond to Belyĭ pairs defined over $\mathbb{Q}(\sqrt{-11})$ [70, p. 163], and complex conjugation transposes each with its mirror-image.

An important tool in studying *dessins d'enfants* is the monodromy group (of $\mathcal{M}$, or equivalently of β): for instance, it is invariant under $\mathrm{Gal}\,\overline{\mathbb{Q}}$ [54]. As we have seen, character theory is effective in determining how a given finite group G can arise in this way as an image of a triangle group Δ, so it has an important role to play in this theory. For surveys on *dessins* see [11, 53], and for recent research see [83, 89].

11 The Inverse Galois Problem

The original problem, closely related to the previous topic, is to prove Hilbert's conjecture that every finite group G is the Galois group (over $\mathbb{Q}$) of an algebraic number field. More generally, one would like to realise G as the Galois group of a regular extension E of the rational function field $\mathbb{Q}(T)$, where 'regular' means that $\overline{\mathbb{Q}} \cap E = \mathbb{Q}$; if G has this property, denoted by Gal_T, then Hilbert's irreducibility theorem implies that by assigning suitable values to T one can realise G as a Galois group over every algebraic number field. (For the details, see [73, 87], and for further general background to this problem, see [39, 86, 96].) Shafarevich [90] has shown that every finite soluble group can be realised as a Galois group over $\mathbb{Q}$, and abelian groups are known to satisfy Gal_T, but in general the problem seems to be very difficult, even for simple groups of low rank over prime fields. For instance, although results of Malle, Matzat and Shih [71, 91] show that $L_2(p)$ satisfies Gal_T for 'most' primes p, there remain infinitely many groups $L_2(p)$ yet to be realised as Galois groups over $\mathbb{Q}$.

Although this problem is algebraic, solutions usually start topologically, realising G as the monodromy group of a branched covering of Σ, and hence as the Galois group of the corresponding extension of the function-field $\mathbb{C}(T)$ of Σ. Rigidity then allows descent to a regular extension of $\mathbb{Q}(T)$ with Galois group G.

Let G be generated by elements $x_1, \ldots, x_r$ satisfying $x_1 \ldots x_r = 1$, and let $P = \{p_1, \ldots, p_r\} \subset \Sigma$. If we apply the ideas in section 8 to the regular representation of G, we obtain a Riemann surface S and a regular covering $\psi : S \to \Sigma$, branched over P, with covering group and monodromy group isomorphic to G; each x_i describes the branching of ψ over p_i. Now the meromorphic functions $f : \Sigma \to \Sigma$ form the rational function field $\mathbb{C}(T)$, with $\mathbb{C}$ and T corresponding to the constant and identity functions, and composition with ψ embeds $\mathbb{C}(T)$ in the field F of meromorphic functions on S. This is a normal extension of $\mathbb{C}(T)$ with G acting as its Galois group [97]. The difficult part of the method is now to choose the points p_i and the generators x_i so that this extension restricts to an extension E of $\mathbb{Q}(T)$ with the same Galois group.

A conjugacy class K in G is *rational* if every character of G takes rational values on K, or equivalently, if $g \in K$ implies $g^i \in K$ for every generator g^i of $\langle g \rangle$. If $K_1, \ldots, K_r$ are conjugacy classes in G, let

$$\Sigma(K_1, \ldots, K_r) = \{\, (g_1, \ldots, g_r) \in K_1 \times \cdots \times K_r \mid g_1 \ldots g_r = 1, \langle g_1, \ldots, g_r \rangle = G \,\}$$

Now G acts by conjugation on this set, and if G has trivial centre then it acts semi-regularly; in [95], Thompson has defined $(K_1, \ldots, K_r)$ to be *rigid* if $\Sigma(K_1, \ldots, K_r)$ is non-empty and is permuted regularly by G, that is, if $|\Sigma(K_1, \ldots, K_r)| = |G|$. Using the character theory in section 2 one can calculate $n(K_1, \ldots, K_r)$, the number of solutions $(g_i) \in K_1 \times \cdots \times K_r$ of $g_1 \ldots g_r = 1$, and section 6 shows how to count those which generate G, so these techniques enable one to find rigid r-tuples in finite groups. The following theorem, due in various forms to Belyĭ, Fried, Matzat, Shih and Thompson, now gives sufficient conditions for descent from $\mathbb{C}$ to $\mathbb{Q}$:

Theorem 1 *If G is a finite group with trivial centre, and $(K_1, \ldots, K_r)$ is a rigid r-tuple of rational conjugacy classes of G, then there is a regular Galois extension E of $\mathbb{Q}(T)$ with Galois group G.*

(This extension is ramified over a set of rational points $p_i \in \Sigma$, and the generators of the corresponding inertia groups belong to the classes K_i.)

For instance S_n has trivial centre (for $n \geq 3$) and all its conjugacy classes are rational; character theory or a simple direct argument shows that the classes consisting of the transpositions, $(n-1)$-cycles and n-cycles form a rigid triple, so S_n satisfies Gal_T. In general, the hypotheses of the theorem are too strong to apply to many groups: thus A_5 has rigid triples (of elements of orders $2, 3$ and 5), but has no rational rigid r-tuples. However, there are less restrictive conditions which are sufficient to imply Gal_T for several classes of groups, including the alternating groups and all the sporadic simple groups except possibly M_{23} [87].

12 Action on homology

So far, I have shown how character theory can be used to count solutions of equations, and the underlying representations have not played a significant role. However, any $G \leq \mathrm{Aut}\, S$ has induced actions on various modules, such as the homology and cohomology groups associated with S, and these representations can be very useful in solving specific problems (see [82, 107] for the general theory).

As in section 3, let $G = \Gamma/\Lambda$ be a finite group of automorphisms of a compact Riemann surface $S = X/\Lambda$ of genus g. The first integer homology group $H_1 = H_1(S; \mathbb{Z})$ of S can be identified with the abelianized fundamental group $\Lambda^{\mathrm{ab}} = \Lambda/\Lambda' \cong \mathbb{Z}^{2g}$, and the natural action of G on H_1 corresponds to the action (induced by conjugation) of Γ/Λ on Λ^{ab}.

Much of the information about the representation of G on H_1 is embodied in its character, which can be computed as follows. Take a triangulation of S/G with vertices including all the non-regular orbits of G. By lifting this to S we obtain a G-invariant triangulation of S, with G acting freely on the i-simplices for $i > 0$. If σ_i and τ_i are the characters of the representations of G on the associated chain-groups C_i and homology groups H_i (with integer coefficients), then the Hopf trace formula [1, §9.4] gives

$$\sum_{i \geq 0} (-1)^i \sigma_i = \sum_{i \geq 0} (-1)^i \tau_i\,,$$

the value of the right-hand side at each $x \in G$ being the Lefschetz number of x. By our choice of triangulation, each non-identity $x \in G$ satisfies $\sigma_i(x) = 0$ for $i > 0$, while $\sigma_0(x)$ is the number $\phi(x)$ of fixed-points of x on S. Since S is connected, H_0 is 1-dimensional, affording the trivial representation of G, and the same applies to H_2 since G preserves the orientation of S. Since $H_i = 0$ for $i > 2$ we therefore have

$$\tau_1(x) = \begin{cases} 2 - \phi(x) & \text{if } x \neq 1 \\ 2g & \text{if } x = 1. \end{cases} \tag{8}$$

In [62] Macbeath has used (8) to find explicit equations for a Riemann surface S of genus 7 with Aut $S \cong L_2(8)$; in general, this passage from group to equations is very difficult. In [65] he proved that

$$\phi(x) = |N_G(\langle x \rangle)| \sum_{i=1}^{r} \frac{\varepsilon_i(x)}{k_i} \qquad (9)$$

for $x \neq 1$, where $\varepsilon_i(x) = 1$ or 0 as x is or is not conjugate to a power of one of the generators x_i of G, and then used this to compute τ_1 for several classes of groups G, such as C_n and $L_2(q)$.

An important corollary of (8) is Hurwitz's theorem that G acts faithfully on H_1 for $g \geq 2$ (there are obvious counterexamples for $g \leq 1$); since G preserves orientation it is therefore embedded in $SL_{2g}(\mathbb{Z})$. In fact G is embedded in the symplectic group $Sp_{2g}(\mathbb{Z})$, since the intersection form, giving the algebraic number of intersections of two 1-cycles in S, is a G-invariant nondegenerate skew-symmetric bilinear form on H_1.

Hurwitz's theorem imposes strong restrictions on the elements $x \in G$. For example, by factorising the characteristic polynomial of x as a product of cyclotomic polynomials, Kirby [56] has shown that $2g \geq \psi(k)$ for a certain function $\psi(k)$ of the order k of x. If $k = p$ is prime then $\psi(p)$ coincides with Euler's function $\phi(p) = p - 1$, so $p \leq 2g + 1$ for $g \geq 2$. The surfaces attaining this bound are the Lefschetz surfaces in Example 2 of section 10; Riera and Rodríguez have analysed them in detail in [81], using some of the ideas in this section.

Another immediate corollary of (8), obtained by considering eigenvalues on H_1, is that each non-identity automorphism of S has at most $2g + 2$ fixed-points, and that any automorphism attaining this bound must be a central involution. The surfaces possessing such an automorphism are the *hyperelliptic surfaces* $y^2 = (x - a_1) \ldots (x - a_{2g+2})$, where the automorphism $(x, y) \mapsto (x, -y)$ has fixed-points $(a_i, 0)$.

Since H_0 is torsion-free, the universal coefficient theorem [28, §29] implies that $H_1(S; R) = H_1 \otimes_{\mathbb{Z}} R$ for any commutative ring of coefficients R, so one can view (8) as giving the character of the representation ρ_1 of G on $H_1(S; R)$. This is particularly useful if $R = \mathbb{C}$, when one can use (8) to calculate the multiplicity

$$(\tau_1, \chi)_G = |G|^{-1} \sum_{x \in G} \tau_1(x)\overline{\chi(x)}$$

of each irreducible character χ of G in τ_1. As we saw in sections 9–10, an important case is that in which Γ is a triangle group. In this situation, Streit [93] has determined an ideal in the group-algebra $\mathbb{C}G$ affording ρ_1, and has used this (with the aid of GAP [84]) to obtain explicit representations of certain automorphism groups G and surfaces S by matrices and polynomials.

One can also study ρ_1 through the natural action of G as a group of automorphisms of the cohomology ring of S. The first de Rham cohomology group $H^1(S; \mathbb{C})$ is the $2g$-dimensional vector space of smooth closed differentials modulo the smooth exact differentials, or equivalently the space of harmonic differentials on S [20, §III.2, §V.3]. Integration of differentials along 1-chains gives a G-invariant non-singular pairing $H^1(S; \mathbb{C}) \times H_1(S; \mathbb{C}) \to \mathbb{C}$ (the cap product), so the representation of G on $H^1(S; \mathbb{C})$ is the dual of ρ_1; since G is finite and τ_1 is real, these representations are in fact equivalent. Now $H^1(S; \mathbb{C})$ is the direct sum of the G-invariant g-dimensional spaces $\Omega(S)$ and $\overline{\Omega}(S)$ of holomorphic and anti-holomorphic differentials, these two summands affording dual (and hence conjugate) representations $\rho_{1,0}$ and $\rho_{0,1} = \overline{\rho_{1,0}}$ of G. If $x \in G$ has order $k > 1$, then at each of its finitely many fixed-points on S its derivative (in local coordinates) is ζ_k^j for some j coprime to k, where $\zeta_k = \exp(2\pi i/k)$; the Eichler trace formula [20, §V.2.9] states that x has trace $1 + \sum(\zeta_k^j - 1)^{-1}$ on $\Omega(S)$, summing over all the fixed-points of x, and from this one can deduce the decomposition of τ_1 into irreducible components. For instance if G acts freely on S then $\rho_{1,0}$ is the sum of the principal representation and $h - 1$ copies of the regular representation, where S/G has genus h.

The G-module $\Omega(S)$ is the special case $q = 1$ of the representation of G on the space of holomorphic q-differentials: when $q > 1$, this has degree $(2q - 1)(g - 1)$ or g as $g \geq 2$ or $g \leq 1$. Chevalley and Weil [8] showed that in this representation, an irreducible representation ρ of G of degree d has multiplicity

$$m_\rho = d(2q - 1)(h - 1) + \sum_{i=1}^{r} \sum_{j=0}^{k_i - 1} N_{i,j}\left((q - 1)(1 - k_i^{-1}) + \left\langle \frac{q - 1 - j}{k_i} \right\rangle\right) + \sigma;$$

here $\sigma = 1$ if $q = 1$ and ρ is the principal representation (so $m_\rho = h$), and $\sigma = 0$ otherwise; $\langle t \rangle = t - \lfloor t \rfloor$ is the fractional part of a real number t, and $N_{i,j}$ is the multiplicity of $\zeta_{k_i}^j$ as an eigenvalue of $\rho(x_i)$. For further details and applications of this representation of G, see [20, 59].

13 Abelian coverings

If, in section 12, we take $R = \mathbb{Z}_n$ as the ring of coefficients, we obtain the reduction of ρ_1 mod n, that is, the representation of G on the homology group $H_1(S; \mathbb{Z}_n) \cong \Lambda/\Lambda'\Lambda^n \cong (\mathbb{Z}_n)^{2g}$; Serre [85] showed that if $g \geq 2$ then this is faithful for all $n \geq 3$, whereas for $n = 2$ its kernel has exponent dividing 2. One can use this representation to construct regular abelian coverings of S: any G-submodule of $\Lambda_n := \Lambda/\Lambda'\Lambda^n$ corresponds to a normal subgroup M of Γ lying between Λ and $\Lambda'\Lambda^n$, so it gives a regular unbranched covering

$X/M \to X/\Lambda = S$, with the finite abelian group Λ/M as the group of covering transformations.

To illustrate this technique, let Γ be the triangle group $\Delta = \Delta(2,3,7)$ and let G be the Hurwitz group $L_2(7)$. Then $n_\Delta(G) = 1$, that is, there is a single normal subgroup Λ of Δ with $\Delta/\Lambda \cong G$, namely the subgroup $\Lambda = K$ associated with Klein's surface $S = \mathcal{U}/K$ of genus $g = 3$ in section 10. The zero submodule of Λ_n lifts to the subgroup $M = \Lambda'\Lambda^n$, which is characteristic in Λ and hence normal in Δ; this gives a Hurwitz group Δ/M of genus $1+2n^6$ and type $n^6.L_2(7)$, that is, an extension of a normal subgroup $(\mathbb{Z}_n)^6$ by $L_2(7)$. Indeed, this is how Macbeath first proved the existence of infinitely many Hurwitz groups in [61].

For certain values of n, further Hurwitz groups can be obtained from proper submodules of Λ_n, and to find these we need to reduce ρ_1 mod n. For simplicity, I will follow Sah [82] and take n to be a prime q. Now (9) gives the character-values $\tau_1(x) = 6, -2, 0, 2, -1$ as $x \in G$ has order $1, 2, 3, 4, 7$ respectively, so $\tau_1 = \chi_2 + \chi_3$ in the ATLAS notation [16], the sum of two conjugate irreducible characters of degree 3 with values in $\mathbb{Q}(\sqrt{-7})$. With knowledge of the ordinary and modular representations of G [5, 16, 44] one can use this to determine the structure of the $\mathbb{Z}_q G$-module Λ_q. (On this point, comment (b) on p. 37 of [82] is incorrect, and in fact there is no Hurwitz group of genus 55; similarly, in comment (e) on p. 38, the value $g = 55$ should be deleted, and likewise 73 since $U_3(3)$ is not a Hurwitz group.)

If $q \equiv 1, 2$ or $4 \bmod 7$ then -7 is a quadratic residue mod q, and Λ_q is a direct sum of two irreducible 3-dimensional G-submodules; these give two normal subgroups M_1, M_2 of Δ of index q^3 in Λ, and hence two non-isomorphic surfaces $\mathcal{U}/M_i$ of genus $1+2q^3$, each acted on by a Hurwitz group of type $q^3.L_2(7)$. If $q = 7$ then Λ_q is reducible but indecomposable, with a single proper G-submodule of dimension 3, so we get a single surface of genus $1 + 2.7^3 = 687$ associated with a Hurwitz group of type $7^3.L_2(7)$. If $q \equiv 3, 5$ or $6 \bmod 7$ then -7 is not a quadratic residue mod q, and Λ_q is irreducible, so we obtain no further Hurwitz groups in this case. In fact, Cohen [9] has determined all the Hurwitz groups having an abelian normal subgroup A with quotient-group $L_2(7)$; those described here are the examples where A is elementary abelian.

14 Applications to maps and hypermaps

The ideas in sections 12 and 13 can be combined with those in section 9 to obtain further results about maps and hypermaps. For instance Biggs [3] and Machì [66] have applied (8) to automorphisms of these structures, with $\phi(x)$ the total number of vertices, edges and faces invariant under x; as before, this gives a faithful symplectic representation on H_1 for $g \geq 2$. Similarly, by finding suitable submodules of $H_1(\mathcal{M}; \mathbb{Z}_n)$ one can construct maps with specific properties as abelian coverings of a given map $\mathcal{M}$. For example, the normal subgroups M of $\Delta(2,3,7)$ constructed in section 13 correspond to regular maps of type $\{7,3\}$ which are abelian coverings of Klein's map $\mathcal{K}$ of genus 3. In particular, the normal subgroups M_1 and M_2 obtained from primes $q \equiv 1, 2$ or $4 \bmod 7$ are conjugate in the extended triangle group $\Delta^*(2,3,7)$ (which contains $\Delta(2,3,7)$ with index 2), so their associated maps form chiral pairs. By Dirichlet's theorem on primes in arithmetic progressions, infinitely many such pairs can be obtained in this way.

15   Isospectral surfaces

A classic problem asks to what extent a permutation representation is determined by its character. This has unexpected applications to algebraic number fields [24, 79], to Riemann surfaces [7], and to Kac's famous question 'Can one hear the shape of a drum?' [55].

Two subgroups H_1 and H_2 of a finite group G are *almost conjugate* if the representations of G on the cosets of H_1 and H_2 have the same character, or equivalently $|H_1 \cap K| = |H_2 \cap K|$ for each conjugacy class K of G. Conjugate subgroups are almost conjugate, but the converse is false. For instance if H_1 and H_2 are non-isomorphic groups with the same numbers of elements of each order (p-groups provide plenty of examples), then they have the same order n, and their regular representations embed them as non-conjugate subgroups of $G = S_n$; however, they are almost conjugate since each element of order k in H_i consists of n/k k-cycles, and such permutations form a single class in S_n. Similarly, if $n \geq 3$ then the stabilizers H_1 and H_2 of a point and a hyperplane in $G = PGL_n(q)$ are not conjugate, but they are almost conjugate: to see this, choose H_2 to be represented by the transposes of the matrices representing H_1, and use the fact that every square matrix over a field is conjugate to its transpose. Guralnick and Saxl [31] give an infinite sequence of alternating groups G in which a maximal subgroup H_1 is almost conjugate to a non-maximal subgroup H_2, so in general a permutation character does not determine primitivity.

If S is a compact Riemann surface of genus $g > 1$ then, as a quotient of the hyperbolic plane, S is a 2-dimensional Riemannian manifold of curvature -1; in this metric, isometries correspond to conformal or anticonformal isomorphisms of Riemann surfaces. The *length spectrum* of S is the increasing sequence of lengths of its primitive closed geodesics (where 'primitive' here means not a proper iterate of another geodesic). Two surfaces are *isospectral* if they have the same length spectra; by Huber's theorem, this is equivalent to their Laplace operators having the same eigenvalue spectra. Isometric surfaces are isospectral, and Gel'fand conjectured that the converse is true.

However, counterexamples can be found as follows (see [7] for the details and
for the background to this problem). If $G \leq \operatorname{Aut} S$ for some compact Riemann surface S, and if H_1, H_2 are almost conjugate subgroups of G acting
freely (without fixed points) on S, then Sunada's theorem [94] implies that
the surfaces S/H_i are isospectral. However, if H_1 and H_2 are not conjugate,
one can choose S so that these two surfaces are not isometric: the basic idea
is to 'thicken out' a Cayley diagram $\mathcal{C}$ for G, replacing its vertices and edges
with spheres and tubes, to obtain a Riemann surface S on which G acts con-
formally and freely; since the two coset diagrams $\mathcal{C}/H_i$ are not isomorphic (as
labelled directed graphs), and since the geometry of this thickening can be
arranged so that $\mathcal{C}/H_i$ can be reconstructed from S/H_i, it follows that S/H_1
and S/H_2 cannot be isometric. For a negative answer to Kac's question, involving an extension of this idea to orbifolds and hence to simply connected
domains in $\mathbb{R}^2$, see [26].

ACKNOWLEDGEMENTS

I am very grateful to Peter Buser, David Singerman, Manfred Streit, Robert
Wilson and Alexander Zvonkin for their helpful comments on early drafts of
this paper.

References

[1] M. A. Armstrong, *Basic topology*, McGraw-Hill, London, 1979.

[2] G. V. Belyĭ, On Galois extensions of a maximal cyclotomic field,
Izv. Akad. Nauk SSSR **43** (1979), 269–276 (Russian); *Math. USSR
Izvestiya* **14** (1980), 247–256 (English transl.).

[3] N. L. Biggs, The symplectic representation of map automorphisms,
Bull. London Math. Soc. **4** (1972), 303–306.

[4] E. Bujalance, J. J. Etayo, J. M. Gamboa and G. Gromadzki, *Automorphism groups of compact bordered Klein surfaces*, Lecture Notes
in Mathematics 1439, Springer-Verlag, Berlin/Heidelberg/New York,
1990.

[5] R. Burkhardt, Die Zerlegungsmatrizen der Gruppen $PSL(2, p^f)$, *J. Algebra* **40** (1976), 75–96.

[6] W. Burnside, *Theory of groups of finite order*, Cambridge University
Press, 1911.

[7] P. Buser, *Geometry and spectra of compact Riemann surfaces*, Birkhäuser, Boston, 1992.

[8] C. Chevalley and A. Weil, Über das Verhalten der Integrale 1. Gattung bei Automorphismen des Funktionenkörpers, *Abh. Math. Sem. Univ. Hamburg* **10** (1934), 358–361.

[9] J. M. Cohen, On Hurwitz extensions by $PSL_2(7)$, *Math. Proc. Cambridge Philos. Soc.* **86** (1979), 395–400.

[10] J. M. Cohen, On non-Hurwitz groups and non-congruence subgroups of the modular group, *Glasgow Math. J.* **22** (1981), 1–7.

[11] P. B. Cohen, C. Itzykson and J. Wolfart, Fuchsian triangle groups and Grothendieck dessins. Variations on a theme of Belyi, *Comm. Math. Phys.* **163** (1994), 605–627.

[12] M. D. E. Conder, Generators for alternating and symmetric groups, *J. London Math. Soc. (2)* **22** (1980), 75–86.

[13] M. D. E. Conder, The symmetric genus of alternating and symmetric groups, *J. Combinatorial Theory (B)* **39** (1985), 179–186.

[14] M. D. E. Conder, Hurwitz groups: a brief survey, *Bull. Amer. Math. Soc. (N. S.)* **23** (1990), 359–370.

[15] M. D. E. Conder, R. A. Wilson and A. J. Woldar, The symmetric genus of sporadic groups, *Proc. Amer. Math. Soc.* **116** (1992), 653–663.

[16] J. H. Conway, R. T. Curtis, S. P. Norton, R. A. Parker and R. A. Wilson, *An* ATLAS *of Finite Groups*, Clarendon Press, Oxford, 1985.

[17] R. Cori, Un code pour les graphes planaires et ses applications, *Astérisque* **27** (1975).

[18] R. Cori and A. Machì, Maps, hypermaps and their automorphisms: a survey I, II, III, *Expositiones Math.* **10** (1992), 403–427, 429–447, 449–467.

[19] H. S. M. Coxeter and W. O. J. Moser, *Generators and relations for discrete groups*, Springer-Verlag, Berlin/Göttingen/Heidelberg/New York, 1965.

[20] H. M. Farkas and I. Kra, *Riemann surfaces*, Springer-Verlag, New York/Heidelberg/Berlin, 1980.

[21] G. Frobenius, Über Gruppencharaktere, *Sitzber. Königlich Preuss. Akad. Wiss. Berlin* (1896), 985–1021.

[22] G. Frobenius, Über einen Fundamentalsatz der Gruppentheorie, II, *Sitzber. Königlich Preuss. Akad. Wiss. Berlin* (1907), 428–437.

[23] G. Frobenius and I. Schur, Über die reellen Darstellungen der endlichen Gruppen, *Sitzber. Königlich Preuss. Akad. Wiss. Berlin* (1906), 186–208.

[24] F. Gassmann, Bemerkungen zur vorstehenden Arbeit von Hurwitz, *Math. Z.* **25** (1926), 665–675.

[25] H. Glover and D. Sjerve, The genus of $PSL_2(q)$, *J. reine angew. Math.* **380** (1987), 59–86.

[26] C. Gordon, D. L. Webb and S. Wolpert, One cannot hear the shape of a drum, *Bull. Amer. Math. Soc. (N.S.)* **27** (1992), 134–138.

[27] D. Gorenstein, *Finite groups*, Harper and Row, New York, 1968.

[28] L. Greenberg, *Lectures on algebraic topology*, Benjamin, New York, 1966.

[29] A. Grothendieck, Esquisse d'un programme, Preprint, Montpellier, 1984.

[30] R. M. Guralnick, The genus of a permutation group, in *Groups, combinatorics and geometry, Durham 1990 (eds. M. W. Liebeck and J. Saxl)*, pp. 351–363. London Math. Soc. Lecture Note Series 165, Cambridge University Press, 1992.

[31] R. M. Guralnick and J. Saxl, Primitive permutation characters, in *Groups, combinatorics and geometry, Durham 1990 (eds. M. W. Liebeck and J. Saxl)*, pp. 364–367. London Math. Soc. Lecture Note Series 165, Cambridge University Press, 1992.

[32] R. M. Guralnick and J. G. Thompson, Finite groups of genus zero, *J. Algebra* **131** (1990), 303–341.

[33] P. Hall, The Eulerian functions of a group, *Quart. J. Math. (Oxford)* **7** (1936), 134–151.

[34] W. R. Hamilton, Letter to John T. Graves 'On the Icosian' (17th October 1856), in *Mathematical papers, Vol. III, Algebra (eds. H. Halberstam and R. E. Ingram)*, pp. 612–625. Cambridge University Press, 1967.

[35] J. Harer and D. Zagier, The Euler characteristic of the moduli space of curves, *Invent. Math.* **85** (1986), 457–485.

[36] W. J. Harvey, Cyclic groups of automorphisms of a compact Riemann surface, *Quart. J. Math. (Oxford) (2)* **17** (1966), 86–97.

[37] A. H. M. Hoare, Subgroups of NEC groups and finite permutation groups, *Quart. J. Math. (Oxford) (2)* **41** (1990), 45–59.

[38] A. Hurwitz, Über algebraische Gebilde mit eindeutigen Transformationen in sich, *Math. Ann.* **41** (1893), 403–442.

[39] Y. Ihara, K. Ribet and J.-P. Serre (eds.), *Galois groups over* **Q**, Springer-Verlag, New York, 1989.

[40] D. M. Jackson, Counting cycles in permutations by group characters, with an application to a topological problem, *Trans. Amer. Math. Soc.* **299** (1987), 785–801.

[41] D. M. Jackson and T. I. Visentin, A character theoretic approach to embeddings of rooted maps in an orientable surface of given genus, *Trans. Amer. Math. Soc.* **322** (1990), 343–363.

[42] D. M. Jackson and T. I. Visentin, Character theory and rooted maps in an orientable surface of given genus: face-colored maps, *Trans. Amer. Math. Soc.* **322** (1990), 365–376.

[43] G. D. James and M. W. Liebeck, *Representations and characters of groups*, Cambridge University Press, 1993.

[44] C. Jansen, K. Lux, R. Parker and R. Wilson, *An* ATLAS *of Brauer characters*, Clarendon Press, Oxford, 1995.

[45] G. A. Jones, Ree groups and Riemann surfaces, *J. Algebra* **165** (1994), 41–62.

[46] G. A. Jones, Enumeration of homomorphisms and surface-coverings, *Quart. J. Math. (Oxford) (2)* **46** (1995), 485–507.

[47] G. A. Jones, Counting normal subgroups of non-euclidean crystallographic groups, preprint, University of Southampton, 1994.

[48] G. A. Jones and S. A. Silver, Suzuki groups and surfaces, *J. London Math. Soc. (2)* **48** (1993), 117–125.

[49] G. A. Jones and S. A. Silver, The monodromy genus of the Suzuki groups, in *Proc. 2nd internat. conf. on algebra (eds. L. A. Bokut', A. I. Kostrikin and K. K. Kutateladze)*, pp. 215–224. American Math. Soc. Contemporary Mathematics Ser. **184** (1995).

[50] G. A. Jones and D. Singerman, Theory of maps on orientable surfaces, *Proc. London Math. Soc. (3)* **37** (1978), 273–307.

[51] G. A. Jones and D. Singerman, *Complex functions, an algebraic and geometric viewpoint*, Cambridge University Press, 1987.

[52] G. A. Jones and D. Singerman, Maps, hypermaps and triangle groups, in *The Grothendieck theory of dessins d'enfants (ed. L. Schneps)*, pp. 115–145. London Math. Soc. Lecture Note Series 200, Cambridge University Press, 1994.

[53] G. A. Jones and D. Singerman, Belyĭ functions, hypermaps and Galois groups, *Bull. London Math. Soc.*, to appear.

[54] G. A. Jones and M. Streit, Galois groups, monodromy groups and cartographic groups, Preprint, Frankfurt and Southampton, 1995.

[55] M. Kac, Can one hear the shape of a drum?, *Amer. Math. Monthly* **73** (1966), 1–23.

[56] D. Kirby, Integer matrices of finite order, *Rendiconti Mat.* **2** (1969), 1–6.

[57] F. Klein, Über die Transformationen siebenter Ordnung der elliptischen Functionen, *Math. Ann.* **14** (1879), 428–471. (*Gesammelte Mathematische Abhandlungen III*, Springer-Verlag, Berlin, 1923.)

[58] K. Kuiken, On the monodromy groups of Riemann surfaces of genus zero, *J. Algebra* **59** (1979), 481–489.

[59] J. Lewittes, Automorphisms of compact Riemann surfaces, *Amer. J. Math.* **85** (1963), 734–752.

[60] M. W. Liebeck and J. Saxl, Minimal degrees of primitive permutation groups, with an application to monodromy groups of Riemann surfaces, *Proc. London Math. Soc. (3)* **63** (1991), 266–314.

[61] A. M. Macbeath, On a theorem of Hurwitz, *Proc. Glasgow Math. Assoc.* **5** (1961), 90–96.

[62] A. M. Macbeath, On a curve of genus 7, *Proc. London Math. Soc. (3)* **15** (1965), 527–542.

[63] A. M. Macbeath, Generators of the linear fractional groups, *Proc. Sympos. Pure Math.* **12** (1967), 14–32.

[64] A. M. Macbeath, The classification of non-euclidean plane crystallographic groups, *Canad. J. Math.* **19** (1967), 1192–1205.

[65] A. M. Macbeath, Action of automorphisms of a compact Riemann surface on the first homology group, *Bull. London Math. Soc.* **5** (1973), 103–108.

[66] A. Machì, Homology of hypermaps, *J. London Math. Soc. (2)* **31** (1985), 10–16.

[67] C. Maclachlan, Abelian groups of automorphisms of compact Riemann surfaces, *Proc. London Math. Soc. (3)* **15** (1965), 699–712.

[68] J. Malgoire and C. Voisin, Cartes cellulaires, *Cahiers Mathématiques de Montpellier* **12** (1977).

[69] G. Malle, Hurwitz groups and $G_2(q)$, *Canadian Math. Bull.* **33** (1990), 349–357.

[70] G. Malle, Fields of definition of some three point ramified field extensions, in *The Grothendieck theory of dessins d'enfants (ed. L. Schneps)*, pp. 115–145. London Math. Soc. Lecture Note Series 200, pp. 147–168. Cambridge University Press, 1994.

[71] G. Malle and B. H. Matzat, Realisierung von Gruppen $PSL_2(\mathbf{F}_p)$ als Galoisgruppen über **Q**, *Math. Ann.* **272** (1985), 549–565.

[72] W. S. Massey, *Algebraic topology: an Introduction*, Harcourt, Brace and World, New York, 1967.

[73] B. H. Matzat, *Konstruktive Galoistheorie*, Lecture Notes in Math. 1284, Springer-Verlag, Berlin, 1987.

[74] A. D. Mednyh, Determination of the number of nonequivalent coverings over a compact Riemann surface, *Dokl. Akad. Nauk SSSR* **239** (1978), 269–271 (Russian); *Soviet Math. Dokl.* **19** (1978), 318–320 (English transl.).

[75] A. D Mednyh, On unramified coverings of compact Riemann surfaces, *Dokl. Akad. Nauk SSSR* **244** (1979), 529–532 (Russian); *Soviet Math. Dokl.* **20** (1979), 85–88 (English transl.).

[76] A. D. Mednyh, On the solution of the Hurwitz problem on the number of nonequivalent coverings over a compact Riemann surface, *Dokl. Akad. Nauk SSSR* **261** (1981), 537–542 (Russian); *Soviet Math. Dokl.* **24** (1981), 541–545 (English transl.).

[77] A. D. Mednykh, On the number of subgroups in the fundamental group of a closed surface, *Comm. Algebra* **16** (1988), 2137–2148.

[78] A. D. Mednykh, Branched coverings of Riemann surfaces whose branch orders coincide with the multiplicity, *Comm. Algebra* **18** (1990), 1517–1533.

[79] R. Perlis, On the equation $\zeta_K(s) = \zeta_{K'}(s)$, *J. Number Theory* **9** (1977), 342–360.

[80] E. Reyssat, *Quelques aspects des surfaces de Riemann*, Birkhäuser, Boston/Basel/Berlin, 1989.

[81] G. Riera and R. Rodríguez, Riemann surfaces and abelian varieties with an automorphism of prime order, *Duke Math. J.* **69** (1993), 199–217.

[82] C.-H. Sah, Groups related to compact Riemann surfaces, *Acta Math.* **123** (1969), 13–42.

[83] L. Schneps (ed.), *The Grothendieck theory of dessins d'enfants*, London Math. Soc. Lecture Note Series 200, Cambridge University Press, 1994.

[84] M. Schönert *et. al.*, GAP—*Groups, Algorithms, and Programming*, Lehrstuhl D für Mathematik, RWTH Aachen, 1992.

[85] J.-P. Serre, Rigidité du foncteur de Jacobi d'echelon $n \geq 3$, *Sém. Henri Cartan* **17–20** (1960/61), 17–18.

[86] J.-P. Serre, Groupes de Galois sur **Q**, *Astérisque* **161–162** (1988), 73–85.

[87] J.-P. Serre, *Topics in Galois theory*, Jones and Bartlett, Boston/London, 1992.

[88] G. B. Shabat and V. A. Voevodsky, Drawing curves over number fields, in *Grothendieck Festschrift III (eds. P. Cartier et al.)*, pp. 199–227. Birkhäuser, Basel, 1990.

[89] G. Shabat and A. Zvonkin, Plane trees and algebraic numbers, in *Jerusalem combinatorics '93 (eds. H. Barcelo and G. Kalai)*, pp. 233–275. Amer. Math. Soc. Contemporary Math. Ser. **178** (1994).

[90] I. R. Shafarevich, Construction of fields of algebraic numbers with given solvable Galois group, *Izv. Akad. Nauk. SSSR* **18** (1954), 525–578 (In Russian); *Amer. Math. Transl.* **4** (1956), 185–237 (English transl.).

[91] K.-Y. Shih, On the construction of Galois extensions of function fields and number fields, *Math. Ann.* **207** (1974), 99–120.

[92] D. Singerman, On the structure of non-Euclidean crystallographic groups, *Proc. Cambridge Philos. Soc.* **76** (1974), 17–32.

[93] M. Streit, Homology, Belyi functions and canonical curves, *Manuscripta Math.*, to appear.

[94] T. Sunada, Riemannian coverings and isospectral manifolds, *Ann. of Math.* **121** (1985), 169–186.

[95] J. G. Thompson, Some finite groups which appear as $\mathrm{Gal}(L/K)$, where $K \subseteq \mathbf{Q}(\mu_n)$, *J. Algebra* **89** (1984), 437–499.

[96] J. G. Thompson, Galois groups, in *Groups—St. Andrews 1989 (eds. C. M. Campbell and E. F. Robertson)*, vol. 2, pp. 455–462. London Math. Soc. Lecture Note Series 160, Cambridge University Press, 1991.

[97] M. Tretkoff, Algebraic extensions of the field of rational functions, *Comm. Pure Appl. Math.* **24** (1971), 491–496.

[98] T. W. Tucker, Finite groups acting on surfaces and the genus of a group, *J. Combinatorial Theory (B)* **34** (1983), 82–98.

[99] A. Weil, The field of definition of a variety, *Amer. J. Math.* **78** (1956) 509–524.

[100] M. C. Wilkie, On non-Euclidean crystallographic groups, *Math. Z.* **91** (1966), 87–102.

[101] R. A. Wilson, The symmetric genus of the Baby Monster, *Quart. J. Math. (Oxford) (2)* **44** (1993), 513–516.

[102] R. A. Wilson, The symmetric genus of the Fischer group Fi_{23}, *Topology* **36** (1996), 379–380.

[103] A. J. Woldar, On Hurwitz generation and genus actions of sporadic groups, *Illinois J. Math.* **33** (1989), 416–437.

[104] A. J. Woldar, Sporadic simple groups which are Hurwitz, *J. Algebra* **144** (1991), 443–450.

[105] J. Wolfart, The 'obvious' part of Belyï's theorem and Riemann surfaces with many automorphisms, Preprint, Frankfurt, 1995.

[106] D. Zagier, On the distribution of the number of cycles of elements in symmetric groups, Preprint, Bonn, 1995.

[107] H. Zieschang, E. Vogt and H-D. Coldewey, *Surfaces and planar discontinuous groups*, Lecture Notes in Mathematics 835, Springer-Verlag, Berlin/Heidelberg/New York, 1980.

On the characterization of finite groups by characters

Wolfgang Kimmerle

Abstract

The first object is a survey on the isomorphism problem of integral
group rings and the Zassenhaus conjectures related to this question.
Then it is shown how automorphisms of character tables may be used
to get information about automorphisms of integral group rings. This
permits the proof of one of the Zassenhaus conjectures for series of
simple or almost simple groups. Finally a short proof is given that
$\{1, 2, 3\}$-characters determine a finite group up to isomorphism.

Introduction

R. Brauer in his famous lectures in 1963 on representations of finite groups
posed more than 40 problems which have had a big influence on this topic
since then [7]. Many of these problems concern the question of which properties of a finite group G are reflected by its character table. The basic problem
is to determine what kind of information about G additional to its character
table is needed in order to determine G up to isomorphism. We shall consider
two aspects of these questions.

The first one is the isomorphism problem of integral group rings of finite
groups. This is an old and still open problem. It was considered for the first
time in G. Higman's thesis in 1939 [38, p. 100]. The question is whether
$\mathbb{Z}G \cong \mathbb{Z}H$ implies that $G \cong H$. It is a result of G. Glauberman that finite
groups G and H with isomorphic integral group rings $\mathbb{Z}G$ and $\mathbb{Z}H$ have the
same character table [16, (3.17)]. Thus the isomorphism problem fits well
into the context of Brauer's problems. Note that K. W. Roggenkamp and
A. Zimmermann recently constructed two non-isomorphic infinite polycyclic
groups G and H whose integral group rings are Morita equivalent. This gives
more or less a counterexample to the isomorphism problem for infinite groups
[36, Corollary 1].

In section 1 we discuss the isomorphism problem under the additional
aspect of conjectures of H. Zassenhaus concerning the structure of torsion

subgroups of the unit group of $\mathbb{Z}G$. At the end of section 1 we give a survey on recent results comparing the knowledge for the case when G is soluble with that when G is a general finite group. This shows that it is necessary to consider the questions in particular for non-abelian simple groups.

Section 2 is devoted to the contribution of character tables to the Zassenhaus conjectures. It will become clear that the ATLAS [8] was of great help in order to find a method to attack the Zassenhaus conjectures for non-abelian simple groups. We sketch this method which was studied first in [5] and has been further developed by F. M. Bleher in [2] and [3]. It will also become transparent that the modular ATLAS [17] is even more helpful and that knowledge of generic character tables permits proof of the results. Thus the computer algebra systems GAP [40] and CHEVIE [11] are extremely useful.

In section 3 we come to the second aspect and prove that knowledge of the 2- and 3-characters added to the character table determines a finite group up to isomorphism. Historically k-characters arose as follows. In 1896 G. Frobenius studied in detail the group determinant $D(G)$ of a finite group G [10], which had been introduced by R. Dedekind in 1886. In particular Frobenius studied the prime factors of this determinant and introduced, via the coefficients of the monomials arising in the prime factors of $D(G)$, functions from G^k into $\mathbb{C}$. These functions were later called k-characters [18] and they coincide for $k = 1$ with the ordinary irreducible characters. It was proved in 1990 by E. Formanek and D. Sibley [9, Theorem 5] that the group determinant of a finite group determines the group up to isomorphism. Later a different—short and constructive—proof was given by R. Mansfield [28]. H. J. Hoehnke and K. W. Johnson [14] prove that the irreducible 1-, 2-, 3-characters already determine the group by showing that these characters determine all k-characters and thus the group determinant, which then, thanks to the result of Formanek and Sibley, determine the group. The aim of section 3 is to give a very short argument which shows directly that the 1-, 2-, 3-characters of a finite group G determine G up to isomorphism. This reports on joint work with K. W. Roggenkamp [24]. Since the 1-, 2-, 3-characters can be derived from $D(G)$, this gives also a short proof of the result of Formanek and Sibley.

1   Torsion units of integral group rings

1.1   Notation

Throughout, R denotes an integral domain of characteristic zero and G is a finite group. We assume that no prime divisor of $|G|$ is invertible in R. Let K denote a field containing R. The map $\varepsilon : RG \longrightarrow R$ defined by $\varepsilon(\sum r_g g) = \sum r_g$ is called the *augmentation map*. A unit u of RG is called *normalized*, if $\varepsilon(u) = 1$. The group of normalized units of RG is denoted by $V(RG)$. A subgroup H of $V(RG)$ is called a *group basis*, if H is an R-basis of RG. Let H be a group basis of RG and let C be a conjugacy class of H. Then the sum taken over all elements of C is called the class sum $\underline{C}$.

1.2 The class sum correspondence

See [29, Theorem C], [37], [35, IV.1]. If H is a group basis of RG, then H is in class sum correspondence to G, i.e. there is a bijection $\sigma : G \to H$ such that the image of a conjugacy class C of G is a conjugacy class of H. Moreover their class sums $\sigma(\underline{C})$ and $\underline{C}$ coincide. In particular the R-linear extension of σ fixes the centre of RG element-wise, since class sums form an R-basis of the centre of RG. Note however that σ is not *a priori* a group homomorphism. By [29, Proposition 2] σ is compatible with the power map, i.e. for each n the conjugacy class of $\sigma(g^n)$ coincides with that one of $(\sigma(g))^n$.

Assume now for a moment that $R = \mathbb{Z}$. It is an open question whether the torsion subgroups of $V(\mathbb{Z}G)$ are determined by G. More precisely the following questions have been studied extensively in the last twenty years (see [42, p. 497–98]).

1.3 The Zassenhaus Conjectures

(ZC1) Let u be a unit of finite order of $V(\mathbb{Z}G)$. Then u is conjugate within $\mathbb{Q}G$ to an element of G.

(ZC2) Let H be a subgroup of $V(\mathbb{Z}G)$ with the same order as G. Then H is conjugate to G by a unit of $\mathbb{Q}G$.

(ZC3) Let U be a finite subgroup of $V(\mathbb{Z}G)$. Then U is conjugate within $\mathbb{Q}G$ to a subgroup of G.

(ZC3), if true for $\mathbb{Z}G$, describes the torsion part of $V(\mathbb{Z}G)$ as G modulo conjugation. For G abelian this means that the elements of G are the only torsion units of $V(\mathbb{Z}G)$. Note that this was proved by G. Higman in his thesis [13, Theorem 3], [38]. It should be remarked that Zassenhaus himself in his lectures called (ZC2) *Brauer and Higman invariance*. By the Noether–Skolem theorem the conjecture (ZC2) may be rephrased—replace $\mathbb{Z}$ by R as in 1.1—in the following way.

(*) For each group basis H of RG there exists an R-algebra automorphism σ of RG such that $\sigma(G) = H$ and such that σ fixes the centre $Z(RG)$ element-wise.

An arbitrary group isomorphism from G to a group basis H extends by R-linear extension to a ring automorphism of RG. This permits the study of conjecture (ZC2) for many groups G by considerations of normalized ring automorphisms (see section 2).

1.4   The p-version

Whereas conjecture (ZC1) is still open, Roggenkamp and Scott constructed a metabelian counterexample to (ZC2) [27], [35, IX,§1], [31]. Thus the central question is, whether there is a suitable replacement or a modification of (ZC2), which is true in general and still implies a positive solution to the isomorphism problem (IP). No counterexample is known to the following local version of (ZC2).

(**ZC2**)$_p$ Let H be a group basis of RG and let p be a rational prime. Then there exists a ring automorphism σ_p such that $\sigma_p(G) = H$ and σ_p restricted to the class sums of p-elements is the identity.

By (*) it is clear that (ZC2)$_p$ is a weaker version of (ZC2) and may be regarded as a Sylow variant of (ZC2). Note that, if (ZC2)$_p$ is true, the isomorphism problem (IP) has a positive answer. For other variations of (ZC2) we refer to [20].

1.5   Some main results

We give a brief survey on results about (IP) and (ZC 2), (ZC2)$_p$ respectively from a point of view which especially regards the chief factor structure of the group G. For finite soluble groups quite a lot is known.

Theorem 1.5.1 [33, Corollary 3]   (ZC2) *is true provided G is nilpotent.*

A. Weiss gave a shorter proof of Theorem 1.5.1 [44] and finally proved that even (ZC3) holds if G is nilpotent [45, Theorem 1].

Theorem 1.5.2 [34] [41, p. 266]   (ZC2) *is true provided the generalized Fitting group $F^*(G)$ is a p-group.*

Theorem 1.5.3 [19, 5.20]   (ZC2)$_p$ *is true for every prime p, if G is supersoluble.*

Call a finite group G an *S-group* if all chief factors of G are isomorphic to one and the same simple group S. If S is non-abelian simple, then S-groups are direct products of copies of S by Schreier's conjecture. Thus, if $G = \prod G_i$ is a direct product of S_i-groups G_i, then G is a direct product of a nilpotent group and a perfect semisimple group. The nilpotent groups are precisely the soluble groups which are direct products of S-groups. Thus the following result is a kind of general analogue to Theorem 1.5.1.

Theorem 1.5.4 [22, 2.8] *If G is a direct product of a nilpotent group and a semisimple group, then* (IP) *has a positive answer for G. Moreover* (ZC2) *holds provided* (ZC2) *holds for every non-abelian composition factor G.*

Note that the second part of Theorem 1.5.4 follows from the general fact that (ZC2), (IP) respectively hold for direct products provided they hold for each factor [20, 2.4], [19, 5.4]. Clearly the second part of 1.5.4 only makes sense if there are non-abelian simple groups for which (ZC2) is valid. So this gives one motivation to check (ZC2) for non-abelian simple groups.

Nilpotent groups are also those soluble groups whose generalized Fitting subgroup is the group itself. This gives a second possibility to generalize the notion nilpotent to the class of all finite groups and leads to quasi-nilpotent groups [15, X,§13]. Note that these are precisely those groups with the property that each element acts as an inner automorphism on each chief factor of the group.

The obvious generalization of the class of finite supersoluble groups is the class $\mathfrak{S}$ of finite groups which consists of those groups whose chief factors are simple. Neither for all quasi-nilpotent groups nor for all groups of $\mathfrak{S}$ is a positive answer to (IP) known. There is also no general analogue to Theorem 1.5.2. It is clear that in order to get results in the direction of such generalizations one has to deal with almost simple groups and quasisimple groups. Some results on such groups are known (see section 2.9).

1.6   Projective limits

As usual let $O_{p'}(G)$ be the largest normal subgroup of G of order not divisible by the prime p. Write G as a finite projective limit (or in other words as a subdirect product) of quotients $G/O_{p'}(G)$, i.e.

$$ G \;\cong\; \{\, (gO_{p'_1}(G),\dots,gO_{p'_k}(G)) \;\in\; \prod_{i=1}^{k} G/O_{p'_i}(G); \; g \in G\}, $$

where the primes p_i are taken such that $\bigcap_{p_i} O_{p'_i}(G) = 1$. Note that it is not necessary to take all primes dividing the order of G. In the case when G is p-constrained, it follows from Theorem 1.5.2 that the conjecture (ZC2) is valid for $G/O_{p'}(G)$. Thus a soluble group may always be regarded as a projective limit of groups for which (ZC2) holds. In section 2.4 we discuss this construction in a little bit more detail with respect to automorphisms of RG. The counterexamples to (ZC2) show that the correctness of (ZC2) in general does not carry over to such a projective limit. But a certain variation like (ZC2)$_p$ might hold at least for classes of groups as discussed in section 1.5. Note that the proof of Theorem 1.5.3 makes use of the projective limit construction [23, 4.15 and 5.4]. It is a natural question whether such a construction holds for a general finite group. In the case when G belongs to the class $\mathfrak{S}$, one sees immediately that this is possible if (ZC2) is true for quasisimple or almost simple groups.

2 Automorphisms

2.1 Normalized group ring automorphisms

An R-algebra automorphism σ of RG is called *normalized* if it preserves the augmentation, i.e. $\varepsilon(\sigma(g)) = \varepsilon(g)$. The normalized R-algebra automorphisms of RG form a normal subgroup $\mathrm{Aut}_n(RG)$ of the group $\mathrm{Aut}(RG)$ of all R-algebra automorphisms of RG. Clearly the unit group $U(RG)$ is a direct product of the normalized units $V(RG)$ and R^*. Thus no information is lost, if one studies only $\mathrm{Aut}_n(RG)$. Moreover $|\mathrm{Aut}(RG) : \mathrm{Aut}_n(RG)|$ coincides naturally with the number of different group homomorphisms from G to R^*. If σ is a group automorphism of a group basis X of RG, then σ induces naturally a normalized R-algebra automorphism of RG by extending σ R-linearly to RG.

2.2 Zassenhaus decomposition

We say that $\sigma \in \mathrm{Aut}_n(RG)$ admits a *Zassenhaus decomposition* with respect to a group basis X of RG, if σ is the composition of a normalized automorphism induced from a group automorphism of X followed by a central automorphism γ (i.e. an automorphism which fixes the centre of RG element-wise). If γ fixes only the class sums of the p-elements of G, we say that σ admits a *Zassenhaus variation decomposition*.

Clearly the conjecture (ZC2) of Zassenhaus follows if the isomorphism problem has a positive answer for RG and each normalized automorphism has a Zassenhaus decomposition with respect to G.

The link between the conjecture and normalized automorphisms is even stronger. In [19, 5.3] it is shown that (ZC2) is true for RG provided each $\sigma \in \mathrm{Aut}_n R(G \times G)$ and each $\tau \in \mathrm{Aut}_n RG$ admits a Zassenhaus decomposition with respect to each group basis of $R(G \times G)$, RG, respectively.

The following is now an immediate consequence. If for a class Γ of finite groups which is closed under taking direct products, each normalized R-algebra automorphism admits a Zassenhaus decomposition with respect to each group basis, then (ZC2) is true for all $G \in \Gamma$.

2.3 Čech style cohomology

As in section 1.6 write G as the projective limit of its quotients $G/O_{p_i'}(G)$. This projective limit of groups corresponds of course to the projective limit

Γ of the group rings $R(G/O_{p'_i}(G))$. Note that Γ does not coincide with RG but it is a characteristic image of RG.

Starting with $\sigma \in \text{Aut}_n(RG)$ one obtains induced automorphisms σ_i, σ_j and σ_{ij} of $R(G/O_{p'_i}(G))$, $R(G/O_{p'_j}(G))$ and $R(G/(O_{p'_i}(G) \cdot O_{p'_j}(G))$. The Zassenhaus decompositions

$$\sigma_i = \rho_i \cdot \gamma_i, \sigma_j = \rho_j \cdot \gamma_j,$$

where ρ_i and ρ_j are induced from group automorphisms of $G/O_{p'_i}(G)$ and $G/O_{p'_j}(G)$, induce on $G/(O_{p'_i}(G) \cdot O_{p'_j}(G))$ central group automorphisms

$$\rho_{ij} = \overline{\rho_i} \cdot \overline{\rho_j}^{-1}.$$

Now σ admits a Zassenhaus variation decomposition (section 2.2), if these ρ_{ij} are induced simultaneously from a group automorphism of G. This obstruction may be interpreted in the form of Čech style cohomology sets comparable to the covering of a sphere by open sets, for details see [23, Section 3], [35, IX,§2].

2.4   Chief series

We say that finite groups G and H have the same chief series, if given a chief series $1 = N_0 < N_1 < \cdots < N_k = G$ of G there is a chief series of H of the form $1 = M_0 < M_1 < \cdots < M_k = G$ such that for each $1 \leq i \leq k$ the ith chief factors M_i/M_{i-1} and N_i/N_{i-1} are isomorphic, and *vice versa*, for each chief series of H there is one of the same length of G such that ith chief factors are isomorphic for each i.

Theorem 2.4.1 [19, Satz 6.3] *Assume that G and H have the same character table. Then G and H have the same chief series, in particular their chief factors are isomorphic and have the same multiplicity.*

The hypothesis of having the same character table always has to be understood in the sense of Brauer's problems, i.e. the headline of the table is not given, in particular the orders of the representatives of the conjugacy classes are not known. Under the stronger hypothesis $\mathbb{Z}G \cong \mathbb{Z}H$, Theorem 2.4.1 has been proved in [22, Theorem 2.3]. The assumptions of the theorem however may even be weakened. It suffices to assume that G and H are in Jordan–Hölder class correspondence. For details we refer to [25, Theorem 5].

The following result contains a positive answer to Brauer's problem 12 [7]. It may be derived from Theorem 2.4.1.

Theorem 2.4.2 [19, Satz 6.2], [25, Theorem 6a] *The character table of a finite group determines abelian Hall subgroups up to isomorphism.*

Again the assumptions may be weakened. It suffices to assume that G and H have the same conjugacy class structure [26].

2.5 Automorphisms of character tables

The aim of this subsection is the reformulation of the conjecture (ZC2) in terms of automorphisms of character tables.

Denote by $\mathrm{Irr}(G)$ the set of the ordinary irreducible characters of G. Extend normalized automorphisms of RG K-linearly to KG and choose K to be sufficiently large. $\mathrm{Aut}_n(RG)$ acts naturally on the set of ordinary irreducible characters $\mathrm{Irr}(G)$. On the other hand by the class sum correspondence a normalized automorphism σ of RG maps class sums into class sums. Thus linking $\underline{C}$ with the conjugacy class C we get an action of σ on the conjugacy classes $\mathrm{Cl}(G)$ of G. Let $\sigma \in \mathrm{Aut}_n(RG)$ and denote its action on $\mathrm{Irr}(G)$ and on $\mathrm{Cl}(G)$ also by σ then we have the equation

$$\sigma(\chi)(\sigma(C)) = \chi(C). \tag{1}$$

Let $\mathrm{CT}(G)$ be the ordinary character table of G. Define its automorphism group $\mathrm{Aut}(\mathrm{CT}(G))$ as the group of all permutations π of $\mathrm{Irr}(G) \cup \mathrm{Cl}(G)$ which map $\mathrm{Irr}(G)$ into $\mathrm{Irr}(G)$ such that for each $\chi \in \mathrm{Irr}(G)$ and for each $C \in \mathrm{Cl}(G)$ the equation

$$\pi(\chi)(\pi(C)) = \chi(C) \tag{2}$$

holds. Then by equation 1 we have a map μ of $\mathrm{Aut}_n(RG)$ into $\mathrm{Aut}(\mathrm{CT}(G))$. Denote the image of μ by $\mathrm{Aut}_{RG}(\mathrm{CT}(G))$. Fixing a group basis isomorphic to G each group automorphism of G extends R-linearly to a normalized R-algebra automorphism of RG. Thus we may regard $\mathrm{Aut}(G)$ in this way as a subgroup of $\mathrm{Aut}_n(RG)$. Denote $\mu(\mathrm{Aut}G)$ by $\mathrm{Aut}_G(\mathrm{CT}(G))$.

Proposition 2.5.1 *Each normalized automorphism of RG admits a Zassenhaus decomposition with respect to G if, and only if, $\mathrm{Aut}_G(\mathrm{CT}(G)) = \mathrm{Aut}_{RG}(\mathrm{CT}(G))$.*

Examples 2.5.2 a) Let G be the automorphism group of A_6. Then $\mathrm{Aut}(\mathrm{CT}(G)) = 1$. Hence by Proposition 2.5.1 (ZC2) is valid for G.

b) Let $G = \mathrm{A}_6$, $G = \mathrm{S}_6 = \mathrm{A}_6.2_1$ or $G = M_{10} = \mathrm{A}_6.2_3$. Then $\mathrm{Aut}(\mathrm{CT}(G)) = \mathrm{Aut}_G(\mathrm{CT}(G))$. Again we can apply 2.5.1 and get that (ZC2) is valid for these groups. If $G = \mathrm{PGL}(2,9) = \mathrm{A}_6.2_2$, then $\mathrm{Aut}_G(\mathrm{CT}(G)) \cong C_2$ whereas $\mathrm{Aut}(\mathrm{CT}(G)) \cong C_2 \times C_2$. So we need in this case additional arguments to conclude that $\mathrm{Aut}_G(\mathrm{CT}(G)) = \mathrm{Aut}_{RG}(\mathrm{CT}(G))$, see Example 2.7.1.

2.6 Autoequivalences of blocks

The object of this subsection is to demonstrate that the study of modular representations, in particular the use of Brauer trees, gives additional information on $\mathrm{Aut}_{RG}(\mathrm{CT}(G))$.

Let (K, R, k) be a p-modular system sufficiently large for G. The basic idea is now to consider via this p-modular system simultaneously the action of a normalized automorphism $\sigma \in \text{Aut}_n(RG)$ on KG and on kG. The automorphism σ gives rise to an autoequivalence of the categories of finitely generated modules of KG, RG and kG sending a module M to its twisted module M^σ by acting via σ.

Let M be a simple KG-module, let M_0 be an R-form of M and denote the reduction of M_0 mod p by $\overline{M}$. If C is a composition factor of $\overline{M}$, then C^σ is a composition factor of $\overline{M}^\sigma$. The multiplicities coincide. Moreover σ induces an autoequivalence of the category of finitely generated kG-modules and also of the category of finitely generated B-modules for any block B fixed by σ. Clearly the trivial simple module is fixed and thus so is the principal block B_0 of kG.

Theorem 2.6.1 [5, Theorem 2] *Assume that the finite group G has cyclic Sylow p-subgroups. Then each non-trivial autoequivalence of the principal p-block B_0 which fixes the trivial simple module preserves the isomorphism classes of all finitely generated B_0-modules.*

The consequence in terms of irreducible characters is that $\chi \in \text{Irr}(G)$ is fixed by σ provided its restriction to the p-regular classes of $\text{Cl}(G)$ is in the principal p-block and χ is non-exceptional. This clearly gives additional information for $\text{Aut}_{RG}(\text{CT}(G))$ and suffices to prove (ZC2) for the groups $\text{PSL}(2, p)$. This was the first infinite series of non-abelian simple groups for which (ZC2) was established [5, Theorem 1].

Example 2.6.2 We consider the ordinary character table of the group $G = \text{Aut}(J_3) = J_3.2$, shown in Table 1. The irreducible characters and the conjugacy classes are ordered as in GAP or in the ATLAS. For the example a part of $\text{CT}(G)$ suffices. Here, and later on, upper case letters denote certain irrationalities whose actual values are not important.

The other irreducible characters $\chi_1, \ldots, \chi_7, \chi_{10}, \chi_{11}, \chi_{16}, \chi_{17}, \chi_{24}, \ldots, \chi_{30}$ are fixed by each $\sigma \in \text{Aut}_n(RG)$. This follows from their degrees and their character values on classes outside J_3. In terms of conjugacy classes, $\text{Aut}(\text{CT}(G))$ is generated by the permutations $(24a, 24b)$, $(17a, 17b)(34a, 34b)$ and $(9a, 9b, 9c)(18a, 18b, 18c)$. Note that in the table all conjugacy classes which are fixed by each character table automorphism are omitted except $1a$ and $19a$. The latter classes will be needed below. Note also that $\text{Out}(G) = 1$. Thus $\text{Aut}_G(\text{CT}(G)) = 1$.

In order to show that $\text{Aut}_{RG}(\text{CT}(G)) = 1$ we consider the prime $p = 19$. Note that $|J_3.2| = 2^8 \cdot 3^5 \cdot 5 \cdot 17 \cdot 19$. The irreducible characters given in the table above are all non-exceptional with respect to $p = 19$. The length of a conjugacy class $\text{Cl}(g)$ of $g \in G$ is divisible by 19 provided g does not belong

Table 1: Part of the character table of $J_3.2$

	1a	9a	9b	9c	17a	17b	19a	18a	18b	18c	24a	24b	34a	34b
χ_8	816	.	.	.	.	.	-1	.	.	.	E	$-E$	.	.
χ_9	816	.	.	.	.	.	-1	.	.	.	$-E$	E	.	.
χ_{12}	1215	.	.	.	A	$*A$	-1	.	.	.	.	.	$-A$	$-*A$
χ_{13}	1215	.	.	.	A	$*A$	-1	.	.	.	.	.	A	$*A$
χ_{14}	1215	.	.	.	$*A$	A	-1	.	.	.	.	.	$-*A$	$-A$
χ_{15}	1215	.	.	.	$*A$	A	-1	.	.	.	.	.	$*A$	A
χ_{18}	1920	B	C	D	-1	-1	1	F	G	H	.	.	-1	-1
χ_{19}	1920	B	C	D	-1	-1	1	$-F$	$-G$	$-H$	.	.	1	1
χ_{20}	1920	D	B	C	-1	-1	1	H	F	G	.	.	-1	-1
χ_{21}	1920	D	B	C	-1	-1	1	$-H$	$-F$	$-G$	.	.	1	1
χ_{22}	1920	C	D	B	-1	-1	1	G	H	F	.	.	-1	-1
χ_{23}	1920	C	D	B	-1	-1	1	$-G$	$-H$	$-F$	.	.	1	1

to the class $19a$. Hence all the irreducible characters χ listed above belong to the principal 19-block because

$$\frac{\chi(g)}{\chi(1)} \cdot |Cl(g)| \equiv |Cl(g)| \bmod 19$$

for all 19-regular elements of G. By Theorem 2.6.1 it follows that all these characters χ are fixed by all $\sigma \in \mathrm{Aut}_n(RG)$ and therefore $\mathrm{Aut}_{RG}(\mathrm{CT}(G)) = 1$.

2.7 Brauer tables and tensor products

The action of $\mathrm{Aut}_n(RG)$ on the category $\mathrm{mod}(kG)$ induces an action on the irreducible Brauer characters of G and together with the action on the conjugacy classes of p-elements, $\sigma \in \mathrm{Aut}_n(RG)$ yields a character table automorphism of the p-Brauer table of G [3, Proposition 2.1], [2, Lemma 2.6].

Example 2.7.1 Consider the character table of $G = \mathrm{PGL}(2,9)$, shown in Table 2. $\mathrm{Aut}(\mathrm{CT}(G))$ is generated by $(\chi_4, \chi_6)(5a, 5b)(\chi_5, \chi_7)(10a, 10b)$ and $(8a, 8b)(\chi_{10}, \chi_{11})$. The only prime for which G has cyclic Sylow subgroups is 5. However for $p = 5$ the characters χ_{10} and χ_{11} have defect zero and χ_4, χ_5 are exceptional with multiplicity 2. Thus the consideration of the 5-blocks gives no new information about $\mathrm{Aut}_{RG}(\mathrm{CT}(G))$. The Brauer table for $p = 3$ is as in Table 3 (see [17, p. 4]).

An automorphism of this table which moves the class $5a$ must move the class $8a$ as well. Consequently $\mathrm{Aut}_{RG}(\mathrm{CT}(G))$ has order 2 and coincides with $\mathrm{Aut}_G(\mathrm{CT}(G))$. By Proposition 2.5.1 it follows that (ZC2) holds for $\mathrm{PGL}(2,9)$.

Table 2: The character table of PGL$(2, 9)$

	1a	2a	3a	4a	5a	5b	2b	8a	8b	10a	10b
χ_1	1	1	1	1	1	1	1	1	1	1	1
χ_2	1	1	1	1	1	1	−1	−1	−1	−1	−1
χ_3	10	2	1	−2	.	.	.	.	.	.	.
χ_4	8	.	−1	.	A	*A	2	.	.	−A	−*A
χ_5	8	.	−1	.	A	*A	−2	.	.	A	*A
χ_6	8	.	−1	.	*A	A	2	.	.	−*A	−A
χ_7	8	.	−1	.	*A	A	−2	.	.	*A	A
χ_8	9	1	.	1	−1	−1	−1	1	1	−1	−1
χ_9	9	1	.	1	−1	−1	1	−1	−1	1	1
χ_{10}	10	−2	1	.	.	.	.	B	−B	.	.
χ_{11}	10	−2	1	.	.	.	.	−B	B	.	.

Table 3: The 3-modular character table of PGL$(2, 9)$

	1a	2a	4a	5a	5b	2b	8a	8b	10a	10b
χ_1	1	1	1	1	1	1	1	1	1	1
χ_2	1	1	1	1	1	−1	−1	−1	−1	−1
χ_3	3	−1	1	A	*A	−1	B	*B	C	*C
χ_4	3	−1	1	A	*A	1	−B	−*B	−C	−*C
χ_5	3	−1	1	*A	A	−1	*B	B	*C	C
χ_6	3	−1	1	*A	A	1	−*B	−B	−*C	−C
χ_7	4	.	−2	−1	−1	.	D	−D	E	−E
χ_8	4	.	−2	−1	−1	.	−D	D	−E	E
χ_9	9	1	1	−1	−1	−1	1	1	−1	−1
χ_{10}	9	1	1	−1	−1	1	−1	−1	1	1

F. Bleher observed that normalized automorphisms act naturally on tensor products of Brauer characters.

Lemma 2.7.2 [3, Proposition 2.1] *Let ζ and ξ be Brauer characters and $\sigma \in \mathrm{Aut}_n(RG)$. Then*

$$\sigma(\zeta \otimes \xi) = \sigma(\zeta) \otimes \sigma(\xi).$$

This allows the application of Steinberg's tensor product theorem and leads to a systematic treatment of series of simple groups of Lie type using the defining characteristic.

Theorem 2.7.3 [2, Section 3] *The Zassenhaus conjecture (ZC2) holds for the following groups of Lie type.*

$$\mathrm{SL}(2, p^f), \mathrm{PSL}(2, p^f), {}^2B_2(2^{2m+1}), {}^2G_2(3^{2m+1}),$$

$${}^2F_4(2^{2m+1}), \mathrm{SL}(3, 3^m), \mathrm{SU}(3, 3^{2m}), \mathrm{Sp}(4, 2^m), G_2(p^m), {}^3D_4(p^{3m}),$$

where p always denotes a rational prime.

The following results follow from Theorem 2.7.3.

Theorem 2.7.4 [3, Theorems 1 and 2], [2, Theorem 3.26] *(ZC2) holds for every minimal simple group, for every simple Zassenhaus group and for every simple group with abelian Sylow 2-subgroups.*

2.8   Alternating and sporadic groups

With respect to the alternating groups A_n it is unknown whether normalized group ring automorphisms admit a Zassenhaus decomposition. It is proved for $n \leq 10$ and $n = 12$. The group $\mathrm{Aut}_n(G)$ does not act only on $\mathbb{C}G$, it acts also on $\mathbb{Q}G$. At the level of the rational group algebra $\mathbb{Q}G$ the following is true.

Theorem 2.8.1 [19, Satz 5.9] *For alternating groups A_n each $\sigma \in \mathrm{Aut}_n(\mathbb{Z}A_n)$ may be modified by $\tau \in \mathrm{Aut}(A_n)$ such that $\tau \cdot \sigma$ fixes each block of the Wedderburn decomposition of $\mathbb{Q}A_n$. In other words each irreducible $\mathbb{Q}$-character of A_n is fixed by $\tau \cdot \sigma$.*

In the proof of Theorem 2.8.1 it is shown that for $n \neq 6$ a bijection of the set of conjugacy classes of A_n which respects the length of the classes as well as the order of a representative fixes each conjugacy class of A_n which is invariant under conjugation in S_n. Note that by 1.2 a normalized automorphism of RG always induces such a bijection on the set of conjugacy classes of G. Since the conjugacy class of an even permutation π of a symmetric group S_n splits on restriction to A_n if and only if the cycle type of π consists of odd cycles of pairwise different length, the following follows.

Corollary 2.8.2 *For* A_n *the variation* $(ZC2)_p$ *holds for each prime p.*

Proof: For a given degree n and a given prime p there is at most one sum $\sum_{j=1}^{k} p^{i_j} = n$ with $0 \le i_1 < \cdots < i_k$. Thus there is at most one conjugacy class of a p-element in S_n which splits into two classes restricted to A_n. These two classes are linked by an outer automorphism τ of A_n. Consequently, if $\sigma \in \mathrm{Aut}_n(RA_n)$ does not fix all conjugacy classes of p-elements, then the composition of σ with τ fixes all these classes. Thus each σ admits a Zassenhaus variation decomposition with respect to p. □

With respect to the sporadic groups the following is known.

Theorem 2.8.3 [5, Section 4], [2, Section 2 and A.2] *(ZC2) is valid for the sporadic groups*

$$M_{11}, M_{12}, M_{22}, M_{23}, M_{24}, \mathrm{HS}, \mathrm{Co}_3, \mathrm{Co}_2, \mathrm{Co}_1, \mathrm{HN}, \mathrm{Th}, J_1, J_2, \mathrm{Ru}, \mathrm{B}.$$

For the other sporadic groups $(ZC2)_p$ is true for most primes p. In particular for the Monster M it holds for each prime [1].

2.9 Other insoluble groups

We close this section with some results for insoluble groups which are not simple.

Theorem 2.9.1 [4] *(ZC2) holds for finite Coxeter groups.*

Note that 2.9.1 contains the result that (ZC2) is valid for the symmetric groups S_n [30]. Another subcase is that (ZC2) is valid for the Weyl groups of type B. That for these groups normalized automorphisms admit a Zassenhaus decomposition follows from results of Giambruno, Sehgal and Valenti [43, (44.17)], [12]. For results on other wreath products of the type $H \wr S_n$ see [43, §44].

The next result is an immediate consequence of the results on chief series (see section 2.4).

Theorem 2.9.2 [22, 2.7] *(IP) holds for the automorphism group of a semi-simple group.*

Theorem 2.9.3 *Let G be a quasisimple group. Assume that the centre $Z(G)$ is cyclic of order p and that the abelianized group $\overline{G}$ is cyclic of order q where q and p denote primes. Assume further that for r different from q the variation $(ZC2)_r$ is valid for $G/Z(G)$ and assume that (IP) is true for the derived group G'. Then $(ZC2)_r$ holds for G.*

As a corollary one obtains that $(ZC2)_p$ has a positive answer for the double cover groups of the symmetric groups. Note that by the theory of Schur multipliers the double covers of S_n have isomorphic complex group algebras. In a recent letter J. Humphreys pointed out to me that the same is true for the modular group algebras of these double covers over a field of characteristic p, where p denotes an odd prime.

Finally for additional information about (IP) or the Zassenhaus conjectures the reader may consult the following references which contain surveys on these topics written under different aspects: [32], [35], [39], [43], [21].

3 k-characters

Definition 3.1 Let K be field and let $\chi \in \mathrm{Irr}_K(G)$. For each $k \in \mathbf{N}$ the k-character $\chi^{(k)}$ is the map $\chi^{(k)} : G^k \longrightarrow K$ defined recursively by $\chi^{(1)} = \chi$ and $\chi^{(k)}(g_1, \ldots, g_k) = \chi(g_1) \cdot \chi^{(k-1)}(g_2, \ldots, g_k) - \chi^{(k-1)}(g_1 \cdot g_2, \ldots, g_k) - \cdots - \chi^{(k-1)}(g_2, \ldots, g_1 \cdot g_k)$.

Proposition 3.2 *For a given $\chi \in \mathrm{Irr}_K(G)$ the 2- and 3-characters are explicitly given by the formulae*

$$\chi^{(2)}(x, y) = \chi(x) \cdot \chi(y) - \chi(x \cdot y)$$

and

$$\begin{aligned} \chi^3(x, y, z) = \ & \chi(x) \cdot \chi(y) \cdot \chi(z) - \chi(x) \cdot \chi(y \cdot z) - \chi(y) \cdot \chi(x \cdot z) \\ & - \chi(z) \cdot \chi(x \cdot y) \chi(x \cdot y \cdot z) + \chi(x \cdot z \cdot y) \end{aligned}$$

Note that the 2-character $\chi^{(2)}$ is just the obstruction for χ to be a homomorphism.

Definition 3.3 Let $\kappa \subset \mathbf{N}$. We say that two finite groups G and H have the same κ-characters over K provided there exist bijections

$$\beta : G \longrightarrow H \text{ and } \tilde{\ } : \mathrm{Irr}_K(G) \longrightarrow \mathrm{Irr}_K(H),$$

such that for each $k \in \kappa$,

$$\chi^{(k)}(x_1, \ldots, x_k) = \tilde{\chi}^{(k)}(\beta(x_1, \ldots, x_k)) \text{ for all } x_1 \in G, \chi \in \mathrm{Irr}_K(G).$$

Remarks 3.4 The group determinant $D_K(G)$ of G over K is defined as the determinant of the matrix $(X_{g_i \cdot g_j^{-1}})$, where the X_{g_i} are indeterminates over K labelled by the group elements g_i. The irreducible factors of $D_K(G)$ are in one-to-one correspondence with the irreducible characters $\chi \in \mathrm{Irr}_K(G)$. The coefficients of the irreducible factors may be calculated from the k-characters of G and conversely they determine the k-characters [10]. Thus the knowledge of the group determinant is equivalent to the knowledge of the k-characters. Note in this context that, if χ corresponds to the irreducible factor Φ, the degree of Φ coincides with the degree d of χ and χ^k is zero provided $k \geq d+1$. Moreover the character value $\chi(g)$ is the coefficient of $X_1^{(d-1)} \cdot X_g$ in Φ. Let m denote the maximum of the degrees of the $\chi \in \mathrm{Irr}_K(G)$. Then finite groups G and H have the same group determinant if, and only if, they have the same $\{1, \ldots, m\}$-characters.

The proof of the following result is joint work with K. W. Roggenkamp [24].

Proposition 3.5 *Let G and H be finite groups with the same $\{1, 2, 3\}$-characters over K and assume that the characteristic of K does not divide $|G|$. Then G and H are isomorphic.*

Proof: By assumption we have bijections $\beta : G \longrightarrow H$ and $\tilde{\ } :$ $\mathrm{Irr}_K(G) \longrightarrow \mathrm{Irr}_K(H)$ such that $\chi(x) = \tilde{\chi}^{(1)}(\beta(x))$ for all $x \in G$, $\chi \in \mathrm{Irr}_K(G)$. An immediate consequence of this is that

$$\beta(1) = 1. \tag{3}$$

Consider now the 2-characters. Since

$$\chi^{(2)}(x, y) = \tilde{\chi}^{(2)}(\beta(x), \beta(y))$$

$\forall x, y \in G$, $\forall \chi \in \mathrm{Irr}_K(G)$ we conclude that for every $1 \leq i \leq n$ and $\forall x, y \in G$,

$$\chi(x) \cdot \chi(y) - \chi(x \cdot y) = \tilde{\chi}(\beta(x)) \cdot \tilde{\chi}(\beta(y)) \tilde{\chi}(\beta(x) \cdot \beta(y)).$$

However, the 1-characters of G and H are isomorphic, and so we get

$$\tilde{\chi}(\beta(x) \cdot \beta(y)) = \chi(x \cdot y) = \tilde{\chi}(\beta(x \cdot y))).$$

Using the isomorphism of the 1-characters, we conclude that for all $\xi \in \mathrm{Irr}_K(H)$, and for all $x, y \in G$, we have

$$\xi(\beta(x \cdot y)) = \xi(\beta(x) \cdot \beta(y)). \tag{4}$$

Since this equation holds for every irreducible character of H, we conclude—using the second orthogonality relation for characters of H—that

$$\beta(x \cdot y) \sim \beta(x) \cdot \beta(y), \tag{5}$$

where $\sim$ denotes as usual conjugacy of group elements. In particular we get from equation 4 that for every irreducible character $\xi(1) = \xi(\beta(x) \cdot \beta(x^{-1}))$, whence

$$\beta(x^{-1}) = \beta(x)^{-1}. \tag{6}$$

We now use the 3-characters. From the definition it follows that for all $\chi \in \mathrm{Irr}_K(G)$, and for all $x, y, z \in G$ we have

$$\chi(x) \cdot \chi(y) \cdot \chi(z) - \chi(x) \cdot \chi(y \cdot z) - \chi(y) \cdot \chi(x \cdot z) - \chi(z) \cdot \chi(x \cdot y)$$
$$+\chi(x \cdot y \cdot z) + \chi(x \cdot z \cdot y) =$$
$$\tilde{\chi}(\beta(x)) \cdot \tilde{\chi}(\beta(y)) \cdot \gamma(\chi))(\beta(z)) - \tilde{\chi}(\beta(x)) \cdot$$
$$\tilde{\chi}(\beta(y) \cdot \beta(z)) - \tilde{\chi}(\beta(y)) \cdot \tilde{\chi}(\beta(x) \cdot \beta(z)) - \tilde{\chi}(\beta(z)) \cdot \tilde{\chi}(\beta(x) \cdot \beta(y))$$
$$+\tilde{\chi}(\beta(x) \cdot \beta(y) \cdot \beta(z)) + \tilde{\chi}(\beta(x) \cdot \beta(z) \cdot \beta(y)).$$

Because G and H have the same $\{1,2\}$-characters the above equation reduces to

$$\chi(x \cdot y \cdot z) + \chi(x \cdot z \cdot y) = \tilde{\chi}(\beta(x) \cdot \beta(y) \cdot \beta(z)) + \tilde{\chi}(\beta(x) \cdot \beta(z) \cdot \beta(y)). \quad (7)$$

Thus $\forall \xi \in \mathrm{Irr}_K(H)$ and $\forall x, y, z \in G$:

$$\xi(\beta(x \cdot y \cdot z)) + \xi(\beta(x \cdot z \cdot y)) = \xi(\beta(x) \cdot \beta(y) \cdot \beta(z)) + \xi(\beta(x) \cdot \beta(z) \cdot \beta(y)). \quad (8)$$

Apply the second orthogonality relations again. Then equation 8 implies

$$\begin{aligned}
\beta(x) \cdot \beta(y) \cdot \beta(z) &\sim \beta(x \cdot y \cdot z) \text{ or} && (9)\\
\beta(x) \cdot \beta(y) \cdot \beta(z) &\sim \beta(x \cdot z \cdot y),\\
\text{and } \beta(x) \cdot \beta(z) \cdot \beta(y) &\sim \beta(x \cdot y \cdot z) \text{ or}\\
\beta(x) \cdot \beta(z) \cdot \beta(y) &\sim \beta(x \cdot z \cdot y).
\end{aligned}$$

We now specialize $\beta(z) = (\beta(x) \cdot \beta(y))^{-1}$, and get from the possibilities listed in equation 9

$$\begin{aligned}
1 &= & \beta(x \cdot y \cdot z) && \sim \beta(x \cdot y) \cdot \beta(z)\\
&= & \beta(x \cdot y) \cdot (\beta(x) \cdot \beta(y))^{-1} &&\\
& & \text{or} &&\\
1 &= & \beta(x \cdot z \cdot y) &&\\
&\sim & \beta(z \cdot y \cdot x) && \sim \beta(z) \cdot \beta(y \cdot x)\\
&= & (\beta(x) \cdot \beta(y))^{-1} \cdot \beta(y \cdot x) && .
\end{aligned}$$

Here we have used that equation 5 implies that

$$\beta((x \cdot y) \cdot z) \sim \beta(x \cdot y) \cdot \beta(z).$$

Because the elements $x \cdot z \cdot y$ and $z \cdot y \cdot x$ are conjugate the elements $\beta(x \cdot z \cdot y)$ and $\beta(z \cdot y \cdot x)$ are conjugate. In fact, we have the following chain:

$$\beta(x \cdot y) \sim \beta(x) \cdot \beta(y) \sim \beta(y) \cdot \beta(x) \sim \beta(y \cdot x).$$

Summarizing, we have for $x, y \in G$ either

$$\beta(x \cdot y) = \beta(x) \cdot \beta(y) \text{ or } \beta(x \cdot y) = \beta(y) \cdot \beta(x).$$

But this implies that β is either an isomorphism or an anti-isomorphism [6, 4, Ex. 26], [28].

Consequently the groups G and H are isomorphic, since even if β is an anti-isomorphism, we can compose it with the natural anti-isomorphism $g \longrightarrow g^{-1}$ of G.   □

We obtain the following theorem of Formanek and Sibley as an immediate consequence of the above proposition and Remark 3.

Corollary 3.6 *The group determinant of a finite group G over a field K of characteristic not dividing the group order determines G up to isomorphism.*

The next consequence may be regarded as one answer to the problems and questions raised by R. Brauer in [7].

Corollary 3.7 *A finite group G is determined by its character table and a map Ψ from G^3 into the set of conjugacy classes $\mathrm{Cl}(G)$ which associates to each triple (x, y, z) the class $C_{x \cdot y \cdot z}$.*

References

[1] F. M. Bleher, *Zassenhaus Vermutung und einfache Gruppen*, Diplomarbeit, Stuttgart, 1993.

[2] F. M. Bleher, *Automorphismen von Gruppenringen und Blocktheorie*, Dissertation, Stuttgart, 1995.

[3] F. M. Bleher, Tensor products and a conjecture of Zassenhaus, *Arch. Math.* **64** (1995), 289–298.

[4] F. M. Bleher, M. Geck and W. Kimmerle, Automorphisms of generic Iwahori–Hecke algebras and integral group rings of finite Coxeter groups, *J. Algebra*, to appear.

[5] F. M. Bleher, G. Hiss and W. Kimmerle, Autoequivalences of blocks and a conjecture of Zassenhaus, *J. Pure Appl. Algebra* **103** (1995), 23–43.

[6] N. Bourbaki, *Algèbre I*, Herman, Paris, 1970.

[7] R. Brauer, Representations of finite groups, in *Lectures on modern mathematics*, Vol. I, pp. 133–175, J. Wiley, 1963. Reprinted in *R. Brauer Collected papers Vol. II*, pp. 183–225, MIT Press, Cambridge, MA, 1980.

[8] J. H. Conway, R. T. Curtis, S. P. Norton, R. A. Parker and R. A. Wilson, ATLAS *of finite groups*, Oxford University Press, 1985.

[9] E. Formanek and D. Sibley, The group determinant determines the group, *Proc. Amer. Math. Soc.* **112** (1991), 649–656.

[10] G. Frobenius, Über die Primfaktoren der Gruppendeterminante, *Sitzungsberichte Akad. Wiss. Berlin* (1896), 1343–1382.

[11] M. Geck, G. Hiss, F. Lübeck, G. Malle and G. Pfeiffer, CHEVIE—A system for computing and processing generic character tables, IWR Preprint 95-05, Heidelberg, 1995.

[12] A. Giambruno, S. K. Sehgal and A. Valenti, Automorphisms of the integral group rings of some wreath products, *Comm. Algebra* **19** (1991), 519–534.

[13] G. Higman, The units of group rings, *Proc. London Math. Soc. (2)* **46** (1940), 231–248.

[14] H.-J. Hoehnke and K. W. Johnson, The 1- , 2- , 3-characters are sufficient to determine a group, Preprint, 1991.

[15] B. Huppert and N. Blackburn, *Finite groups III*, Springer, Berlin, 1982.

[16] I. M. Isaacs, *Character theory of finite groups*, Academic Press, New York, 1976.

[17] C. Jansen, K. Lux, R. A. Parker, R. A. Wilson, *An* ATLAS *of Brauer characters*, London Math. Soc. Monographs, New Series 11, Oxford University Press, 1995.

[18] K. W. Johnson, On the group determinant, *Math. Proc. Cambridge Philos. Soc.* **109** (1991), 299–311.

[19] W. Kimmerle, Beiträge zur ganzzahligen Darstellungstheorie endlicher Gruppen, *Bayreuther Math. Schr.* **36** (1991), 1–139.

[20] W. Kimmerle, Variations of the Zassenhaus conjecture, in [35], pp. 117–124.

[21] W. Kimmerle, On automorphisms of ℤG and the Zassenhaus conjectures, *Proceedings of the ICRA VII, Cocoyoc, Mexico (eds. R. Bautista, R. Martinez-Villa and J. A. de la Peña)*, Canad. Math. Soc. Conf. Proc., vol. 18, 1996.

[22] W. Kimmerle, R. Lyons, R. Sandling and D. Teague, Composition factors from the group ring and Artin's theorem on orders of simple groups, *Proc. London Math. Soc. (3)* **60** (1990), 89–122.

[23] W. Kimmerle and K. W. Roggenkamp, Projective limits of group rings, *J. Pure Appl. Algebra* **88** (1993), 119–142.

[24] W. Kimmerle and K. W. Roggenkamp, Group determinants and 3-characters, Preprint, 1991.

[25] W. Kimmerle and R. Sandling, Group theoretic and group ring theoretic determination of certain Sylow and Hall subgroups and the resolution of a question of R. Brauer, *J. Algebra* **171** (1995), 329–346.

[26] W. Kimmerle and R. Sandling, The determination of Abelian Hall subgroups by a conjugacy class structure, *Publicacions Mat.* **36** (1992), 685–691.

[27] L. Klingler, Construction of a counterexample to a conjecture of Zassenhaus, *Comm. Algebra* **19** (1991), 2303–2330.

[28] R. Mansfield, A group determinant determines its group, *Proc. Amer. Math. Soc.* **116** (1992), 939–941.

[29] D. S. Passman, Isomorphic groups and group rings, *Pacific J. Math. (2)* **35** (1965), 561–583.

[30] G. L. Peterson, Automorphisms of the integral group ring of S_n, *Proc. Amer. Math. Soc.* **59** (1976), 14–18.

[31] K. W. Roggenkamp, Observations to a conjecture of H. Zassenhaus, in *Groups—St. Andrews 1989 (eds. C. M. Campbell and E. F. Robertson)*, Vol. 2, pp. 427–444. London Math. Soc. Lecture Note Series 160, Cambridge University Press, 1991.

[32] K. W. Roggenkamp, The isomorphism problem for integral group rings of finite groups, in *Proceedings of the international congress of mathematicians, Kyoto 1990 (ed. I. Satake)*, pp. 369–380. Springer, 1991.

[33] K. W. Roggenkamp and L. L. Scott, Isomorphisms of p-adic group rings, *Ann. of Math.* **126** (1987), 593–647.

[34] K. W. Roggenkamp and L. L. Scott, A strong answer to the isomorphism problem for finite p-solvable groups with a normal p-subgroup containing its centralizer, Preprint, 1987.

[35] K. W. Roggenkamp and M. Taylor, *Group rings and class groups*, DMV-Seminar 18, Birkhäuser, Basel/Boston/Berlin, 1992.

[36] K. W. Roggenkamp and A. Zimmermann, A counterexample for the isomorphism problem of polycyclic groups, *J. Pure Appl. Algebra* **103** (1995), 101–103.

[37] A. I. Saksonov, On the group ring of finite groups I, *Publ. Math. Debrecen* **18** (1971), 187–209.

[38] R. Sandling, Graham Higman's thesis 'Units in group rings', in *Integral representations and applications (ed. K. W. Roggenkamp)*, pp. 93–116. Lecture Notes in Math. 882, Springer, 1981.

[39] R. Sandling, The isomorphism problem for group rings; a survey, in *Orders and their applications (eds. I. Reiner and K. W. Roggenkamp)*, pp. 256–288. Lecture Notes in Math. 1142, Springer, 1985.

[40] M. Schönert *et al.*, *GAP—Groups, Algorithms, and Programming*, Lehrstuhl D für Mathematik, Rheinisch-Westfälische Technische Hochschule, Aachen, 1992.

[41] L. L. Scott, Recent progress on the isomorphism problem, in *Proc. Symposia in Pure Math. (ed. P. Fong)*, Vol. 47, Part I, pp. 259–274. Amer. Math. Soc., 1987.

[42] S. K. Sehgal, Torsion units in group rings, in *Proceedings Nato Institute on methods in ring theory, Antwerp (ed. F.van Oystaeyen)*, pp. 497–504. D. Reidel, Dordrecht, 1983.

[43] S. K. Sehgal, *Units of group rings*, Longman, 1993.

[44] A. Weiss, *p*-adic rigidity of *p*-torsion, *Ann. of Math.* **127** (1987), 317–332.

[45] A. Weiss, Torsion units in integral group rings, *J. reine angew. Math.* **415** (1991), 175–187.

Finite linear groups of small degree

A. S. Kondratiev

Abstract

This paper is a survey of results in the study of finite subgroups of low-dimensional classical groups. The author discusses mainly his classification of the conjugacy classes and the normalizers of finite absolutely irreducible quasisimple linear groups over finite and algebraically closed fields up to degree 27.

Our aim is to discuss recent developments in the following classical problem.

Problem 1 *Describe the finite linear groups of small degree, i.e. finite subgroups in $GL_n(K)$ for every field K and small n.*

Beginning in the middle of the last century, this problem attracted the attention of many mathematicians. By the seventies of our century it was solved for $K = \mathbb{C}$ and $n \leq 9$ in the papers by Jordan, Klein, Valentiner, Blichfeld (see [4]), Brauer [7], Lindsey [41], Huffman and Wales [16], [17], and Feit [13]. The case $\mathrm{char}\, K = p > 0$ and $n \leq 5$ of Problem 1 was considered before 1982 in the papers by Moore [44], Burnside [8], Wiman [52], Dickson [10], [11], Mitchell [43], Hartley [15], Bloom [5], Mwene [45], [46], Wagner [50], Di Martino and Wagner [12], Zalesskii [53], [54], Suprunenko [49], and Zalesskii and Suprunenko [57]. For finite K this problem stands as Problem 40 in the list of important problems of group representation theory formulated by Brauer in his lectures [6]. Results related to Problem 1 (see, for example, the surveys of Zalesskii [55], [56] and the author [26]) have found numerous applications, in particular, in the classification of finite simple groups (CFSG). After the announcement of the completion of CFSG (see [14]) progress in solving Problem 1 is related to intensive development of investigations modulo CFSG of the following problem (see surveys of Kleidman and Liebeck [22], Liebeck and Saxl [42] and Seitz [48]).

Problem 2 *Determine the maximal subgroups of finite almost simple groups, i.e. groups G with nonabelian simple socle $Soc(G)$.*

Aschbacher [1] and Kleidman and Liebeck [23] reduced the problem of finding the maximal subgroups M of a finite almost simple classical group G having the natural projective module $V = V(n, q)$ of dimension n over the field $GF(q)$ to the case when $S = Soc(M)$ is a nonabelian simple group, V is an absolutely irreducible $GF(q)\widehat{S}$-module for some covering group $\widehat{S}$ of S, and V is not realized over a proper subfield of $GF(q)$. Thus the problem of describing the subgroup structure of finite classical groups is reduced in considerable degree to the study of modular representations of finite quasisimple groups, i.e. coverings of finite simple nonabelian groups (see [33]). We shall mainly use the standard notations and terminology as given in [9] and [19].

In his unpublished monograph [21] Kleidman determined all maximal subgroups of finite classical almost simple groups of dimension ≤ 12. A list of the maximal subgroups of the simple classical groups of dimension at most 11 is contained in [20]. Note that we independently determined the conjugacy classes and the normalizers of absolutely irreducible quasisimple subgroups in $GL_6(q)$ [31] and the irreducible subgroups in $GL_n(2)$ for $7 \leq n \leq 10$ [24], [25], [27], [28]. So far the result of Kleidman was the strongest one on Problem 1. Now we solve Problem 1 up to the degree 27. In the future we hope to extend Kleidman's classification [21] up to dimension 27. The number 27 is taken deliberately, as this is the minimal degree of faithful representations of the group $E_6(q)$.

Aschbacher [2] classified all irreducible FL-modules of dimension ≤ 27 for every quasisimple group L of Lie type over a finite or algebraically closed field F. He uses this result in the investigation of the subgroup structure of the group $E_6(q)$ considered as an isometry group of a symmetric 3-linear form on the 27-dimensional module (see [3]).

In [32], we obtained a complete classification of absolutely irreducible p-modular representations of degree ≤ 27 of quasisimple groups of Lie type defined over a finite field of characteristic $\neq p$. In particular, we have the following theorem.

Theorem 1 *A finite nonabelian simple group G of Lie type defined over a field of characteristic p has a nontrivial absolutely irreducible projective representation of degree ≤ 27 over a field of characteristic $\neq p$ if and only if G is isomorphic to one of the following groups: $L_2(q)$ for $4 \leq q \leq 53$ and $q \neq 32$, $L_3(q)$ for $q \in \{3, 4\}$, $U_3(q)$ for $q \in \{3, 4, 5\}$, $L_4(2)$, $L_4(3)$, $PSp_4(q)$ for $q \in \{3, 4, 5, 7\}$, $U_4(3)$, $U_5(2)$, $Sz(8)$, $PSp_6(2)$, $PSp_6(3)$, $U_6(2)$, $\Omega_8^+(2)$, $P\Omega_7(3)$, $G_2(3)$, $G_2(4)$, $^3D_4(2)$, $^2F_4(2)'$.*

In [34], we obtained the corresponding classification for the remaining case of the finite quasisimple groups of alternating or sporadic types. In particular, we have the following theorem.

Theorem 2 *A finite simple nonabelian alternating or sporadic group G has a nontrivial absolutely irreducible projective (modular) representation of degree ≤ 27 if and only if G is isomorphic to one of the following groups: A_n for $5 \leq n \leq 29$, M_{11}, M_{12}, M_{22}, M_{23}, M_{24}, J_1, J_2, J_3, HS, McL, Suz, Co_3, Co_2, Co_1, Fi_{22}.*

In the proof of Theorems 1 and 2 we use the results of Landazuri and Seitz [40] and James [18] which give some lower bounds for the degrees of projective representations of the finite simple groups of Lie type in non-defining characteristic, and of the alternating groups, respectively. The big sporadic groups are rejected by using known information about their subgroup structure (see [9]) and the following elementary lemma.

Lemma 1 *Let X be a finite subgroup of $GL_n(F)$, where F is an algebraically closed field of characteristic p, and let A be a normal elementary abelian r-subgroup in X, $p \neq r$, and m be the minimal length of X-orbits in $A - \{1\}$. Then $m \leq n$.*

Further, we calculated irreducible modular characters for the many concrete "small" finite quasisimple groups from the lists in the conclusions of Theorems 1 and 2. A typical situation in these concrete calculations is that the modular characters are found up to several nonnegative integer parameters. But for Problems 1 and 2 even such special results are often enough. For example, often we need only a few irreducible modular characters of "small" degrees (of minimal degree or of some following degrees). We used also Parker's collection of modular characters [47].

Remark 1 So far even the following problem is not fully solved.

Problem 3 *Determine all faithful p-modular absolutely irreducible representations of the minimal degree for each finite quasisimple group and each prime p.*

Remark 2 We have also some complete computer free results on calculating p-modular irreducible characters for concrete finite quasisimple groups (see [39]). For example, $\widehat{J_2}$ [29], [30], $\widehat{P\Omega_7(3)}$ ($p \neq 3$) [32], [37], $\widehat{U_6(2)}$ ($p \neq 2$) [32], [35]. Recently we used the computer system GAP for calculating modular characters. In particular, we considered the cases Co_3 for $p = 2, 5$ [36], [38]. Of course, in all our results we use the ATLAS [9].

On the basis of a complete classification of absolutely irreducible modular representations of degree ≤ 27 of all finite quasisimple groups we proved recently the following result.

Table 1: Absolutely irreducible quasisimple subgroups of $GL_{13}(p^m)$

G	Restrictions on p	Restrictions on m	Number of conj. classes	$N_{PGL_{13}(q)}(\overline{G})$
$L_2(p^k)$	$p > 11$	$k\|m$	1	$PGL_2(p^k)$
$S_4(5^k)$	$p = 5$	$k\|m$	1	$S_4(5^k){:}2$
$S_6(3^k)$	$p = 3$	$k\|m$	1	$S_6(3^k){:}2$
$SL_{13}(p^k)$		$k\|m$	1	$PGL_{13}(p^k)$
$\Omega_{13}(p^k)$	$p \neq 2$	$k\|m$	1	$PGO_{13}(p^k)$
$SU_{13}(p^k)$		$2k\|m$	1	$PGU_{13}(p^k)$
$L_2(13)$	$2 \neq p \neq 7$		1	$PGL_2(13)$
$L_2(25)$	$2 \neq p \neq 5$		1	$L_2(25){:}2_2$
$L_2(27)$	$p \neq 3$	$2\|m$ by $p \equiv 2(3)$	1	$L_2(27){:}3$
$L_3(3)$	$p > 3$		1	$AutL_3(3)$
$U_3(4)$	$2 \neq p \neq 5$	$4\|m$ by $p \equiv 2,3(5)$	1	$U_3(4)$
$S_4(5)$	$2 \neq p \neq 5$	$2\|m$ by $p \equiv 2,3(5)$	1	$S_4(5)$
$S_6(3)$	$p \neq 3$	$2\|m$ by $p \equiv 2(3)$	1	$S_6(3)$
A_7	$p = 3,5$		1	S_7
A_8	$p = 3,5$		1	S_8
A_{14}	$2 \neq p \neq 7$		1	S_{14}
A_{15}	$p = 3,5$		1	S_{15}
J_2	$p = 3$	$2\|m$	1	J_2

Theorem 3 *The conjugacy classes and the normalizers of absolutely irreducible quasisimple subgroups in $GL_n(q)$ for $13 \leq n \leq 27$ are known.*

So, we have solved Problem 1 up to degree 27 for finite and algebraically closed fields K.

The exact formulation of Theorem 3 contains a table which is too long to give here, therefore we give only an example of this result for $n = 13$, i.e. a fragment of the table.

Example Let G denote a finite quasisimple absolutely irreducible subgroup in $GL_n(q)$, where $q = p^m$ for a prime p and a natural number m. Denote by $\overline{G}$ the image of G in $PGL_n(q)$. Then Table 1 gives the classification of the conjugacy classes and the normalizers of subgroups G in $GL_n(q)$ for $n = 13$.

In the proof of Theorem 3 the following two problems arise.

(a) Finding the field of definition of a (p-modular) irreducible representation $T : G \longrightarrow GL_n(K)$ ($\overline{K} = K$) with the given Brauer character $\varphi = \beta_T$.

The field of definition of T is defined as the least subfield of $K = \overline{GF(p)}$ over which T may be realized. It is well known that the field of definition of T is equal to $GF(p)(\varphi(x_j)^\mu | 1 \leq j \leq r)$, where $x_1, \ldots, x_r$ form a complete system of representatives of conjugacy classes of p'-elements of the group G and $\mu : \mathbb{Z}[U] \longrightarrow K$ is a ring homomorphism, U is a suitable set of roots of unity in $\mathbb{C}$ of order prime to p.

(b) Counting conjugacy classes and normalizers of absolutely irreducible quasisimple subgroups in $GL_n(q)$.

Let G be a finite quasisimple group, $T : G \longrightarrow GL_n(q)$ be a faithful absolutely irreducible representation of G, and N be the normalizer of $T(G)$ in $GL_n(q)$. For any subgroup X of $GL_n(q)$, let $\overline{X}$ be the image of X in $PGL_n(q)$. It is easy to prove that $\overline{T(G)} \leq \overline{N} \leq Aut(T(G))$, and the conjugacy classes of absolutely irreducible subgroups of $GL_n(q)$ isomorphic to G are in one-to-one correspondence with the $Aut(G)$-orbits of faithful irreducible p-modular Brauer characters of degree n of G, and, consequently, $|Aut(T(G)) : \overline{N}|$ is the order of the $Aut(G)$-orbit containing φ.

Therefore, if we know the set of the irreducible p-modular Brauer characters of degree n of G and the action of $Aut(G)$ on this set then we solve the problems (a) and (b). The method of proof of Theorems 1 and 2 permits calculation of the Brauer characters of representations considered in these theorems.

ACKNOWLEDGEMENTS

The author wishes to thank the London Mathematical Society for financial support of his participation in the conference "The ATLAS ten years on". The research described in this publication was made possible in part by Grants 93-01-01529 from RFFI, NMC000 from ISF and NMC300 from ISF and the Russian Government.

References

[1] M. Aschbacher, On the maximal subgroups of the finite classical groups, *Invent. Math.* **76** (1984), 469–514.

[2] M. Aschbacher, Small degree representations of groups of Lie type, Preprint, Calif. Inst. Technol., Pasadena, 1987.

[3] M. Aschbacher, The 27-dimensional module for E_6, I, *Invent. Math.* **89** (1987), 159–195; II, *J. London Math. Soc.* **37** (1988), 275–293; III, *Trans. Amer. Math. Soc.* **321** (1990), 45–84; IV, *J. Algebra* **131** (1990), 23–39.

[4] H. F. Blichfeld, *Finite collineation groups*, University of Chicago Press, 1917.

[5] D. M. Bloom, The subgroups of $PSL(3, q)$ for odd q, *Trans. Amer. Math. Soc.* **127** (1967), 150–158.

[6] R. Brauer, Representations of finite groups, in *Lectures on modern mathematics (ed. T. L. Saaty)*, Vol. 1, pp. 133–175. Wiley, 1963.

[7] R. Brauer, Über endliche lineare Gruppen von Primzahlgrad, *Math. Ann.* **169** (1967), 73–96.

[8] W. Burnside, On a class of groups defined by congruences, *Proc. London Math. Soc.* **25** (1894), 113–139; **26** (1895), 58–106.

[9] J. H. Conway, R. T. Curtis, S. P. Norton, R. A. Parker and R. A. Wilson, ATLAS *of finite groups*, Clarendon Press, Oxford, 1985.

[10] L. E. Dickson, *Linear groups with an exposition of the Galois field theory*, Teubner, Leipzig, 1901; reprinted, Dover, New York, 1958.

[11] L. E. Dickson, Determination of the ternary modular linear groups, *Amer. J. Math.* **27** (1905), 189–202.

[12] L. Di Martino and A. Wagner, The irreducible subgroups of $PSL(V_5, q)$, where q is odd, *Resultate Math.* **2** (1979), 54–61.

[13] W. Feit, On finite linear groups in dimension at most 10, in *Proceedings of conference on group theory (eds. W. R. Scott and F. Gross)*, pp. 397–407. Academic Press, 1976.

[14] D. Gorenstein, *Finite simple groups. An introduction to their classification*, Plenum Press, New York/London, 1982.

[15] R. W. Hartley, Determination of the ternary collineation groups whose coefficients lie in $GF(2^n)$, *Ann. Math.* **27** (1926), 140–158.

[16] W. C. Huffman and D. B. Wales, Linear groups of degree eight with no elements of order seven, *Illinois J. Math.* **20** (1976), 519–527.

[17] W. C. Huffman and D. B. Wales, Linear groups of degree nine with no elements of order seven, *J. Algebra* **51** (1978), 149–163.

[18] G. D. James, On the minimal dimensions of irreducible representations of symmetric groups, *Math. Proc. Cambridge Philos. Soc.* **94** (1983), 417–424.

[19] C. Jansen, K. Lux, R. Parker and R. Wilson, *An* ATLAS *of Brauer characters*, Clarendon Press, Oxford, 1995.

[20] P. B. Kleidman, *The subgroup structure of some finite simple groups*, Ph. D. Thesis, Cambridge, 1986.

[21] P. B. Kleidman, *The low-dimensional finite classical groups and their subgroups*, Longman Research Notes Series, to appear.

[22] P. B. Kleidman and M. W. Liebeck, A survey of the maximal subgroups of the finite simple groups, *Geom. Dedicata* **25** (1988), 375–390.

[23] P. B. Kleidman and M. W. Liebeck, *The subgroup structure of the finite classical groups*, LMS Lecture Note Series 129, Cambridge University Press, 1990.

[24] A. S. Kondratiev, Irreducible subgroups of the group $GL(7,2)$, *Math. Notes* **37** (1985), no. 3–4, 178–181.

[25] A. S. Kondratiev, Irreducible subgroups of the group $GL(9,2)$, *Math. Notes* **39** (1986), no. 3–4, 173–178.

[26] A. S. Kondratiev, Subgroups of finite Chevalley groups, *Russian Math. Surveys* **41** (1986), no. 1, 65–118.

[27] A. S. Kondratiev, Linear groups of small degrees over the field of order 2, *Algebra and Logic* **25** (1986), 343–357.

[28] A. S. Kondratiev, The irreducible subgroups of the group $GL_8(2)$, *Commun. Algebra* **15** (1987), 1039–1093 (In Russian).

[29] A. S. Kondratiev, Decomposition numbers of the group J_2, *Algebra and Logic* **27** (1988), 333–349.

[30] A. S. Kondratiev, Decomposition numbers of the groups $\widehat{J_2}$ and $Aut(J_2)$, *Algebra and Logic* **27** (1988), 429–444.

[31] A. S. Kondratiev, On finite linear groups of degree 6, *Algebra and Logic* **28** (1989), 122–138.

[32] A. S. Kondratiev, Modular representations of small degrees of finite groups of Lie type, Preprint, UrO AN SSSR, Sverdlovsk, 1989 (In Russian).

[33] A. S. Kondratiev, Subgroups and modular representations of finite quasisimple groups, in *Topics in algebra. Part 2 (eds. S. Balcerzyk, T. Jósefiak, J. Krempa, D. Simson and W. Vogel)*, pp. 357–364. Banach Center Publications, Vol. 26, Pt. 2, PWN, 1990.

[34] A. S. Kondratiev, Modular representations of degree ≤ 27 of finite quasisimple groups of alternating and sporadic types, *Trudy IMM UrO RAN* **1** (1992), 21–49 (In Russian).

[35] A. S. Kondratiev, The decomposition matrices of the group $\widehat{U_6(2)}$, in *Problems in group theory and homological algebra (ed. A. L. Onishchik)*, pp. 44–62. Yaroslavl' State University, 1992 (In Russian).

[36] A. S. Kondratiev, On the 2-modular characters of the group Co_3, in *Algebra and analysis. Abstracts of reports of the international conference in honour of N. G. Chebotarev. Part I (eds. F. G. Avkhadiev et al.)*, pp. 52–53. Kazan' University Press, 1994 (In Russian).

[37] A. S. Kondratiev, The 2-modular characters of the group $P\Omega_7(3)$, *Trudy IMM UrO RAN* **3** (1995), 50–59 (In Russian).

[38] A. S. Kondratiev, On the 5-modular characters of the Conway group Co_3, in *All-Ukraine conference on "Developments and application of mathematical methods in scientific-technical investigations". Abstracts. Part I. (ed. O. B. Chernigenko)*, pp. 58–59. State University "Lviv Polytechnic", 1995.

[39] A. S. Kondratiev, Decomposition matrices of finite quasisimple groups, in *Algebra. Proceedings of the IIIrd international conference on algebra (eds. Yu. L. Ershov, E. I. Khukhro, V. M. Levchuk and N. D. Podufalov)*, pp. 133–144. Walter de Gruyter, 1996.

[40] V. Landazuri and G. M. Seitz, On the minimal degree of projective representations of the finite Chevalley groups, *J. Algebra* **32** (1974), 418–443.

[41] J. H. Lindsey II, Finite linear groups of degree six, *Canad. J. Math.* **23** (1971), 771–790.

[42] M. W. Liebeck and J. Saxl, Maximal subgroups of finite simple groups and their automorphism groups, *Contemp. Math.* **131** (1992), 243–259.

[43] H. H. Mitchell, Determination of the ordinary and modular ternary linear groups, *Trans. Amer. Math. Soc.* **12** (1911), 207–242.

[44] E. H. Moore, Determination of the ordinary and modular ternary linear groups, Decennial Publ. Univ. Chicago (1904), 141–190.

[45] B. Mwene, On some subgroups of the group $PSL_4(2^m)$, *J. Algebra* **41** (1976), 79–107.

[46] B. Mwene, On some subgroups of $PSL(4, q)$, q odd, *Geom. Dedicata* **12** (1982), 189–199.

[47] R. A. Parker, A collection of modular characters, Preprint, Cambridge, 1989.

[48] G. M. Seitz, Subgroups of finite and algebraic groups, in *Groups, combinatorics and geometry (eds. M. W. Liebeck and J. Saxl)*, pp. 316–326. LMS Lecture Note Series 165, Cambridge University Press, 1992.

[49] I. D. Suprunenko, Finite irreducible groups of degree 4 over fields of characteristic 3 and 5, *Vesci Akad. Navuk BSSR Ser. Fiz.-Mat. Navuk* 1981, no. 3, 16–22 (In Russian).

[50] A. Wagner, The subgroups of $PSL(5, 2^a)$, *Resultate Math.* **1** (1978), 207–226.

[51] D. B. Wales, Finite linear groups of degree seven, I, *Canad. J. Math.* **21** (1969), 1025–1056; II, *Pacific J. Math.* **34** (1970), 207–235.

[52] A. Wiman, Bestimmung aller Untergruppen einer doppelt unendlichen Reihe von endlichen Gruppen, *Bihang. till K. Svenska Vet.-Akad. Handlingar* **25**(1) (1899), 1–47.

[53] A. E. Zalesskii, The classification of finite linear groups of degree 5 over a field of characteristic not equal to 0, 2, 3, 5, *Dokl. Akad. Nauk BSSR* **20** (1976), 773–775 (In Russian).

[54] A. E. Zalesskii, The classification of finite linear groups of degrees 4 and 5 over a field of characteristic 2, *Dokl. Akad. Nauk BSSR* **21** (1977), 389–392 (In Russian).

[55] A. E. Zalesskii, Linear groups, *Russian Math. Surveys* **36** (1981), no. 5, 63–126.

[56] A. E. Zalesskii, Linear groups, in *Progress in science and technology. Ser. Current problems in mathematics. Fundamental directions. Vol. 37 (ed. R. W. Gamkrelidze)*, pp. 114–228. Akad. Nauk SSSR, VINITI, 1989 (In Russian). Translation in *Algebra. IV. Encyclopedia Math. Sci. Vol. 37*, pp. 97–196. Springer, 1993.

[57] A. E. Zalesskii and I. D. Suprunenko, The classification of finite irreducible linear groups of degree 4 over a field of characteristic $p > 5$, *Vesci Akad. Navuk BSSR Ser. Fiz.-Mat. Navuk*, 1978, no. 6, 9–15; Corrigendum, *ibid.*, 1979, no. 3, 136 (In Russian).

Minimal parabolic systems for the symmetric and alternating groups

Wolfgang Lempken,
Christopher Parker
and
Peter Rowley

Abstract

We determine the minimal parabolic subgroups and minimal parabolic systems of the alternating and symmetric groups with respect to a Sylow 2-subgroup.

1 Introduction

Using the finite groups of Lie type as a rôle model Ronan and Smith [5] and Ronan and Stroth [6] studied certain subgroups of the sporadic finite simple groups. The underlying idea was to obtain geometric structures analogous to buildings via suitable systems of subgroups in each of these groups. Here we investigate the alternating and symmetric groups for the prime 2 from a perspective similar to [6], the emphasis being on minimal parabolic subgroups, which we now define.

Let G be a finite group, p a prime dividing the order of G and $T \in \mathrm{Syl}_p(G)$. Put $B = N_G(T)$. Then a subgroup P of G which properly contains B is called a *minimal parabolic subgroup* (of G) with respect to B if B is contained in a unique maximal subgroup of P. Note that, unlike Ronan and Stroth [6], here we do not require that $O_p(P) \neq 1$. We denote the collection of minimal parabolic subgroups of G with respect to B by $\mathcal{M}(G, B)$. In the case when G is a finite group of Lie type defined over a field of characteristic p, the subgroups in $\mathcal{M}(G, B)$ are indeed minimal parabolic subgroups of G and our nomenclature coincides with the standard one.

Let $P_1, \ldots, P_m \in \mathcal{M}(G, B)$. Then we call $\mathcal{S} = \{P_1, \ldots, P_m\}$ a *minimal parabolic system of rank m* provided that $G = \langle \mathcal{S} \rangle$ and no proper subset of $\mathcal{S}$ generates G.

Throughout this paper all sets considered are finite (in particular, all groups are finite) and $p = 2$. Let Ω be a set of cardinality $n > 2$. We fix the following notation for the 2-adic decomposition of n:

$$n = 2^{n_1} + 2^{n_2} + \cdots + 2^{n_r} \text{ where } n_1 > n_2 > \cdots > n_r \geq 0.$$

Set $G = \text{Sym}(\Omega)$ and $G^* = \text{Alt}(\Omega)$. Let T be a fixed Sylow 2-subgroup of G and put $T^* = G^* \cap T$. Recall that $N_G(T) = T$ and, provided $n > 5$, we also have that $N_{G^*}(T^*) = T^*$. The distinct orbits of T on Ω will be denoted by $\Omega_1, \Omega_2, \ldots, \Omega_r$ where $|\Omega_i| = 2^{n_i}$. Put $I = \{1, \ldots, r\}$. Note also that $T = T_1 \times T_2 \times \ldots \times T_r$ where, for $i \in I$, $T_i \in \text{Syl}_2(\text{Sym}(\Omega_i))$. Recall that each T_i is a certain iterated wreath product of cyclic groups of order 2, see Huppert [2, Satz 15.3]. Let $i \in I$. Then for $j \in \{1, \ldots, n_i - 1\}$ let $\Sigma_{i;j}$ be the collection of T-invariant block systems of Ω_i consisting of sets of order 2^k where $k \in \{0, \ldots, n_i\} \setminus \{j\}$, and define

$$P(i; j) = \text{Stab}_{\text{Sym}(\Omega_i)} \Sigma_{i;j} \times \left(\prod_{k \in I \setminus \{i\}} T_k \right).$$

Put

$$\mathcal{L}(G, T) = \{P(i; j) \mid i \in I, j \in \{1, \ldots, n_i - 1\}\}.$$

For $i, j \in I$, with $i < j$ (so $n_j < n_i$) set $\Lambda_{i+j} = \Omega_i \cup \Omega_j$. Let Γ_i be the collection of all block systems for T on Ω_i and Γ_j the collection of all block systems of T on Ω_j. We define Σ_{i+j} to be the collection of T-invariant systems of subsets of Λ_{i+j} which are the union of one block system from Γ_i and one from Γ_j with the proviso that the blocks of the two chosen block systems have equal numbers of elements. Then

$$P(i + j) = \text{Stab}_{\text{Sym}(\Lambda_{i+j})}(\Sigma_{i+j}) \times \left(\prod_{k \in I \setminus \{i,j\}} T_k \right)$$

and

$$\mathcal{F}(G, T) = \{P(i + j) \mid i, j \in I, i < j\}.$$

Finally we put

$$\mathcal{P}(G, T) = \mathcal{L}(G, T) \cup \mathcal{F}(G, T)$$

and

$$\mathcal{P}(G^*, T^*) = \{P \cap G^* \mid P \in \mathcal{P}(G, T)\}.$$

In Proposition 3.1 we verify that $\mathcal{P}(G, T) \subseteq \mathcal{M}(G, T)$. Our first theorem, proved in section 4, asserts that these sets are in fact equal.

Theorem 1.1 *Assume that Ω is a set with $|\Omega| > 2$, $G = \mathrm{Sym}(\Omega)$ and $T \in$ $\mathrm{Syl}_2(G)$. Then $\mathcal{M}(G,T) = \mathcal{P}(G,T)$.*

The situation for G^*, given in our next result, is similar, with some small exceptional cases which we shall discuss in section 5.

Theorem 1.2 *Assume that Ω is a set with $|\Omega| > 9$, $G^* = \mathrm{Alt}(\Omega)$ and $T^* \in$ $\mathrm{Syl}_2(G^*)$. Then $\mathcal{M}(G^*,T^*) = \mathcal{P}(G^*,T^*)$.*

Theorems 1.1 and 1.2 then give us a concrete description of the minimal parabolic subgroups of G and G^*, and with this information to hand we prove

Theorem 1.3 *Suppose that Ω is a set with $|\Omega| = n > 2$, $G = \mathrm{Sym}(\Omega)$, $G^* = \mathrm{Alt}(\Omega)$, $T \in \mathrm{Syl}_2(G)$ and $T^* = G^* \cap T$.*

(i) *If $n = 2^{n_1}$, then there is a unique minimal parabolic system $\mathcal{S}$ of G, $\mathcal{S} = \mathcal{M}(G,T) = \mathcal{L}(G,T)$ and $\mathcal{S}$ has rank $n_1 - 1$.*

(ii) *If n is odd, then there are r^{r-2} distinct minimal parabolic systems of G with respect to T, each of which has rank $r - 1$. Furthermore, if $\mathcal{S}$ is a minimal parabolic system of G, then there is a spanning tree $\Gamma = (V, E)$ in the complete labelled graph on r vertices for which*

$$\mathcal{S} = \{P(i + j) \in \mathcal{F}(G,T) \mid (i,j) \in E\}$$

and, moreover, every distinct choice of Γ gives rise to a distinct minimal parabolic system of G with respect to T.

(iii) *If n is even and $r > 1$, then there are $r^{n_r + r - 3}(r - 1)$ distinct minimal parabolic systems of G with respect to T each of which has rank $n_r + r - 1$. If $\mathcal{S}$ is a minimal parabolic system of G with respect to T, then $|\mathcal{S} \cap \mathcal{F}(G,T)| = r - 1$, $|\mathcal{S} \cap \mathcal{L}(G,T)| = n_r$ and there is a spanning tree $\Gamma = (V, E)$ in the complete labelled graph on r vertices and a set $\mathcal{T} = \{(i_1, 1), \ldots, (i_{n_r}, n_r) \mid 1 \le i_j \le r, i_{n_r} \ne r\}$ for which*

$$\mathcal{S} = \{P(i + j) \in \mathcal{F}(G,T) \mid (i,j) \in E\} \cup \{P(i;j) \in \mathcal{L}(G,T) \mid (i,j) \in \mathcal{T}\}$$

and, moreover, every distinct choice of Γ and $\mathcal{T}$ gives rise to a distinct minimal parabolic system of G with respect to T.

(iv) *If $n > 9$, then intersection of minimal parabolic subgroups of G with G^* gives a one to one correspondence between minimal parabolic systems of G with respect to T and minimal parabolic systems of G^* with respect to T^*.*

An unexpected, and perhaps surprising, feature of these minimal parabolic systems is

Corollary 1.4 *Suppose the hypothesis of Theorem 1.3 holds. Then all minimal parabolic systems of G (respectively G^*) with respect to T (respectively T^*) have the same rank.*

We also remark that whenever the rank of a minimal parabolic system is greater than or equal to 2 (that is $n \neq 2^{n_1} + 1$), then the subgroups in the system are all 2-local subgroups. Additionally, all these 2-local subgroups P are 2-constrained (meaning that $C_P(O_2(P)) \leq O_2(P)$) if and only if n is even.

Minimal parabolic systems for symmetric groups with respect to the normalizer of a Sylow p-subgroup, $p \neq 2$, are being investigated by Sandra Covello at Birmingham.

2   Preliminary results

Our first lemma, which is a corollary of a theorem of Jordan [3] and Marggraf [4], plays an important rôle in the proofs of Theorems 1.1 and 1.2. We recall that a primitive permutation group is a transitive permutation group which preserves only trivial systems of blocks.

Lemma 2.1 (Jordan, Marggraf) *Suppose that Σ is a set and $L \leq \mathrm{Sym}(\Sigma)$ is primitive.*

(i) *If L contains a transposition, then $L = \mathrm{Sym}(\Sigma)$.*

(ii) *Suppose L contains a fours-group which is transitive on 4 points and fixes all the other points of Σ. If $|\Sigma| > 9$, then $L \geq \mathrm{Alt}(\Sigma)$.*

Proof: See Wielandt [8, Theorems 13.3 and 13.5]. □

Lemma 2.2 *Let Σ be a set of cardinality not equal to 4, $H_1 = \mathrm{Sym}(\Sigma)$, $H_0 = \mathrm{Alt}(\Sigma)$ and for $i = 0, 1$, $S_i \in \mathrm{Syl}_2(H_i)$.*

(i) *$H_1 \in \mathcal{M}(H_1, S_1)$ if and only if $|\Sigma| = 2^m + 1$ for some $m \geq 1$.*

(ii) *$H_0 \in \mathcal{M}(H_0, S_0)$ if and only if $|\Sigma| = 2^m + 1$ for some $m \geq 2$.*

Proof: We prove (ii), part (i) being similar. For $|\Sigma| \leq 9$ we use the ATLAS [1] to verify the result. Thus we now assume that $|\Sigma| > 9$.

Suppose that $|\Sigma| = 2^m + 1$ for some $m \geq 1$. Then we know that S_0 fixes a unique point $\sigma \in \Sigma$ and operates transitively on $\Sigma \setminus \{\sigma\}$. Suppose $L \not\leq \mathrm{Stab}_{H_0}(\sigma)$ and $H_0 \geq L \geq S_0$. Then L is 2-transitive on Σ, and, as S_0 contains a fours-group which is transitive on four points and fixes the remaining points of Σ, Lemma 2.1(ii) implies that $L = H_0$. Thus all proper subgroups of H_0 which contain S_0 are contained in $\mathrm{Stab}_{H_0}(\sigma)$ and so $H_0 \in \mathcal{M}(H_0, S_0)$.

Conversely, suppose that $|\Sigma| \neq 2^m + 1$ for any $m \geq 1$. If $|\Sigma| = 2^k$, $k \neq 2$, then let L_1 (respectively, L_2) be the stabilizer in H_0 of the S_0-invariant block system with block size 2 (respectively, 4). If $|\Sigma| \neq 2^k$, then S_0 has at least 2 orbits on Σ, say Λ_1, Λ_2. If $\Lambda_1 \cup \Lambda_2 \neq \Sigma$, then set $L_i = \mathrm{Stab}_{H_0}(\Lambda_i)$, $i = 1, 2$. Otherwise we take L_1 to be the stabilizer in H_0 of the S_0-invariant block system with block size 2 and let $L_2 = \mathrm{Stab}_{H_0}(\Lambda_2)$. Then, using Lemma 2.1(ii), we see that $H_0 = \langle L_1, L_2 \rangle$ and so $H_0 \notin \mathcal{M}(H_0, S_0)$. $\qquad\square$

Lemma 2.3 *Suppose that Σ is a set, $H_1 = \mathrm{Sym}(\Sigma)$, $H_0 = \mathrm{Alt}(\Sigma)$ and, for $i = 0, 1$, $S_i \in \mathrm{Syl}_2(H_i)$. Then $N_{H_1}(S_1) = S_1$ and, provided $|\Sigma| > 5$, $N_{H_0}(S_0) = S_0$.*

Lemma 2.4 *Suppose that $|\Sigma| = 2^m$, $H = \mathrm{Sym}(\Sigma)$ and $S \in \mathrm{Syl}_2(H)$. Let Θ be the collection of all block systems of Σ which are preserved by S. If $K \leq H$ leaves invariant each of the block systems in Θ, then $K \leq S$.*

Lemma 2.5 *Suppose that $H = X \times Y$ is a direct product of groups X and Y and suppose that $S \in \mathrm{Syl}_r(H)$ where r is a prime which divides the order of both X and Y. Assume that L is a subgroup of H which contains $B := N_H(S)$. Then $L = \langle (B \cap X)^L \rangle \times \langle (B \cap Y)^L \rangle$. In particular, if $L \in \mathcal{M}(H, B)$, then either $L = \langle (B \cap X)^L \rangle \times (B \cap Y)$ or $L = \langle (B \cap Y)^L \rangle \times (B \cap X)$.*

Proof: Set $B_X = B \cap X$ and $B_Y = B \cap Y$. Since overgroups of the normalizer of a Sylow r-subgroup are self-normalizing we get $L = \langle B_Y^L \rangle \langle B_X^L \rangle$, which, as $\langle B_Y^L \rangle \leq Y$ and $\langle B_X^L \rangle \leq X$, gives $L = \langle B_Y^L \rangle \times \langle B_X^L \rangle$. Now suppose that $L \in \mathcal{M}(H, B)$ and let M be the unique maximal subgroup of L which contains B. Then, if L normalizes neither B_X nor B_Y, M must contain both of $\langle B_Y^L \rangle$ and $\langle B_X^L \rangle$, which is a contradiction. Therefore, L normalizes one of B_X and B_Y and hence either $L = \langle B_X^L \rangle \times B_Y$ or $L = \langle B_Y^L \rangle \times B_X$. $\qquad\square$

Applying induction, Lemma 2.5 gives

Corollary 2.6 *Suppose that $\mathcal{K}$ is a set and $H = \prod_{k \in \mathcal{K}} X_k$ where each X_k, $k \in \mathcal{K}$, has order divisible by the prime r. If L is a subgroup of H which contains $B := N_H(S)$, where $S \in \mathrm{Syl}_r(H)$, then $L = \prod_{k \in \mathcal{K}} \langle (B \cap X_k)^L \rangle$.*

Lemma 2.7 *Suppose that H is a group, r is a prime which divides the order of H, $S \in \mathrm{Syl}_r(H)$ and that $B = N_H(S)$. If $P \in \mathcal{M}(H, B)$ and K is normal in P, then either*

(i) $KB = P$, or

(ii) $S \cap K$ is normal in P.

In particular, if $N_K(S \cap K) = B \cap K$, then either $K \leq B$ or $P = KB$.

Proof: Let M be the unique maximal subgroup of P containing B and set $S_0 = S \cap K$. Then $S_0 \in \mathrm{Syl}_r(K)$. By the Frattini argument we have $P = N_P(S_0)K$. Notice that S_0 is normal in B and so $B \leq N_P(S_0)$. Therefore, either (ii) holds or $N_P(S_0) \leq M$. If the latter case holds, then, as $P \neq M$, we must have $K \not\leq M$. But then $KB \not\leq M$ and, as $P \in \mathcal{M}(H, B)$, we conclude that (i) holds.

Now suppose that $N_K(S_0) = B \cap K$. Then if (i) does not hold, then S_0 is normal in P and, in particular, S_0 is normal in K. But then $K = N_K(S_0) = B \cap K \leq B$. $\qquad\square$

3   Minimal parabolic subgroups of $\mathrm{Sym}(\Omega)$ and $\mathrm{Alt}(\Omega)$

In this section we resume the notation set forth in the introduction. When examining the intersection of subgroups of G with G^* we will encounter the following special situation: $\Omega = \Lambda \dot\cup \Theta$ with $|\Lambda| > 1 < |\Theta|$, $L \leq \mathrm{Sym}(\Lambda) \times \mathrm{Sym}(\Theta)$ and $T \leq L$. So $L = L_\Lambda \times L_\Theta$ where $L_\Lambda = \mathrm{Sym}(\Lambda) \cap L$ and $L_\Theta = \mathrm{Sym}(\Theta) \cap L$, by Lemma 2.5. In such circumstances, we denote $L \cap G^*$ by $L_\Lambda \lambda L_\Theta$. Of course, $L_\Lambda \lambda L_\Theta$ can be alternatively described as

$$((\mathrm{Alt}(\Lambda) \cap L) \times (\mathrm{Alt}(\Theta) \cap L)) \rtimes \langle t \rangle$$

where $t \in T$ is a product of two transpositions, one from $T \cap \mathrm{Sym}(\Lambda)$ and one from $T \cap \mathrm{Sym}(\Theta)$.

The purpose of this section is to demonstrate that the members of $\mathcal{P}(G, T)$ (respectively $\mathcal{P}(G^*, T^*)$ when $n > 9$) are in fact minimal parabolic subgroups of G (respectively G^*) with respect to T (respectively T^*).

Proposition 3.1 *We have that $\mathcal{P}(G, T) \subseteq \mathcal{M}(G, T)$. Furthermore, if $P \in \mathcal{L}(G, T)$, then T is a maximal subgroup of P.*

Proof: Suppose first that $P \in \mathcal{F}(G, T)$. From the definition of P, to verify $P \in \mathcal{M}(G, T)$ we may plainly assume that $\Omega = \Omega_1 \cup \Omega_2$. Let $\mathcal{D}_{n_2}$ be the block system left invariant by P which contains blocks of order 2^{n_2} and let K be the kernel of the operation of P on $\mathcal{D}_{n_2}$. Then P/K operates primitively on $\mathcal{D}_{n_2}$ and, as P/K contains elements which act as transpositions on $\mathcal{D}_{n_2}$, Lemma 2.1(i) implies that $P/K \cong \mathrm{Sym}(\mathcal{D}_{n_2})$. Thus, by Lemma 2.2, $P/K \in \mathcal{M}(\mathrm{Sym}(\mathcal{D}_{n_2}), TK/K)$. To complete the proof of the proposition in this case we only have to note that K is a 2-group. This follows as, again by the definition of P, K preserves all the block systems in Σ_{1+2} and individually fixes each of the blocks in $\mathcal{D}_{n_2}$. Thus K embeds into a direct product of $2^{n_1 - n_2} + 1$ copies of the symmetric group on 2^{n_2} points and in each of the direct factors operates as does a Sylow 2-subgroup.

Suppose now that $P \in \mathcal{L}(G,T)$. Then, by the definition of P, it suffices to deal with the case when $\Omega = \Omega_1$. Let $\mathcal{D}_{n_1}, \ldots, \mathcal{D}_0$ be the distinct systems of blocks of Ω which are preserved by T and assume notation chosen so that, for $k \in \{1, \ldots, n_1\}$, the blocks of $\mathcal{D}_k$ have 2^k elements. We will prove that T is a maximal subgroup of P arguing by induction on n_1. This then shows that $P \in \mathcal{M}(G,T)$. If $n_1 = 2$, then, by definition, $P = P(1;1)$ only stabilizes $\mathcal{D}_0$ and $\mathcal{D}_2$ which is no restriction at all. Thus $P = G \cong \mathrm{Sym}(4)$ and so a Sylow 2-subgroup has index 3 in P and is thus maximal. Therefore, we suppose that $n_1 > 2$. If $P \neq P(1;1)$, then we consider the action of P on $\mathcal{D}_1$. The kernel, K, of this action is a 2-group. Since $P/K \in \mathcal{L}(\mathrm{Sym}(\mathcal{D}_1), T/K)$ it follows by induction that T/K is a maximal subgroup of P/K. Hence T is maximal in P in this case. Therefore, we assume that $P = P(1;1)$. Then P preserves the block system $\mathcal{D}_{n_1-1} = \{\Delta_1, \Delta_2\}$ and so embeds into $A := (\mathrm{Sym}(\Delta_1) \times \mathrm{Sym}(\Delta_2)) \rtimes \langle t \rangle$ where t is an involution in T interchanging Δ_1 and Δ_2. Let $A_i = \mathrm{Sym}(\Delta_i)$, $i = 1, 2$ and set $A_0 = A_1 \times A_2$. Then $P \cap A_0 = (P \cap A_1) \times (P \cap A_2)$, by Lemma 2.5 and, as $A_1^t = A_2$, induction implies that $T \cap A_i$ is a maximal subgroup of $P \cap A_i$ for $i = 1, 2$. Now suppose that M is a subgroup of P strictly containing T. Then, by Lemma 2.5

$$M = (M \cap A_0)T = ((M \cap A_1) \times (M \cap A_2)) \rtimes \langle t \rangle.$$

Since $M \neq T$, $M \cap A_0 \neq T \cap A_0$ and so the maximality of $T \cap A_i$ in $P \cap A_i$ for $i = 1, 2$, and the fact that $(M \cap A_1)^t = M \cap A_2$ implies that $M \cap A_0 = P \cap A_0$. Thus $M = P$, whence T is a maximal subgroup of P, as desired. $\qquad\square$

Proposition 3.2 *Suppose that $|\Omega| > 9$. Then $\mathcal{P}(G^*, T^*) \subseteq \mathcal{M}(G^*, T^*)$.*

Proof: Suppose that $P \in \mathcal{P}(G,T)$ and set $P^* = G^* \cap P$. Assume that P is intransitive Ω and let Λ be a P-orbit of length at least 2. Put $\Theta = \Omega \setminus \Lambda$, $H_\Lambda = \mathrm{Sym}(\Lambda)$ and $H_\Theta = \mathrm{Sym}(\Theta)$. Assume now that $|\Theta| \geq 2$. Then $P \leq H_\Lambda \times H_\Theta$ and, by Lemma 2.5, we may suppose that $P = (P \cap H_\Lambda) \times (T \cap H_\Theta)$ and, by Proposition 3.1, that $P \cap H_\Lambda \in \mathcal{M}(H_\Lambda, T \cap H_\Lambda)$. Now $P^* \leq H_\Lambda \lambda(T \cap H_\Theta)$ and $(H_\Lambda \lambda(T \cap H_\Theta))/((T \cap H_\Theta) \cap G^*) \cong H_\Lambda$. Therefore, as $P^*/((T \cap H_\Theta) \cap G^*) \cong P \cap H_\Lambda$, we conclude that $P^* \in \mathcal{M}(G^*, T^*)$. Suppose now that $|\Theta| = 1$. Then, as Θ is a T-orbit, it follows that n is odd and thus that $n - 1 \geq 10$. Furthermore, we have $P \leq \mathrm{Sym}(\Lambda)$ and $P^* \leq \mathrm{Alt}(\Lambda)$. Hence, as $|\Lambda| > 9$ and $|\Lambda|$ is even, without loss of generality we may suppose that P is transitive on Ω. Suppose that $\mathcal{D}$ is a system of blocks of P with blocks of size 2. Let K be the kernel of the action of P on $\mathcal{D}$. Then K is a 2-group and P contains a complement, P_1, to K which consists only of even permutations. Also, P_1 embeds into $\mathrm{Sym}(\mathcal{D})$ and, by Proposition 3.1, is a minimal parabolic subgroup therein. As $P^* = P_1(K \cap G^*)$ we conclude that $P^* \in \mathcal{M}(G^*, T^*)$ in this case. Thus P preserves no block system with blocks of size 2. If $P \in \mathcal{F}(G,T)$, we infer from the definition of P that $n = 2^m + 1$ and that $P = G$. But then $P^* = G^*$ and Lemma 2.2 implies that $P^* \in \mathcal{M}(G^*, T^*)$. So we may assume that $P \in \mathcal{L}(G, T)$. Then, because P preserves no block system with members of size 2, it does, by definition, leave a block system with blocks, Δ_1 and Δ_2, of size $n/2$ invariant. So $P \le \mathrm{Sym}(\Delta_1) \wr \mathrm{Sym}(2)$. Let $P_1 = P \cap \mathrm{Sym}(\Delta_1)$ and $P_2 = P \cap \mathrm{Sym}(\Delta_2)$. Then $P = (P_1 \times P_2) \rtimes \langle t \rangle$ and $P^* = R \rtimes \langle t \rangle$ where $t \in T^*$ is an involution which interchanges Δ_1 and Δ_2 and $R = P^* \cap (P_1 \times P_2)$. Moreover, for $i = 1, 2$, the definition of P shows that $P_i \in \mathcal{L}(\mathrm{Sym}(\Delta_i), T \cap \mathrm{Sym}(\Delta_i))$. It follows from Proposition 3.1 that for $i = 1, 2$, $T \cap P_i$ is maximal in P_i. We will prove that T^* is a maximal subgroup of P^*; this plainly suffices to show that $P^* \in \mathcal{M}(G^*, T^*)$. Let M_1 be a subgroup of P^* which properly contains T^*. Then, by the Dedekind Modular Law, $M := M_1 \cap R$ is also not a 2-group. Since n is a 2-power and $n > 9$, $n/2 > 5$ and so, by Lemma 2.3, for $i = 1, 2$, $T \cap \mathrm{Alt}(\Delta_i)$ is self-normalizing in $\mathrm{Alt}(\Delta_i)$. Hence, by Lemma 2.5, $M \cap (\mathrm{Alt}(\Delta_1) \times \mathrm{Alt}(\Delta_2)) = (M \cap \mathrm{Alt}(\Delta_1)) \times (M \cap \mathrm{Alt}(\Delta_2))$ with $(M \cap \mathrm{Alt}(\Delta_1))^t = M \cap \mathrm{Alt}(\Delta_2)$. We now consider $M \mathrm{Alt}(\Delta_1)/\mathrm{Alt}(\Delta_1)$. If this is not a 2-group, then as $R \mathrm{Alt}(\Delta_1)/\mathrm{Alt}(\Delta_1)$ is a minimal parabolic subgroup with a maximal Sylow 2-subgroup we see that $M \mathrm{Alt}(\Delta_1)/\mathrm{Alt}(\Delta_1) = R \mathrm{Alt}(\Delta_1)/\mathrm{Alt}(\Delta_1)$ and thus $M \cap \mathrm{Alt}(\Delta_1) = P_1 \cap \mathrm{Alt}(\Delta_1)$. Since $(M \cap \mathrm{Alt}(\Delta_1))^t = M \cap \mathrm{Alt}(\Delta_2)$ it follows that $M_1 = P^*$ and so T^* is maximal in P^*, as desired.   □

4   Proof of the main theorems

In this section we continue the notation of section 3.

Proof of Theorem 1.1: We prove the theorem by induction on n. For $n = 3$ we clearly have $\{G\} = \mathcal{M}(G, T) = \mathcal{P}(G, T) = \mathcal{F}(G, T)$. Thus we suppose that $n > 3$. Let P be a minimal parabolic subgroup of G with respect to T. Suppose that P is not transitive on Ω and let Λ be an orbit of P of length greater than 1. Set $\Theta = \Omega \setminus \Lambda$. Then $P \le \mathrm{Sym}(\Theta) \times \mathrm{Sym}(\Lambda)$. By Lemma 2.5 we may assume without loss of generality that $P = (T \cap \mathrm{Sym}(\Theta)) \times \langle (T \cap \mathrm{Sym}(\Lambda))^P \rangle$ and so $P_\Lambda := \langle (T \cap \mathrm{Sym}(\Lambda))^P \rangle$ is a minimal parabolic subgroup of $\mathrm{Sym}(\Lambda)$ with respect to $T \cap \mathrm{Sym}(\Lambda)$. But then, $P_\Lambda \in \mathcal{P}(\mathrm{Sym}(\Lambda), T \cap \mathrm{Sym}(\Lambda))$ by induction and thus $P = P_\Lambda \times (T \cap \mathrm{Sym}(\Theta)) \in \mathcal{P}(G, T)$. Henceforth we assume that P operates transitively on Ω.

Let $\mathcal{D}$ be a block system for P on Ω which contains more than one block. We claim that every block, Δ, of $\mathcal{D}$ must be contained in some Ω_i, $i \in I$. Suppose that this is false. Then there exists $\Delta \in \mathcal{D}$ and $i, j \in I$, $i \ne j$, such that $\Delta \cap \Omega_i \ne \emptyset$ and $\Delta \cap \Omega_j \ne \emptyset$. Since T_i (respectively T_j) operates transitively on Ω_i (respectively Ω_j) and fixes all the elements in Ω_j (respectively Ω_i), and Δ is a block for P, we conclude that $\Omega_j \cup \Omega_i \le \Delta$. Therefore, we deduce that Δ is a union of at least two orbits of T. As $|\Delta|$ divides n it now follows that

$\Omega_r \leq \Delta$. On the other hand, $|\Delta|$ is not a 2-power and so no block of $\mathcal{D}$ is contained in a T-orbit. Hence, by the preceding argument, every block of $\mathcal{D}$ must contain Ω_r and so $|\mathcal{D}| = 1$, which is against our original assumption. This proves the claim. In particular we note that there is a member of $\mathcal{D}$ which is contained in Ω_r.

Suppose, for the moment, that $r > 1$. Additionally we assume that $\mathcal{D}$ is chosen so that P acts primitively on the blocks of $\mathcal{D}$. Let K be the kernel of the operation of P on $\mathcal{D}$. Then P/K embeds into $\mathrm{Sym}(\mathcal{D})$. Because P/K operates primitively on $\mathcal{D}$ and TK/K contains a transposition we get $P/K \cong \mathrm{Sym}(\mathcal{D})$ by Lemma 2.1. Since TK/K is contained in a unique maximal subgroup of P and $r > 1$, Lemma 2.2 implies $|\mathcal{D}| = 2^m + 1$ for some m. Thus as some member of $\mathcal{D}$ is contained in Ω_r, we conclude that $\Omega_r \in \mathcal{D}$ and $r = 2$. We now determine the structure of K. Let M be the unique maximal subgroup of P which contains T. Then, as TK is not transitive on Ω, while P is, we must have $KT \leq M$. Let L be the subgroup of G which fixes all the blocks in $\mathcal{D}$. Then, since L is a direct product of symmetric groups of degree 2^{n_r} and $K \cap T$ is a Sylow 2-subgroup of L, Lemmas 2.3 and 2.7 imply that $T \cap K = K$. Hence $P \in \mathcal{F}(G, T)$.

Finally we assume that $r = 1$ (so $n = 2^{n_1}$ is a 2-power). As $n \geq 3$, we have $n_1 \geq 2$. If $n_1 = 2$, then $G \cong \mathrm{Sym}(4) \in \mathcal{P}(G, T) = \mathcal{L}(G, T)$ and we are done. Thus we assume $n_1 \geq 3$. Then, by Lemma 2.2, $P \neq G$ and so, by Lemma 2.1, P is not primitive on Ω. Let $\mathcal{D}$ be a non-trivial block system for P on Ω and K be the kernel of the action of P on $\mathcal{D}$. Then either $P/K \in \mathcal{L}(\mathrm{Sym}(\mathcal{D}), TK/K)$ or $P/K = TK/K$. In the former case it follows that $K = T \cap K$ as in the earlier case and so $P \in \mathcal{L}(G, T)$. Thus the latter case pertains. Let L be the subgroup of G which fixes all the blocks of $\mathcal{D}$. So $L = L_1 \times \cdots \times L_s$ where each $L_i \cong \mathrm{Sym}(\Delta)$ for $\Delta \in \mathcal{D}$. Moreover, T acts transitively on the set $\{L_i \mid 1 \leq i \leq s\}$. Notice that $K \leq L$. Let $K_1 = K \cap L_1$ and $T_1 = T \cap K_1 (= T \cap L_1)$. Then Lemma 2.6 implies that $K_1 = \langle T_1^K \rangle > T_1$. We claim that K_1 is a minimal parabolic subgroup of L_1 with respect to T_1. Let M_1 be a maximal subgroup of K_1 containing T_1. Suppose that $M_1^t \leq K_1$ for some $t \in T$, then $K_1^t \cap K_1 \neq 1$ and so $K_1 = K_1^t$ and $L_1^t = L_1$. As $L_1 \cong \mathrm{Sym}(\Delta)$ for some $\Delta \in \mathcal{D}$ with $|\Delta| = 2^d$ it follows that t operates on L_1 as an inner automorphism (see Huppert [2, Satz 5].) Since $T_1^t \leq K_1 \cap T = T_1$, t operates on L_1 like an element of T_1 and thus normalizes M_1. It follows that $|\langle M_1^T \rangle| = |M_1|^s < |K|$ and thus $\langle M_1^T \rangle T \leq M$ where M is the unique maximal subgroup of P which contains T. As $K_1 \not\leq M$ it follows that M_1 is the unique maximal subgroup of K_1 containing T_1. Therefore, as $|\Delta|$ is a 2-power, $K_1 \in \mathcal{L}(L_1, T_1)$ follows by induction. Finally $\langle K_1^T \rangle T \in \mathcal{L}(G, T)$ and the theorem is complete. $\square$

Lemma 4.1 *Suppose that Λ and Θ are sets with $|\Lambda| > 5$ and $|\Theta| > 1$.*

Set $X = \text{Sym}(\Lambda) \times \text{Sym}(\Theta)$, $X^* = \text{Sym}(\Lambda) \wr \text{Sym}(\Theta)$, $S \in \text{Syl}_2(X)$ *and* $S^* = S \cap X^*$. *If* $Y \leq X^*$ *contains* S^*, *then* $S \leq N_X(Y)$.

Proof: Set $Z = \text{Alt}(\Lambda) \times \text{Alt}(\Theta)$ and let $s = s_\Lambda s_\Theta$ be a product of the transpositions $s_\Lambda \in S \cap \text{Sym}(\Lambda)$ and $s_\Theta \in S \cap \text{Sym}(\Theta)$. Since S^* clearly normalizes $Y \cap \text{Alt}(\Lambda)$ and s_Θ centralizes $Y \cap \text{Alt}(\Lambda)$, $S = \langle S^*, s_\Theta \rangle$ normalizes $Y \cap \text{Alt}(\Lambda)$. Similarly, S normalizes $Y \cap \text{Alt}(\Theta)$. Thus we may suppose that $Y \cap Z > (Y \cap \text{Alt}(\Lambda)) \times (Y \cap \text{Alt}(\Theta))$. But, then Lemmas 2.3 and 2.5 together with our hypothesis imply that $|\Theta| = 4$ or 5 and that $Y \cap Z \not\geq N_Z(S \cap Z)$. On the other hand, $(Y \cap \text{Alt}(\Lambda)) \times (Y \cap \text{Alt}(\Theta))$ is a normal subgroup of Y and so the Frattini argument implies that $Y = N_Y(S \cap Z)((Y \cap \text{Alt}(\Lambda)) \times (Y \cap \text{Alt}(\Theta)))$, which, as $[N_Z(S \cap Z) : S \cap Z] = 3$, implies that $N_Z(S \cap Z) \leq Y \cap Z$, a contradiction. □

To prove Theorem 1.2 we prove a stronger result which will imply it.

Theorem 4.2 *Suppose that* $n > 9$ *and* P *is any subgroup of* G^* *which contains* T^*. *Then* $T \leq N_G(P)$.

Proof: Let M be a counterexample to the theorem with $|P| + |\Omega|$ minimal. Suppose that $P \notin \mathcal{M}(G^*, T^*)$. Then there exist distinct maximal subgroups M_1 and M_2 of P which properly contain T. Since P is chosen minimal subject to not being normalized by T, M_1 and M_2 are both normalized by T. Thus $P = \langle M_1, M_2 \rangle$ is normalized by T which is a contradiction. Therefore, we may assume that $P \in \mathcal{M}(G^*, T^*)$.

If P fixes a point, $\sigma \in \Omega$, then P embeds into $\text{Alt}(\Omega \setminus \{\sigma\})$ and then, as T has the same orbit structure as T^*, $T \leq \text{Sym}(\Omega \setminus \{\sigma\})$ and we have a contradiction to the minimal choice of $|P| + |\Omega|$. Thus P fixes no point of Ω. Now assume that P is intransitive. Let Λ be a P-orbit and set $\Theta = \Omega \setminus \Lambda$. Since $|\Omega| > 9$, we may assume that $|\Lambda| > 5$ and so we can apply Lemma 4.1 to find that P is normalized by T, which is a contradiction. Thus we may assume that P is transitive on Ω. Since $n > 9$, Lemma 2.1(ii) implies that P is imprimitive. We may deal with the case when n is not a 2-power exactly as in the proof of Theorem 1.1. So we assume that $n = 2^{n_1}$. Let $\mathcal{D}$ be a system of imprimitivity for P on which P acts primitively and let K be the kernel of the operation of P on $\mathcal{D}$. If K is a 2-group, then, as T preserves all the block systems preserved by T^*, K is normalized by T and, as P operates primitively on $\mathcal{D}$ and contains a transposition, $P/K \cong \text{Sym}(\mathcal{D}) \cong \text{Sym}(4)$ by Lemma 2.2. Thus we see that $P \leq P(1; n_1 - 1) \in \mathcal{L}(G, T)$ and so P is normalized by T. Therefore we may now assume that K is not a 2-group. As P operates primitively on $\mathcal{D}$ we have $|P : K| = 2$ and $K \leq \text{Sym}(\Lambda) \wr \text{Sym}(\Theta)$ where $\Omega = \Lambda \dot\cup \Theta$ and $|\Lambda| = |\Theta| \geq 8$. Let $T_1 = T \cap (\text{Sym}(\Lambda) \times \text{Sym}(\Theta))$. Then Lemma 4.1 implies that T_1 normalizes K. But $P = KT^*$ and so P is normalized by $T^*T_1 = T$, and we are done. □

Proof of Theorem 1.2: Suppose the hypothesis of Theorem 1.2 holds. Let $P \in \mathcal{M}(G^*, T^*)$. Then Theorem 4.2 implies that P is normalized by T, whence PT is a subgroup of G. If $P \in \mathcal{M}(G, T)$, then the theorem holds. Let M_1 and M_2 be distinct maximal subgroups of PT which contain T and let M be the unique maximal subgroup of P which contains T^*. Then $(M_1 \cap P)(M_2 \cap P) \leq M$. Thus, using Theorem 4.2, $MT \geq M_1 M_2 = P$ and we conclude that $P = M$, a contradiction. Hence Theorem 1.2 holds. $\square$

Proof of Theorem 1.3: Because of Theorem 1.1 the question now is, how do we generate G with a minimal set of subgroups from $\mathcal{P}(G, T)$? First of all we observe that the members of $\mathcal{L}(G, T)$ preserve each of the T-orbits Ω_i, $i \in I$, while a member $P(i + j)$ of $\mathcal{F}(G, T)$ is transitive on $\Omega_i \cup \Omega_j$ and preserves all the systems of imprimitivity on $\Omega_i \cup \Omega_j$ with blocks of size dividing $|\Omega_j|$. Now to generate G we must, by Lemma 2.1,

(a) fuse all the T-orbits on Ω; and

(b) destroy any imprimitivity.

We begin with (a). To do this we use the elements in $\mathcal{F}(G, T)$. Since T has exactly r orbits, and $P(i + j) \in \mathcal{F}(G, T)$ fuses the orbits Ω_i and Ω_j we can form the complete labelled graph $\Gamma = (V, E)$ where $V = \{\Omega_j \mid j \in I\}$ and $E = \mathcal{F}(G, T)$ and Ω_i, Ω_j are the vertices on the edge $P(i + j)$. To reach the goal of (a) with a minimal number of subgroups from $\mathcal{P}(G, T)$ it is necessary and sufficient to select the edges from any spanning tree of Γ. Thus a minimal transitive system of minimal parabolic subgroups of G with respect to T contains exactly $r - 1$ of the members of $\mathcal{F}(G, T)$ and, by Cayley's Theorem [7, Theorem 2.1], the number of ways of selecting $r - 1$ suitable subsets of $\mathcal{F}(G, T)$ is r^{r-2}.

Now let $\{P_1, P_2, \ldots, P_{r-1}\}$ be a set of $r - 1$ subgroups of $\mathcal{F}(G, T)$ which generate a transitive subgroup of G and put $H = \langle P_1, P_2, \ldots, P_{r-1} \rangle$. Then H only preserves block systems which have a block contained in Ω_r. Therefore, if $n_r = 0$, H is primitive and so, by Lemma 2.1, $H = G$ and part (ii) holds.

We now use induction on n_r. Notice that H preserves blocks of imprimitivity of every size dividing 2^{n_r}. Let $\mathcal{D}$ be a block system for H in which the blocks have order 2. Then $\mathcal{D}$ is preserved by the subgroups $P(i; j) \in \mathcal{L}^* := \{P(i; j) \in \mathcal{L}(G, T) \mid i \in I, j \geq 2\}$. We consider the embedding of H and the subgroups in $\mathcal{L}^*$ into $\mathrm{Sym}(\mathcal{D})$ and let K be the corresponding kernel. We see that the image of $\mathcal{L}^*$ is $\mathcal{L}(\mathrm{Sym}(\mathcal{D}), T/K)$ (which if $n_r = 1$ is empty). We next apply induction to obtain minimal parabolic systems (and their descriptions) for $\mathrm{Sym}(\mathcal{D})$. Each such system then gives a set of subgroups in G which generate $\mathrm{Stab}_G(\mathcal{D})$ ($\cong 2 \wr \mathrm{Sym}(\mathcal{D})$) in G. We now investigate the various cases of the theorem. If $r = 1$, then, as $n \geq 3$ and $n_r \geq 2$, induction indicates that there is a unique system for $\mathrm{Sym}(\mathcal{D})$. This is then extended to the unique system for G by adding the unique remaining element in $\mathcal{L}(G,T) \setminus \mathcal{L}^*$. Hence part (i) also holds. Next assume that $n_r = 1$. Then we have $|\mathcal{D}|$ is odd and so we have, by part (ii), r^{r-2} systems for $\text{Sym}(\mathcal{D})$, and adding any of the $r-1$ members of $\mathcal{L}(G,T) \setminus \mathcal{L}^*$ will give a minimal parabolic system for G. Thus in this case we obtain $r^{r-2}(r-1)$ systems. If $n_r > 1$, then we have $r^{n_r-1+r-3}(r-1)$ systems generating H and r choices for the extra minimal parabolic subgroup in $\mathcal{L}(G,T) \setminus \mathcal{L}^*$ giving a grand total of $r^{n_r+r-3}(r-1)$ systems in this case, so giving (iii). This completes the verification of parts (i)–(iii). From Proposition 3.2 we deduce that intersection does indeed define a function from the parabolic systems of G to the parabolic systems of G^*. Theorems 1.3 and 4.2 show that this function is a bijection. Hence part (iv) holds and this completes the proof of the theorem.   □

5   Some examples

To illustrate the constructive aspects of Theorems 1.1 and 1.3 we now work out all the minimal parabolic subgroups and parabolic systems for $\text{Sym}(\Omega)$ with $n = |\Omega| = 26$. We maintain the notation introduced in sections 3 and 4. So $n = 16 + 8 + 2 = 2^4 + 2^3 + 2$ with $\Omega = \Omega_1 \cup \Omega_2 \cup \Omega_3$ where $|\Omega_1| = 16$, $|\Omega_2| = 8$ and $|\Omega_3| = 2$. Also $T = T_1 \times T_2 \times T_3$ with $|T_1| = 2^{15}$, $|T_2| = 2^7$ and $|T_3| = 2$, and $|T| = 2^{23}$. Since $r = 3$ and $n_r = 1$, by Theorem 1.3(iii), G possesses $3^{1+3-3}(3-1) = 6$ distinct minimal parabolic systems with respect to T. We begin by listing the subgroups in $\mathcal{F}(G,T)$

$$P(1+2) \cong (((C_2 \wr C_2) \wr C_2) \wr \text{Sym}(3)) \times T_3$$
$$P(1+3) \cong (C_2 \wr \text{Sym}(9)) \times T_2$$
$$P(2+3) \cong (C_2 \wr \text{Sym}(5)) \times T_1$$

In the above the symmetric groups $\text{Sym}(3)$, $\text{Sym}(9)$ and $\text{Sym}(5)$ are acting in their natural permutation representations on the systems of imprimitive blocks given by (respectively) $\Omega_2^{P(1+2)}$, $\Omega_3^{P(1+3)}$ and $\Omega_3^{P(2+3)}$.

Turning to the set $\mathcal{L}(G,T)$, we first examine the $P(1;j)$ and suppose that our notation is chosen so that the block systems preserved by T in $\Omega_1 := \{1, 2, \ldots, 16\}$ are

$$\mathcal{D}_1 = \{\{1,2\},\{3,4\},\{5,6\},\{7,8\},\{9,10\},\{11,12\},\{13,14\},\{15,16\}\}$$
$$\mathcal{D}_2 = \{\{1,2,3,4\},\{5,6,7,8\},\{9,10,11,12\},\{13,14,15,16\}\} \text{ and}$$
$$\mathcal{D}_3 = \{\{1,2,3,4,5,6,7,8\},\{9,10,11,12,13,14,15,16\}\}$$

We then have $\Sigma_{1;1} = \{\mathcal{D}_2, \mathcal{D}_3\}$, $\Sigma_{1;2} = \{\mathcal{D}_1, \mathcal{D}_3\}$ and $\Sigma_{1;3} = \{\mathcal{D}_1, \mathcal{D}_2\}$. Then the groups $P(1;j)$, $(j = 1, 2, 3)$ are as follows:

$$P(1;1) \cong ((\text{Sym}(4) \wr C_2) \wr C_2) \times T_2 \times T_3$$

(where $\mathrm{Sym}(4)$ operates in its natural permutation representation on the set $\{1,2,3,4\}$),

$$P(1;2) \cong ((C_2 \wr \mathrm{Sym}(4)) \wr C_2) \times T_2 \times T_3$$

(where $\mathrm{Sym}(4)$ operates in its natural permutation representation on the set $\{\{1,2\},\{3,4\},\{4,5\},\{5,6\}\}$),

$$P(1;3) \cong ((C_2 \wr C_2) \wr \mathrm{Sym}(4)) \times T_2 \times T_3$$

(where $\mathrm{Sym}(4)$ operates in its natural permutation representation on the set $\{\{1,2,3,4\},\{5,6,7,8\},\{9,10,11,12\},\{13,14,15,16\}\}$).

The remaining members of $\mathcal{L}(G,T)$ are

$$P(2;1) = T_1 \times (\mathrm{Sym}(4) \wr C_2) \times T_3$$

and

$$P(2;2) = T_1 \times (C_2 \wr \mathrm{Sym}(4)) \times T_3.$$

Finally using Theorem 1.3 (iii) we have that the minimal parabolic systems for G with respect to T are as follows.

$$\{P(1+2), P(1+3), P(1;1)\}$$
$$\{P(1+2), P(1+3), P(2;1)\}$$
$$\{P(1+2), P(2+3), P(1;1)\}$$
$$\{P(1+2), P(2+3), P(2;1)\}$$
$$\{P(1+3), P(2+3), P(1;1)\}$$
$$\{P(1+3), P(2+3), P(2;1)\}$$

We conclude this section, and the paper, with a list of the minimal parabolic subgroups and systems for $\mathrm{Alt}(n)$, $5 \le n \le 9$. Among these exceptional cases $\mathrm{Alt}(7)$ stands out as being particularly well endowed with minimal parabolic systems.

$$
\begin{aligned}
n = 5 \quad & \mathcal{M}(G^*, T^*) = \{G^*\} \\
n = 6 \quad & \mathcal{M}(G^*, T^*) = \{2^2\mathrm{Sym}(3), 2^2\mathrm{Sym}(3)\} \\
n = 7 \quad & \mathcal{M}(G^*, T^*) = \{P_0 \cong \mathrm{Sym}(5), P_1 \cong \mathrm{Sym}(4), P_2 \cong \mathrm{Sym}(3) \wedge D_8, \\
& \qquad P_3 \cong \mathrm{Sym}(4), P_1^\sigma \cong \mathrm{Sym}(4), P_4 \cong \mathrm{Sym}(4)\} \\
n = 8 \quad & \mathcal{M}(G^*, T^*) = \{2^5\mathrm{Sym}(3), 2^5\mathrm{Sym}(3), 2^5\mathrm{Sym}(3)\} \\
n = 9 \quad & \mathcal{M}(G^*, T^*) = \{2^5\mathrm{Sym}(3), 2^5\mathrm{Sym}(3), 2^5\mathrm{Sym}(3), G^*\}
\end{aligned}
$$

In all of the above cases except for $n = 7$, $\mathrm{Alt}(n)$ has a unique minimal parabolic system, whereas $\mathrm{Alt}(7)$ has eight minimal parabolic systems as detailed below (see [6, Appendix to Section 2]).

$$\{P_1, P_3, P_4\}, \; \{P_1^\sigma, P_3, P_4\}, \; \{P_1, P_3, P_1^\sigma\},$$
$$\{P_2, P_3\}, \; \{P_0, P_1\}, \; \{P_0, P_2\}, \; \{P_0, P_3\}, \; \{P_0, P_1^\sigma\}.$$

As a final observation, we point out that our theorems all hold for Alt(n) when $n = 5$ or 6, while Theorem 1.2 fails for $n = 7, 8$ and 9, Theorem 1.3 fails for $n = 7$ and 8 and Corollary 1.4 fails for $n = 7$.

Acknowledgements

Parker and Rowley were partially supported by an LMS scheme 3 grant.

References

[1] J. H. Conway, R. T. Curtis, S. P. Norton, R. A. Parker and R. A. Wilson, *An* ATLAS *of finite groups*, Clarendon Press, Oxford, 1985.

[2] B. Huppert, *Endliche Gruppen I*, Springer, Berlin, 1967.

[3] C. Jordan, Théorèmes sur les groupes primitifs, *J. Math. Pures Appl.* **16** (1871), 383–408.

[4] B. Marggraf, *Über primitive Gruppen mit transitiven Untergruppen geringeren Grades*, Dissertation, Giessen, 1892.

[5] M. A. Ronan and S. D. Smith, 2-local geometries for some sporadic groups, in *The Santa Cruz conference on finite groups (eds. B. Cooperstein and G. Mason)*, pp. 283–289. Proceedings of Symposia in Pure Mathematics, Vol. 37, American Mathematical Society, 1980.

[6] M. A. Ronan and G. Stroth, Minimal parabolic geometries for the sporadic groups, *European J. Combin.* **5** (1984), 59–91.

[7] D. Stanton and D. White, *Constructive combinatorics*, Springer Berlin, 1986.

[8] H. Wielandt, *Finite permutation groups*, Academic Press, New York, 1964.

Probabilistic methods in the generation of finite simple groups

Martin W. Liebeck
and
Aner Shalev

Abstract

We survey recent progress, made using probabilistic methods, on several problems concerning generation of finite simple groups. For example, we outline a proof that all but finitely many classical groups different from $PSp_4(q)$ ($q = 2^a$ or 3^a) can be generated by an involution and an element of order 3.

1 Results

In this survey we present some new methods and results in the study of generating sets for the finite (nonabelian) simple groups. The results are largely taken from the three papers [16], [17], [18]. We shall present the results in this first section, and outline some proofs in sections 2 and 3. We begin by describing some of the basic questions and work in the area.

It is a well known consequence of the classification that every finite simple group can be generated by two elements. This result was established early this century for the alternating groups by Miller [23] and for the groups $PSL_2(q)$ by Dickson [8]. Various other simple groups were handled by Brahana [5] and by Albert and Thompson [1], but it was not until 1962 that Steinberg [26] showed that all finite simple groups of Lie type can be generated by two elements. To complete the picture, in 1984 Aschbacher and Guralnick [2] established the same conclusion for sporadic groups.

A refinement of the two element generation question asks whether every finite simple group can be generated by an involution and a further element. Partial results on this question were obtained in the above-mentioned papers [23], [5], [1], [2], but only recently has the question been answered completely, in the affirmative, by Malle, Saxl and Weigel [22].

Another, much-studied, refinement asks which finite simple groups are (2,3)-generated (that is, are generated by an element of order 2 and an element of order 3). This question has attracted wide attention throughout this century, one reason being that $(2,3)$-generated groups are precisely the images of $PSL_2(\mathbb{Z})$. (Notice that some simple groups are definitely *not* $(2,3)$-generated; for example, the Suzuki groups ${}^2B_2(q)$ do not possess elements of order 3.) The question is far from being completely answered, although there has been considerable recent progress, mainly for classical groups. For example, as the culmination of work of many authors, it is now known that for all odd prime powers q and for all n (with the exception of $(n,q) = (2,9)$), the simple group $PSL_n(q)$ is $(2,3)$-generated (see [21] for $n = 2$, [10] for $n = 3$, [27] for $n = 4$, [25] for $n \geq 13$ (where the restriction on q is not needed), and [6], [7] for the remaining cases). Some other classical groups of large rank (such as $PSp_{2n}(q)$ for $n \geq 37$) are dealt with in [28]. All this work is deterministic in nature; suitable pairs of elements of the groups in question are constructed, and are shown to generate the group. A wide variety of tools and methods is employed in these papers, and the proofs tend to become increasingly complicated. The basic conjecture in the field is that all finite simple classical groups, except for some groups of low rank in characteristic 2 and 3, are $(2,3)$-generated.

In fact, each of the above questions can profitably be looked at from a probabilistic point of view. The story starts in 1892, with a conjecture of Netto [24], that most pairs of elements of the symmetric group generate the symmetric group or the alternating group. This conjecture was proved by Dixon [9] in 1969. In his paper (written prior to the classification), Dixon makes the following rather general conjecture.

Dixon's Conjecture *Two elements chosen at random from a finite simple group G generate G with probability $\to 1$ as $|G| \to \infty$.*

In other words, the conjecture states that if

$$P(G) = \frac{|\{(x,y) \in G \times G : \langle x, y \rangle = G\}|}{|G|^2}$$

(so that $P(G)$ is the probability that a randomly chosen pair of elements generates G), then $P(G) \to 1$ as $|G| \to \infty$. Dixon's paper [9] contains a proof of his conjecture for the alternating groups $\mathrm{Alt}(n)$. Twenty years later Babai [4] was able to find the exact generation probability (up to a small error term): he showed that $P(\mathrm{Alt}(n)) = 1 - n^{-1} + O(n^{-2})$. (The n^{-1} term comes from the fact that this is the probability that two randomly chosen permutations fix a point.) For groups of Lie type, an important contribution was made by Kantor and Lubotzky [13] in 1990; they confirmed Dixon's conjecture for the classical groups, and for some small rank exceptional groups of Lie type (see also [12]). Unlike the work of Dixon, which uses only elementary arguments, the proofs of Babai and of Kantor and Lubotzky rely on the Classification Theorem, and on suitable information on the subgroup structure of the simple groups in question.

Our first result is as follows.

Theorem 1 [16] *Dixon's conjecture holds.*

We outline our proof of this result in sections 2 and 3.

Precise estimates for the generation probabilities $P(G)$ for groups G of Lie type are obtained in [18], 6.1 and 6.2 (see also [12], 3.3). Roughly speaking, these estimates imply that if two randomly chosen elements do not generate G, then with probability tending to 1, they lie in one of a few specified "large" subgroups of G. With a few exceptions, these specified subgroups are (1) the parabolic subgroups of largest order, and (2) the stabilizers in certain classical groups of nonsingular 1-spaces. For example, if $n > 8$ then

$$P(L_n(q)) = 1 - \frac{2}{p(n,q)} + O(p(n,q)^{-5/4}),$$

where $p(n,q) = (q^n - 1)/(q - 1)$, and

$$P(E_8(q)) = 1 - \frac{1}{p(q)} + O(p(q)^{-16/15}),$$

where $p(q)$ is the index in $E_8(q)$ of a parabolic subgroup of type E_7.

As a refinement of Dixon's conjecture, Kantor and Lubotzky ([13], Conjecture 1) suggested that the result subsequently proved in [22] (that every simple group is generated by an involution and a further element) might have a probabilistic version.

Kantor–Lubotzky Conjecture *Let G be a finite simple group. Then the probability that a randomly chosen involution of G and a randomly chosen additional element generate G tends to 1 as $|G| \to \infty$.*

Our next result is:

Theorem 2 ([17], 1.1 and [18], 1.1) *The Kantor–Lubotzky conjecture holds.*

This is proved for classical groups in [17] and for exceptional groups in [18]; we outline the proof for classical groups in section 2, and make some remarks on the proof for exceptional groups at the end of section 3.

Just to show that not all conjectures in this field are true, we note that a stronger conjecture made by Kantor and Lubotzky (see [13], Conjecture 2) has recently been refuted by Guralnick, Kantor and Saxl [11]. This conjecture states that if G is a simple group, x is a fixed non-identity element of G, and y is randomly chosen in G, then $\langle x, y \rangle = G$ with probability tending to 1 as $|G| \to \infty$. In [11] it is shown that the conjecture is false when G is alternating and x is a 3-cycle, and also when G is a classical group over $\mathbb{F}_q$ with q fixed, and x is a long root element of G. On the other hand, in [11] the conjecture is proved to be true if $G = G(q)$ is of Lie type of bounded rank over $\mathbb{F}_q$, and $q \to \infty$.

Our method of proving Theorem 2 has some additional by-products.

Theorem 3 ([17], 1.2) *Let G be a finite simple classical group.*

(i) The probability that three randomly chosen involutions x, y, z of G generate G tends to 1 as $|G| \to \infty$.

(ii) The same is true if we let x, y, z be randomly chosen elements of a largest conjugacy class of involutions in G.

It follows from part (i) above that all but finitely many simple classical groups can be generated by three involutions, a result which already follows from [22]. However, the problem of finding the finite simple (classical) groups which can be generated by three *conjugate* involutions is still very much open. Part (ii) of Theorem 3 gives rise to the following new result.

Corollary 4 *All but finitely many of the finite simple classical groups can be generated by three conjugate involutions.*

The next result is the analogue of the Kantor–Lubotzky conjecture, with elements of order 3 replacing involutions. Note that the Suzuki groups are of course genuine exceptions in the statement, as they do not contain elements of order 3.

Theorem 5 ([17], 7.1(iii) and [18], 1.2) *Let G be a finite simple group which is not a Suzuki group. Then the probability that a randomly chosen element of order 3 and a randomly chosen additional element of G generate G tends to 1 as $|G| \to \infty$.*

Consequently, apart from the Suzuki groups and finitely many other possible exceptions, all finite simple groups can be generated by two elements, one of which has order 3.

Again, this is proved for classical groups in [17] and for exceptional groups in [18]; we sketch the proof for classical groups in section 2.

Next we move on to the (2, 3)-generation problem discussed above. Our main contribution is the next result; the proof is sketched in section 2.

Theorem 6 ([17], 1.4) *Let G be a finite simple classical group different from $PSp_4(q)$. Then the probability that a randomly chosen involution of G and a randomly chosen element of order 3 generate G tends to 1 as $|G| \to \infty$.*

If $G = PSp_4(p^a)$, where p is a prime exceeding 3, then the above probability tends to $\frac{1}{2}$ as $|G| \to \infty$.

Corollary 7 *All but finitely many of the finite simple classical groups which are different from $PSp_4(q)$ ($q = 2^a$ or 3^a) are $(2,3)$-generated.*

Since no restrictions are made on the characteristic of the underlying field, this result seems to be new even for groups such as $PSL_n(q)$ for $5 \leq n \leq 12$, if q is even. For many other families of classical groups the result is also new in odd characteristic. It would be interesting to know whether the deterministic methods used so far in the investigation of the $(2,3)$-generation problem could yield results such as Corollary 7.

The next result settles the remaining case of 4-dimensional symplectic groups in characteristic 2 and 3.

Theorem 8 ([17], 1.6) *Let $p = 2$ or 3, and let $q = p^a$. Then $PSp_4(q)$ is not $(2,3)$-generated.*

We therefore obtain two infinite families of simple classical groups which are not images of the modular group. This is rather unexpected. The only previously known infinite family of non $(2,3)$-generated finite simple groups was that of the Suzuki groups, which simply do not possess elements of order 3. Of course, it follows from Theorem 6 that there are no more exceptional families among the classical groups.

It must be admitted that our probabilistic approach has some disadvantages. First, it does not yield an explicit list of exceptions in Corollary 7 (though in principle such a list could be worked out by a careful analysis of the proof). Secondly, even when we conclude that a certain group is $(2,3)$-generated, we do not obtain an explicit pair of generators x, y demonstrating this. However, the next result shows that at least one of the required generators can be determined rather easily.

The point is that the proof of Theorem 6 actually shows a bit more, namely: if $C_2, C_3 \subset G$ are conjugacy classes of largest size of elements of order 2 and 3 respectively, then a random element of C_2 and a random element of C_3 will almost certainly generate G. This gives rise to the following.

Corollary 9 [17] *There exists a constant c with the following property: let $G \neq PSp_4(q)$ be a finite simple classical group of order at least c, and let $C_2, C_3 \subset G$ be as above; then for each element $x \in C_2$ there exists an element $y \in C_3$ such that $\langle x, y \rangle = G$.*

Conversely, we can fix $y \in C_3$ and conclude that for some $x \in C_2$ we have $\langle x, y \rangle = G$.

The estimates obtained in the proof of our $(2,3)$-generation results also establish the following.

Theorem 10 ([17], 1.7) *Let $G \neq PSp_4(q)$ be a finite simple classical group. Then the probability that two randomly chosen elements x, y of order 3 in G generate G tends to 1 as $|G| \to \infty$. Furthermore, the same holds if we choose x, y from a largest conjugacy class of elements of order 3 in G.*

Corollary 11 *All but finitely many of the finite simple classical groups which are different from $PSp_4(q)$ can be generated by two conjugate elements of order 3.*

It is worth mentioning that all the results above also hold for alternating groups (with much easier proofs). More general results for alternating groups will appear in a forthcoming paper of Pyber (see also Luczak and Pyber [20]). It should also be mentioned that our techniques can probably handle more general (a, b)-generation problems (a group is said to be (a, b)-generated if it can be generated by an element of order a and an element of order b).

2   Proofs: classical groups

In this section we make some remarks on the proofs of some of the results stated in section 1 concerning classical groups.

Let G be a finite simple group. For a finite group H and a positive integer s, let $i_s(H)$ denote the number of elements of H of order s. If s, t are positive integers, let $P_{s,t}(G)$ be the probability that two randomly chosen elements of orders s and t generate G, Thus

$$P_{s,t}(G) = \frac{|\{(x, y) \in G \times G : |x| = s, |y| = t, \langle x, y \rangle = G\}|}{i_s(G) i_t(G)}.$$

Let $Q_{s,t}(G) = 1 - P_{s,t}(G)$ be the complementary probability. Define $P_{s,t,r}(G)$, $Q_{s,t,r}(G)$ for positive integers s, t, r in similar fashion. We use $*$ to denote unspecified order; thus $i_*(G) = |G|$, and the quantity $P(G)$ defined in section 1 is the same as $P_{*,*}(G)$.

With the above notation, Dixon's conjecture states that $P_{*,*}(G) \to 1$ (as $|G| \to \infty$); the Kantor–Lubotzky conjecture that $P_{2,*}(G) \to 1$; Theorem 3(i) that $P_{2,2,2}(G) \to 1$ (for G classical); Theorem 5 that $P_{3,*}(G) \to 1$ (for G not a Suzuki group); Theorem 6 that $P_{2,3}(G) \to 1$ (for G classical, $G \neq PSp_4(q)$); and Theorem 10 that $P_{3,3}(G) \to 1$ (for G classical, $G \neq PSp_4(q)$).

Assume now that G is classical. We describe a uniform approach to all the above results. Let $s, t \in \{2, 3, *\}$, and choose at random elements $x, y \in G$ of order s, t respectively. If $\langle x, y \rangle \neq G$, then $x, y \in M$ for some maximal subgroup M of G. The probability of this event, given M, is $\frac{i_s(M)}{i_s(G)} \cdot \frac{i_t(M)}{i_t(G)}$. Consequently

$$Q_{s,t}(G) \leq \sum_{M \text{ max } G} \frac{i_s(M)}{i_s(G)} \cdot \frac{i_t(M)}{i_t(G)}.$$

Letting $\mathcal{M}$ be a set of representatives for the conjugacy classes of the maximal subgroups of G, we see that

$$Q_{s,t}(G) \leq \sum_{M \in \mathcal{M}} \frac{i_s(M)}{i_s(G)} \cdot \frac{i_t(M)}{i_t(G)} \cdot |G : M|. \tag{1}$$

To show that $P_{s,t}(G) \to 1$ in the various cases, it suffices to show that the right hand side of the inequality (1) tends to 0 as $|G| \to \infty$. Clearly one of the keys to doing this lies in estimating the quantities $\frac{i_s(M)}{i_s(G)}$ for $s = 2, 3$.

To do this, we first obtain estimates for $i_2(G)$ and $i_3(G)$. Since the conjugacy classes in G of involutions and of elements of order 3 are known, this can be carried out quite routinely. For groups of large rank we find that $i_2(G)$ is roughly $|G|^{1/2}$ and $i_3(G)$ is roughly $|G|^{2/3}$.

Having done this, we next estimate $i_2(M)$ and $i_3(M)$ for each maximal subgroup M of G. This is the hard core of the proof. The results on the maximal subgroups of G given in [3], [14] play a crucial role here. For maximal subgroups M which are close to simple groups the estimates of the previous paragraph are applicable, but in general (for example when M is parabolic) much extra work needs to be done. It ultimately follows from our estimates that there exists a constant c such that

$$\frac{i_2(M)}{i_2(G)} \leq c|G : M|^{-2/5} \tag{2}$$

and, with a few exceptions (such as $G = PSp_4(q)$),

$$\frac{i_3(M)}{i_3(G)} \leq c|G : M|^{-8/13}. \tag{3}$$

Putting together (1), (2) and (3) we obtain

$$Q_{s,t}(G) \leq \sum_{M \in \mathcal{M}} |G : M|^{-\alpha_{s,t}} \tag{4}$$

where $\alpha_{*,*} = 1$, $\alpha_{2,*} = 2/5$, and with a few exceptions, $\alpha_{3,*} = 8/13$, $\alpha_{2,3} = 1/65$, $\alpha_{3,3} = 3/13$. (And similarly $Q_{2,2,2}(G) \leq \sum |G : M|^{-1/5}$.) In particular, $\alpha_{s,t}$ is positive in each case.

Thus we are led to define a "zeta function" $\zeta_G(s)$, encoding the indices of maximal subgroups, as follows:

$$\zeta_G(s) = \sum_{M \in \mathcal{M}} |G : M|^{-s}.$$

The final step in the proof is to show that if $s > 0$ then $\zeta_G(s) \to 0$ as $|G| \to \infty$ (for classical groups G). This is carried out in [17], Theorem 2.1. Theorems 1, 3, 5, 6 and 10 (for classical groups G) follow from this fact, combined with (4).

3   Proofs: exceptional groups

The approach described in section 2 for the classical groups does not work for exceptional groups (at least those of rank 4 or more), because of insufficient knowledge about the maximal subgroups of such groups. Indeed, for our proofs of Theorems 1, 2 and 5 in the case where G is exceptional, we find it necessary first to obtain new information on maximal subgroups (although for the most part this follows quickly from the recent work in [15], [19] on such matters).

To illustrate, we sketch our proof of Theorem 1 for G exceptional. As in section 2, we know that

$$Q(G) = 1 - P(G) \leq \sum_{M \ max \ G} |G : M|^{-2}. \qquad (5)$$

Thus the aim is to show that the right hand side tends to 0 as $|G| \to \infty$. We may as well assume that G is of type F_4, E_6, $^2E_6$, E_7 or E_8 (as the maximal subgroups of the other (small rank) exceptional groups are known).

To estimate the right hand side of (5), we partition the set of maximal subgroups of G into two sets, $\mathcal{K}$ and $\mathcal{U}$: the set $\mathcal{U}$ consists of all almost simple maximal subgroups of G of order at most $|G|^{5/13}$, and $\mathcal{K}$ consists of the remaining maximal subgroups. In [16], Theorem 1.2, we show that $\mathcal{K}$ consists of a small number of conjugacy classes of subgroups, all of which are known. From this it follows easily that

$$\sum_{M \in \mathcal{K}} |G : M|^{-2} \to 0. \qquad (6)$$

It remains to estimate $\sum_{M \in \mathcal{U}} |G : M|^{-2}$. Let $M \in \mathcal{U}$, so M has simple socle, S say, of order s. By [22], S is generated by an involution and a further element of G, so taking account of multiplicities we find that the number of choices for $M \in \mathcal{U}$ with socle isomorphic to S is at most $i_2(G)|G|/s$. Hence the contribution of these subgroups to the sum in question is, roughly speaking, at most

$$\frac{i_2(G)|G|}{s} \cdot \frac{s^2}{|G|^2} = \frac{i_2(G)s}{|G|}.$$

Now by definition of $\mathcal{U}$, we know that $s \leq |G|^{5/13}$; and we check from the literature on involutions in exceptional groups that $i_2(G) \leq c|G|^{7/13}$ for some constant c. Consequently the above contribution is at most $c|G|^{-1/13}$. Since the number of divisors s of $|G|$ is at most $|G|^{o(1)}$, and there are at most two simple groups of any given order s (up to isomorphism), we conclude that

$$\sum_{M \in \mathcal{U}} |G : M|^{-2} \to 0. \qquad (7)$$

Theorem 1 now follows from (5), (6) and (7).

We note that the proof of Theorem 2 for exceptional groups of Lie type involves more delicate arguments. In fact, in counting "unknown" almost simple maximal subgroups M in an exceptional group G, we use the fact (established in Corollary 7) that the classical groups are usually $(2,3)$-generated. Now, if the socle S of the maximal subgroup M is $(2,3)$-generated, then by counting generating pairs consisting of elements of orders 2 and 3 (and taking account of multiplicities) we find that G has at most $i_2(G)i_3(G)/|S|$ maximal subgroups which are isomorphic to M (or whose socle is isomorphic to S). While this upper bound is probably still far from best possible, it improves the bound obtained in the previous paragraph, and it is strong enough for the purpose of proving Theorem 2. We therefore see that $(2,3)$-generation of classical groups yields new results on random generation of exceptional groups of Lie type.

References

[1] A. A. Albert and J. G. Thompson, Two element generation of the projective unimodular group, *Illinois J. Math.* **3** (1959), 421–439.

[2] M. Aschbacher and R. Guralnick, Some applications of the first cohomology group, *J. Algebra* **90** (1984), 446–460.

[3] M. Aschbacher, On the maximal subgroups of the finite classical groups, *Invent. Math.* **76** (1984), 469–514.

[4] L. Babai, The probability of generating the symmetric group, *J. Combin. Theory Ser. A.* **52** (1989), 148–153.

[5] H. R. Brahana, Pairs of generators of the known simple groups whose orders are less than one million, *Ann. of Math. (2)* **31** (1930), 529–549.

[6] L. Di Martino and N. Vavilov, (2,3)-generation of SL(n,q), I, cases $n = 5,6,7$, *Comm. Algebra* **22** (1994), 1321–1347.

[7] L. Di Martino and N. Vavilov, (2,3)-generation of SL(n,q), II, cases $n \geq 8$, *Comm. Algebra* **24** (1996), 487–515.

[8] L. E. Dickson, *Linear groups with an exposition of the Galois field theory*, Teubner, Leipzig, 1901; reprinted, Dover, 1958.

[9] J. D. Dixon, The probability of generating the symmetric group, *Math. Z.* **110** (1969), 199–205.

[10] D. Garbe, Über eine Klasse von arithmetisch definierbaren Normalteilern der Modulgruppe, *Math. Ann.* **235** (1978), 195–215.

[11] R. M. Guralnick, W. M. Kantor and J. Saxl, The probability of generating a classical group, *Comm. Algebra* **22** (1994), 1395–1402.

[12] W. Kantor, Some topics in asymptotic group theory, in *Groups, combinatorics and geometry (eds. M.W. Liebeck and J. Saxl)*, pp. 403–421. London Math. Soc. Lecture Note Series **165**, Cambridge University Press, 1992.

[13] W. M. Kantor and A. Lubotzky, The probability of generating a finite classical group, *Geom. Dedicata* **36** (1990), 67–87.

[14] P. B. Kleidman and M. W. Liebeck, *The subgroup structure of the finite classical groups*, London Math. Soc. Lecture Note Series **129**, Cambridge University Press, 1990.

[15] M. W. Liebeck and G. M. Seitz, Maximal subgroups of exceptional groups of Lie type, finite and algebraic, *Geom. Dedicata* **36** (1990), 353–387.

[16] M. W. Liebeck and A. Shalev, The probability of generating a finite simple group, *Geom. Dedicata* **56** (1995), 103–113.

[17] M. W. Liebeck and A. Shalev, Classical groups, probabilistic methods, and the $(2,3)$-generation problem, *Ann. of Math.* **144** (1996), 77–125.

[18] M. W. Liebeck and A. Shalev, Simple groups, probabilistic methods, and a conjecture of Kantor and Lubotzky, *J. Algebra* **184** (1996), 31–57.

[19] M. W. Liebeck, J. Saxl and D. M. Testerman, Simple subgroups of large rank in groups of Lie type, *Proc. London Math. Soc.* **72** (1996), 425–457.

[20] T. Luczak and L. Pyber, On random generation of the symmetric group, *Combin. Probab. Comput.* **2** (1993), 505–512.

[21] A. M. Macbeath, Generators of the linear fractional groups, *Proc. Sympos. Pure Math.* **12** (1967), 14–32.

[22] G. Malle, J. Saxl and T. Weigel, Generation of classical groups, *Geom. Dedicata* **49** (1994), 85–116.

[23] G. A. Miller, On the groups generated by two operators, *Bull. Amer. Math. Soc.* **7** (1901), 424–426.

[24] E. Netto, *The theory of substitutions*, Chelsea, New York, 1964 (reprint of the 1892 edition).

[25] P. Sanchini and M. C. Tamburini, Constructive (2, 3)-generation: a permutational approach, *Rend. Sem. Mat. Fis. Milano* **LXIV** (1994), 141–158.

[26] R. Steinberg, Generators for simple groups, *Canad. J. Math.* **14** (1962), 277–283.

[27] M. C. Tamburini and F. Vassalo, (2,3)-generazione di SL(4, q) in caratteristica dispari e problemi collegati, *Boll. Un. Mat. Ital.* *(7)* **8-B** (1994), 121–134.

[28] M. C. Tamburini, J. S. Wilson and N. Gavioli, The (2,3)-generation of some classical groups, *J. Algebra* **168** (1994), 353–370.

Condensing tensor product modules

Klaus Lux
and
Markus Wiegelmann

Abstract

We describe an algorithm that performs the condensation of arbitrary tensor product modules for group algebras over finite fields. This algorithm is very helpful in studying the submodule structure and determining subquotients of tensor products.

Introduction

Let G be a finite group, p be a prime number, F be a finite field of characteristic p, FG be the group algebra, H be a subgroup of G of order prime to p, and $e_H := \frac{1}{|H|} \sum_{h \in H} h$. The objective of this paper is to describe an algorithm which, given two FG-modules V and W, constructs the condensed tensor product module $(V \otimes_F W)e_H$ for the condensation algebra $e_H FG e_H$. This algorithm plays an important role in determining p-modular character tables of finite simple groups, see for example [4]. Its main application is the analysis of large tensor product modules for the group algebra FG using the condensation method, see [13]. The condensation method is based on a close relationship between the category of finitely generated, right FG-modules and the category of finitely generated, right $e_H FG e_H$-modules for the subalgebra $e_H FG e_H$ of FG, see for example [3].

If one wants to exploit this relationship one is faced with the following task: Given a matrix representation of FG on the FG-module V, determine a matrix representation of $e_H FG e_H$ on the condensed $e_H FG e_H$-module $V e_H$. This process is called *condensation* of the FG-module V, and the matrix representation on $V e_H$ will be called the *condensed matrix representation*. Observe that the $e_H FG e_H$-module $V e_H$ is an F-subspace of the FG-module V. Given an $e_H FG e_H$-submodule W of $V e_H$, the calculation of the embedding of W into V is called *uncondensing* W. This can be done if the explicit embedding of $V e_H$ into V has been determined beforehand. Since the condensed module of the FG-submodule WFG is again W, see [7], we can use the process of

174

uncondensing to construct FG-submodules of V from $e_H FGe_H$-submodules of Ve_H.

The first programs for calculating condensed modules were written by Parker and Thackray, see [15], for the class of FG-permutation modules. Note that in this case the condensed matrix representation can be obtained without first constructing a matrix representation of FG on the FG-permutation module.

In [13], A. Ryba gave a general description of condensation including an algorithm for the condensation of antisymmetric powers of a given FG-module.

Here we prove by giving an explicit algorithm that the condensation of arbitrary tensor products can be performed without constructing a representation for the tensor product module itself. For the reader familiar with the MeatAxe, see [10], we describe how useful this method is for analysing module structures of large tensor product modules and thus for determining p-modular character tables of finite simple groups, for example.

In section 1, we state some basic results about the condensation of tensor products. In addition to the theoretical description, we give a first rough outline of our condensation method.

The main program is described in section 2. It relies on a very efficient way of storing a basis for the condensed module and of performing the projection onto the condensed module. Here we require special bases with respect to which the input matrix representations of FG on the FG-modules are given. In addition, we describe a program for embedding elements of the condensed tensor product module into the tensor product module. It can be used to uncondense a given submodule of the condensed tensor product module and hence can be used to construct matrix representations on submodules of the tensor product module.

In section 3, we describe two methods for calculating the special bases, reflecting the semisimplicity of the restriction of a given FG-module to FH. These special bases are required for the condensation algorithm stated in section 2. The first is based on peakwords, which were introduced by Lux, Müller and Ringe in [7], and the second method is based on the generalized Schur relations.

Finally, section 4 demonstrates the use of our programs. We construct a new matrix representation of dimension 517 over $GF(5)$ for the sporadic simple Lyons group. It appears—as predicted by A.Ryba—in the tensor product of the known 111-dimensional representation with itself. A table containing the running times and some references to further applications complete this section.

1   Preliminary results

Throughout this paper let G be a finite group and FG be the group algebra of G over a finite field F of characteristic p. Furthermore, let H be a subgroup of G of order prime to p and let $e_H := \frac{1}{|H|} \sum_{h \in H} h$. Note that e_H is an idempotent and that for a given FG-module V the space Ve_H is the subspace of elements of V fixed by H. Moreover, we denote the restriction of the FG-module V to FH by V_{FH}. If V and W are two FG-modules then we use the shorthand notation $V \otimes W$ for the tensor product module $V \otimes_F W$. Furthermore, if V is an FG-module then $V^* = \mathrm{Hom}_F(V, F)$ denotes the contragredient FG-module.

Definition 1.1 We define the *condensation algebra* corresponding to e_H as the subalgebra $e_H FG e_H$ of FG. For an FG-module V, the $e_H FG e_H$-module Ve_H is called the *condensed module*.

The aim of this section is to give a theoretical, basis independent description of an algorithm which given FG-modules V and W in the form of matrix representations of FG computes the condensed representation of $e_H FG e_H$ on the condensed tensor product module. Before we give an outline of the algorithm, we first mention the following reduction.

Proposition 1.2 *Let V and W be FG-modules. Their restrictions to the semisimple subalgebra FH can be decomposed into direct sums of simple FH-submodules, say*

$$V_{FH} = \bigoplus_{S} \bigoplus_{i=1}^{m_S} S_i \quad and \quad W_{FH} = \bigoplus_{S} \bigoplus_{j=1}^{n_{S^*}} T_j,$$

where S runs through a set of representatives for the isomorphism types of simple FH-modules, m_S and n_{S^*} are the multiplicities of S respectively S^* as a constituent of V_{FH} respectively W_{FH}, and $S \cong S_i$, $S^* \cong T_j$ for $1 \le i \le m_S$ and $1 \le j \le n_{S^*}$. Note that we allow m_S and n_{S^*} to be zero. Then

$$(V \otimes W)e_H = \bigoplus_{S} \bigoplus_{i=1,j=1}^{m_S, n_{S^*}} (S_i \otimes T_j)e_H.$$

Proof: The statement follows immediately from the distributivity of the tensor product and the observation that for given simple FH-modules S and T we have $(S \otimes T) \cong_{FH} \mathrm{Hom}_F(S^*, T)$, see [2], and therefore $(S \otimes T)e_H \cong \mathrm{Hom}_{FH}(S^*, T)$ is non-zero if and only if $S^* \cong T$.   □

We now translate the above reduction into an algorithm for condensing tensor products.

Algorithm 1.3 (Conceptual) Let the notation be as in Proposition 1.2. Then the condensed $e_H FGe_H$-module $(V \otimes W)e_H$ can be determined by the following steps.

1. Determine the pairwise nonisomorphic constituents $S_1, \ldots, S_s$ of V_{FH} and $T_1, \ldots, T_t$ of W_{FH} together with their multiplicities m_i of S_i and n_j of T_j. Sort them such that $S_i \cong T_i^*$ for $1 \leq i \leq s$. This implies that we have to allow the multiplicity of T_j in W to be zero.

2. For all i with $1 \leq i \leq s$ and both m_i and n_i non-zero, determine a basis Q_i of $(S_i \otimes T_i)e_H$. Then a basis Q of $(V \otimes W)e_H$ can be easily derived from the Q_i by a "concatenation" procedure. See section 2.

3. For $g \in G$ determine the action of $e_H ge_H$ on $(V \otimes W)e_H$ with respect to the basis Q. This is done by projecting the elements qg for $q \in Q$ by e_H onto $(V \otimes W)e_H$ and expressing the result as a linear combination of the elements of Q.

Of course the matrix of $e_H ge_H$ on $(V \otimes W)e_H$ with respect to Q should be calculated without building the large tensor product matrices explicitly. It will be shown in the next section, that this is indeed possible, provided the bases for the matrix representations of V and W are chosen carefully.

2   The condensation algorithm

2.1   Notations and basic definitions

Throughout this section we will assume that we are given two FG-modules V of dimension m and W of dimension n with bases B of V and C of W. We denote the matrix representation of FG on V with respect to B by M and the matrix representation of FG on W with respect to C by N. Moreover, we define a basis $B \otimes C$ of $V \otimes W$ by

$$(b_1 \otimes c_1, b_1 \otimes c_2, \ldots, b_1 \otimes c_n, b_2 \otimes c_1, \ldots, b_2 \otimes c_n, \ldots, b_m \otimes c_1, \ldots, b_m \otimes c_n).$$

The matrix of $g \in G$ with respect to the basis $B \otimes C$ is then given by the Kronecker product of the matrices $M(g)$ with $N(g)$.

We are now going to describe an algorithm that derives a matrix representation for the $e_H FGe_H$-module $(V \otimes W)e_H$ directly from M and N.

Definition 2.1 Let A be a finite-dimensional F-algebra and let V be a semisimple A-module. Moreover, let the constituents of V be the pairwise nonisomorphic simple A-modules $S_1, \ldots, S_s$, where S_i occurs with multiplicity $m_i \geq 0$. Furthermore, let M_i be a matrix representation for the A-module S_i, where $1 \leq i \leq s$. A basis B of V is called a *symmetry basis* of V with respect to the sequence $[M_1, \ldots, M_s]$, see also [14], if the matrix representation M of A on the A-module V with respect to B satisfies the following condition: the matrices $M(a)$ for all $a \in A$ are block diagonal matrices, the sequence on the diagonal being $[M_1(a), \ldots, M_1(a), M_2(a), \ldots, M_s(a), \ldots, M_s(a)]$, where the matrix $M_i(a)$ appears m_i times for $1 \leq i \leq s$, that is

$$
M(a) = \begin{pmatrix}
M_1(a) & & & & & & & \\
 & \ddots & & & & & 0 & \\
 & & M_1(a) & & & & & \\
 & & & M_2(a) & & & & \\
 & & & & \ddots & & & \\
 & & & & & M_2(a) & & \\
 & & & & & & \ddots & \\
 & & & & & & & M_s(a) \\
 & 0 & & & & & & & \ddots \\
 & & & & & & & & & M_s(a)
\end{pmatrix}.
$$

In the next section, we will give two algorithms that compute a symmetry basis for a semisimple A-module given by a matrix representation. In this section, we will assume that we are given symmetry bases for the semisimple FH-modules V_{FH} and W_{FH}.

More precisely and in order to fix the notation, let $S_1, \ldots, S_s$ be the pairwise nonisomorphic constituents of V_{FH} with multiplicities $m_1, \ldots, m_s$ and let $T_1, \ldots, T_t$ be the pairwise nonisomorphic constituents of W_{FH}. For $1 \leq i \leq s$, let $T_i \cong S_i^*$ and $n_i \geq 0$ be the multiplicity of T_i, $1 \leq j \leq t$, as a constituent of W_{FH}. Finally, let M_i and N_j be matrix representations for S_i and T_j for $1 \leq i \leq s$ and $1 \leq j \leq t$, respectively. We also assume that the chosen bases B and C of V and W are symmetry bases for V_{FH} and W_{FH} with respect to the sequence of M_i's and the sequence of N_j's, respectively.

2.2   Reduction to simple blocks

By the assumptions made in the preceding subsection, we can restrict the analysis to a fixed i, with $1 \leq i \leq s$.

From $S_i^* \cong T_i$ for $1 \leq i \leq s$ we get that $0 \neq E_i := \frac{1}{|H|} \sum_{h \in H} M_i(h) \otimes N_i(h)$ is the matrix of e_H with respect to the basis $B_i \otimes C_i$. Here B_i and C_i denote bases of S_i and T_i such that M_i and N_i are the matrix representations of FH on S_i and T_i with respect to B_i and C_i. Then a chosen basis of the row space of E_i can be interpreted as the coefficient matrix $q_i = m(Q_i, B_i \otimes C_i)$ of the decomposition of a basis Q_i for $(S_i \otimes T_i)e_H$ in terms of the basis $B_i \otimes C_i$.

Since the rows of E_i can be written uniquely as linear combinations of q_i, there is a unique matrix p_i with $E_i = p_i q_i$. As $E_i^2 = E_i$ it follows that $q_i p_i$ is the identity matrix. From these properties we can deduce the following projection formula for an element of $S_i \otimes T_i$.

Proposition 2.2 *For a given $v_i \in S_i \otimes T_i$ with coefficient row $m(v_i, B_i \otimes C_i)$ the coefficient vector $m(v_i e_H, Q_i)$ is given by $m(v_i e_H, Q_i) = m(v_i, B_i \otimes C_i) p_i$.*

2.3 The complete basis

Since Proposition 1.2 implies that only the subproducts of the above type $S_i \otimes T_i$ contain elements of the condensed module, the next step is to construct a basis Q of $(V \otimes W) e_H$ from the bases Q_i of the $(S_i \otimes T_i) e_H$. We reorder the basis $B \otimes C$ of the tensor product $V \otimes W$ in the following way. According to the decomposition of V_{FH} and W_{FH} into simple FH-submodules as in Proposition 1.2, which is reflected in the block structure of M and N on FH, we partition the bases B and C as follows:

$$B = B_{11} \sqcup B_{12} \sqcup \ldots \sqcup B_{1m_1} \sqcup B_{21} \sqcup \ldots \sqcup B_{s1} \sqcup \ldots \sqcup B_{sm_s}. \qquad (1)$$

Here $B_{i\alpha}$ is a basis of the simple submodule of type S_i of V_{FH} corresponding to the block number α of type M_i in M. Similarly

$$C = C_{11} \sqcup C_{12} \sqcup \ldots \sqcup C_{1n_1} \sqcup C_{21} \sqcup \ldots \sqcup C_{s1} \sqcup \ldots \sqcup C_{tn_t}. \qquad (2)$$

We choose a new basis P for $V \otimes W$ in the following way

$$P = \sqcup_{j=1}^{t} \sqcup_{i=1}^{s} \sqcup_{\beta=1}^{n_i} \sqcup_{\alpha=1}^{m_i} B_{i\alpha} \otimes C_{j\beta},$$

where $\sqcup$ stands for appending the sequences $B_{i\alpha} \otimes C_{j\beta}$ in the increasing order of the indices. Note that P and $B \otimes C$ are equal as sets but differ in the ordering unless all blocks have size one, and that the matrix for $h \in H$ on $V \otimes W$ with respect to P is a block diagonal matrix, the blocks being the Kronecker product matrices $M_i(h) \otimes N_j(h)$.

By Proposition 1.2 all elements of Q have non-zero coefficients only for the parts $B_{i\alpha} \otimes C_{i\beta}$ for $i \leq s$. Choosing the Q_i for those parts we get the following result.

Lemma 2.3 *With q_i as above, a basis Q of $(V \otimes W) e_H$ is given by*

$$m(Q, P) := \sqcup_{i=1}^{s} \sqcup_{\beta=1}^{n_i} \sqcup_{\alpha=1}^{m_i} q_i m(B_{i\alpha} \otimes C_{i\beta}, P),$$

where $\sqcup$ stands for pasting matrices together.

Observe that the basis Q respects the decomposition of P into the parts $B_{i\alpha} \otimes C_{j\beta}$. Therefore, we can apply Proposition 2.2 in order to determine the decomposition of an element $v \in (V \otimes W) e_H$ as a linear combination of the elements of Q for each part of Q in turn.

2.4   The partial action

From the previous subsection it follows that the projection with e_H onto $(V \otimes W)e_H$ and the decomposition as a linear combination into the elements of Q can be done part by part by multiplying with the p_i's. Hence, the action of a group element $g \in G$ can be determined part by part, too. More precisely, we subdivide the matrix $M(g)$ with respect to the partition of B given in equation (1) as

$$
\begin{pmatrix}
M_{1111}(g) & \cdots & M_{111m_1}(g) & \cdots & M_{11s1}(g) & \cdots & M_{11sm_s}(g) \\
M_{1211}(g) & \cdots & M_{121m_1}(g) & \cdots & M_{12s1}(g) & \cdots & M_{12sm_s}(g) \\
\vdots & & \vdots & & \vdots & & \vdots \\
M_{1m_111}(g) & \cdots & M_{1m_11m_1}(g) & \cdots & M_{1m_1s1}(g) & \cdots & M_{1m_1sm_s}(g) \\
\vdots & & \vdots & & \vdots & & \vdots \\
M_{s111}(g) & \cdots & M_{s11m_1}(g) & \cdots & M_{s1s1}(g) & \cdots & M_{s1sm_s}(g) \\
\vdots & & \vdots & & \vdots & & \vdots \\
M_{sm_s11}(g) & \cdots & M_{sm_s1m_1}(g) & \cdots & M_{sm_ss1}(g) & \cdots & M_{sm_ssm_s}(g)
\end{pmatrix}
$$

and $N(g)$ with respect to the partition of C given in equation (2) as

$$
\begin{pmatrix}
N_{1111}(g) & \cdots & N_{111n_1}(g) & \cdots & N_{11t1}(g) & \cdots & N_{11tn_t}(g) \\
N_{1211}(g) & \cdots & N_{121n_1}(g) & \cdots & N_{12t1}(g) & \cdots & N_{12tn_t}(g) \\
\vdots & & \vdots & & \vdots & & \vdots \\
N_{1n_111}(g) & \cdots & N_{1n_11n_1}(g) & \cdots & N_{1n_1t1}(g) & \cdots & N_{1n_1tn_t}(g) \\
\vdots & & \vdots & & \vdots & & \vdots \\
N_{t111}(g) & \cdots & N_{t11n_1}(g) & \cdots & N_{t1t1}(g) & \cdots & N_{t1tn_t}(g) \\
\vdots & & \vdots & & \vdots & & \vdots \\
N_{tn_t11}(g) & \cdots & N_{tn_t1n_1}(g) & \cdots & N_{tn_tt1}(g) & \cdots & N_{tn_ttn_t}(g)
\end{pmatrix}
$$

and work with the Kronecker products $M_{i\alpha k\gamma}(g) \otimes N_{j\beta l\delta}(g)$ during all the calculations. This is possible because of the following result.

Proposition 2.4 *Let x be an element of the basis Q whose non-zero coefficients expressed in terms of the basis P belong to the part $B_{i\alpha} \otimes C_{i\beta}$ of the basis P. Then the coefficient row*

$$(y_{1111}, \ldots, y_{k\gamma l\delta}, \ldots, y_{sm_stn_t}) := m(xg, P)$$

is given by $y_{k\gamma l\delta} = x_{i\alpha i\beta} \cdot (M_{i\alpha k\gamma}(g) \otimes N_{i\beta l\delta}(g))$.

2.5   The algorithm

The results of subsections 2.2 and 2.3 yield the following explicit method for determining the matrix of a given element $e_H g e_H$ with $g \in G$ on $(V \otimes W)e_H$ with respect to the basis Q.

Algorithm 2.5 (Tensorcondense)
INPUT:

- For $g \in G$ matrices $M(g)$ and $N(g)$ with respect to symmetry bases.

- The basis matrices $q_1, \ldots, q_s$ and projection matrices $p_1, \ldots, p_s$.

- The integers $s, t, m_i, n_i, \dim_F(S_i), \dim_F(\mathrm{End}_{FH}(S_i))$.

CALCULATION:

Define $c := \sum_{i=1}^s m_i \cdot n_i \cdot \dim_F(\mathrm{End}_{FH}(S_i))$;
Initialize $R := 0 \in F^{c \times c}$;
For i from 1 to s, α from 1 to m_i, β from 1 to n_i
For j from 1 to s, γ from 1 to m_i, δ from 1 to n_i
  Define $O(g) := M_{i\alpha j\gamma}(g) \otimes N_{i\beta j\delta}(g)$;
  Determine $R_{i\alpha\beta j\gamma\delta} := q_i \cdot O(g) \cdot p_j$;
  Insert $R_{i\alpha\beta j\gamma\delta}$ into R;

OUTPUT:

- The matrix R representing $e_H g e_H$ on $(V \otimes W)e_H$ with respect to the basis Q.

2.6 Uncondense

From the description of the last section, it follows that we can solve the following problem computationally: given the coefficient row $m(v, Q)$ of an element $v \in (V \otimes W)e_H$, compute the coefficient row $m(v, B \otimes C)$. This solves the problem of calculating the embedding of $(V \otimes W)e_H$ into $V \otimes W$ mentioned in the introduction.

More precisely, the program which implements this idea is given the matrices q_i and, in addition, the multiplicities and dimensions of the constituents of V_{FH} and W_{FH}. Note that these data encode the decomposition of the basis Q as a linear combination of the basis P of $V \otimes W$. The base change matrix from the basis P to the basis $B \otimes C$ is a permutation matrix and can be derived from the given multiplicities and dimensions. We refer to [16] for a more detailed description of the program.

Lemma 2.6 *Let G, V, W be as above. Let $B = [b_1, \ldots, b_m]$ be a basis of V, and $C = [c_1, \ldots, c_n]$ be a basis of W, let M be the matrix representation of FG on V with respect to B and let N be the matrix representation of FG on W with respect to C. Then the vector space of $m \times n$ matrices $F^{m \times n}$ is an FG-module with the action of $g \in G$ given by $Xg := M(g)^T X N(g)$ for all $X \in F^{m \times n}$. Moreover, the linear map $\Phi: V \otimes W \to F^{m \times n}$ which maps $b_i \otimes c_j$ to the i, j-th elementary matrix E_{ij} for $1 \leq i \leq m$, $1 \leq j \leq n$, is an FG-isomorphism.*

We apply this in the following way, see also [6]. Suppose we want to determine the action of the Kronecker product of $M(g)$ and $N(g)$ on a given vector v in the tensor product module $V \otimes W$. Then take the matrix $\Phi(v)$, multiply by $M(g)^T$ from the left and by $N(g)$ from the right, and use Φ^{-1} to get the answer in the original space $V \otimes W$. This approach avoids calculating explicitly with the Kronecker product matrix $M(g) \otimes N(g)$. So, using the uncondense program and the method described above for determining the action of elements in G on $V \otimes W$, we can construct efficiently the submodule WFG of a given $e_H FGe_H$ submodule W in $(V \otimes W)e_H$. In the last section, we will use this method to construct a 517-dimensional submodule of the sporadic simple Lyons group in a tensor product module of dimension 12321.

3 Calculating symmetry bases

3.1 Preliminaries

In Algorithm 2.5, the input representations were expected to be given with respect to symmetry bases. The two algorithms presented here for calculating such bases determine appropriate vectors lying in simple submodules and then form the span of these vectors under the action of FH. In order to calculate suitable vectors, the first method uses kernels of so-called *deltawords* and the second one uses the images of certain weighted sums over group elements in FH, both associated with every isomorphism type of simple submodules.

Given a vector $v \in V$ we use the following well-known procedure, see [10], to calculate the span of v under the action of the F-algebra A with respect to a fixed ordered generating system E of A.

Algorithm 3.1 (Spinning Algorithm)
INPUT:

- A finite ordered generating system E of A.

- An element $0 \neq v \in V$.

CALCULATION:

Initialize $B := [v]$;
For all $x \in B$:
    For all $a \in E$:
        If $xa \notin \langle B \rangle_F$:
            Append xa to B;

OUTPUT:

- An F-Basis B of $vA \leq_A V$.

In the following, we denote by $B_E(v)$ the basis of $vA \leq V$ produced by the above algorithm. Note that the algorithm commutes with any A-monomorphism ϕ, that is: $B_E(\phi(v)) = \phi(B_E(v))$.

3.2   The deltaword method

The method described in this subsection works for an arbitrary finite-dimensional F-algebra A and any semisimple A-module V. Let S_i, $i = 1, \ldots, s$, be the pairwise nonisomorphic constituents of V. For a given $a \in A$ denote by a_V the linear map on V induced by a. A *deltaword* for S_i is an element a_i of A satisfying the following properties: $\mathrm{Ker}((a_i)_{S_j}) = 0$ if $j \neq i$ and $\dim_F(\mathrm{Ker}((a_i)_{S_i})) = \dim_F(\mathrm{End}_A(S_i))$. If it also satisfies $\mathrm{Ker}(((a_i)^2)_{S_i}) = \mathrm{Ker}((a_i)_{S_i})$ it is called a *peakword* for S_i. For results on the existence and properties of peakwords we refer to [7]. We will make use of the following Lemma.

Lemma 3.2 *If $v_i \in \mathrm{Kernel}((a_i)_V)$, then $V_i := \langle B_E(v_i) \rangle_F$ is a simple submodule of V isomorphic to S_i. Furthermore, the matrix representation X_i of A on V_i with respect to the basis $B_E(v_i)$ is independent of the chosen v_i.*

Proof: All we have to prove is that the endomorphism algebra $\mathrm{End}_A(V)$ acts transitively on the non-zero vectors of $\mathrm{Ker}((a_i)_V)$. But, since $\mathrm{End}_A(V)$ is isomorphic to a direct sum of full matrix algebras over $\mathrm{End}_A(S_i)$ and since $\mathrm{Ker}((a_i)_V)$ is an irreducible $\mathrm{End}_A(V)$-module, there is an element of $\mathrm{End}_A(V)$ mapping a given non-zero vector of $\mathrm{Ker}((a_i)_V)$ to any other vector of $\mathrm{Ker}((a_i)_V)$. $\square$

The following algorithm computes a symmetry basis of V with respect to the sequence of matrix representations $[X_1, X_2, \ldots, X_s]$ defined by Lemma 3.2.

Algorithm 3.3 (Deltaword method for a symmetry basis)
INPUT:

- A finite ordered generating system E of A.

- For each constituent $S_i, 1 \leq i \leq s$, of V, a deltaword $a_i \in A$.

CALCULATION:

Initialize $B := [\,]$;
For i from 1 to s:
    Determine a basis σ_i for $\mathrm{Ker}((a_i)_V)$;
    For all vectors z in σ_i:
        If $z \notin \langle B \rangle$:
            Append $B_E(z)$ to B;

OUTPUT:

- A symmetry basis B of V.

Proof: By the definition of a deltaword, the dimension of $\mathrm{Ker}((a_i)_V)$ is n_i times the dimension of $\mathrm{Kernel}((a_i)_{S_i})$. Since the subspace generated by one element of σ_i under the action of A contains a subspace of dimension $\dim_F(\mathrm{Ker}((a_i)_{S_i})) = [\mathrm{End}_A(S_i) : F]$ of $\mathrm{Ker}((a_i)_V)$, one gets at least n_i simple submodules of type S_i for every i. Since the intersection of two different simple submodules is trivial, the concatenated sequence B stays linearly independent during the whole procedure. A counting argument immediately implies that we finally get a basis for V at the end of the algorithm.

The matrix of $a \in A$ with respect to the basis B is clearly a block diagonal matrix, the blocks being $X_i(a)$ by Lemma 3.2.   □

Note that this algorithm can be generalized to an algorithm that computes the socle series of an A-module for an arbitrary finite-dimensional algebra A, see [8].

3.3   The operator method

We now describe a second method for determining appropriate vectors generating simple submodules. In contrast to the deltaword method, we need to be able to run through the elements of H. The operators we are interested in are given as the following weighted sums over the elements of H.

Definition 3.4 Let X be an irreducible matrix representation of FH. By $X(h)_{ij}$ we denote the entry at position (i, j) in the matrix $X(h)$. Then the element $\gamma_X := \sum_{h \in H} X(h^{-1})_{11} h \in FH$ is called the γ-operator with respect to X.

In order to prove the correctness of the method, we need the following property of the γ-operator.

Proposition 3.5 Let S be a simple FH-module, B be a basis of S, s_1 be the first basis vector of B and let X be the irreducible matrix FH-representation on S with respect to B. Assume that we are given an arbitrary FH-module V. Then the following statement holds

$$V\gamma_X = s_1 \cdot \mathrm{Hom}_{FH}(S, V).$$

Proof: It suffices to prove the proposition for a simple module $T = V$. The general result follows immediately, since an arbitrary FH-module is semisimple.

Let Y be a matrix representation for T and define for a matrix $P \in F^{\dim(S) \times \dim(T)}$ the trace matrix

$$\mathrm{Tr}_H(P) := \sum_{h \in H} X(h^{-1}) P Y(h).$$

Recall that $X(g)\mathrm{Tr}_H(P) = \mathrm{Tr}_H(P)Y(g)$ for all $g \in H$. This means that $\mathrm{Tr}_H(P)$ is an FH-homomorphism from S to T.

Taking P to be an elementary matrix E_{1j} yields as the first row of $\mathrm{Tr}_H(E_{1j})$ exactly row j of $Y(\gamma_X)$. Now in case $S \not\cong T$ we get that $\mathrm{Tr}_H(E_{1j})$ is the zero homomorphism and $Y(\gamma_X)$ is zero, too. Hence the statement holds in case $S \not\cong T$.

Furthermore, if $T \cong S$ we get that $T\gamma_X \subseteq s_1 \cdot \mathrm{Hom}_{FH}(S,T)$. To prove equality it suffices to compare the dimensions of both vector spaces.

As $T\gamma_X \neq 0$ for $T \cong S$ (since otherwise γ_X would have to be 0), the dimension of $T\gamma_X$ is at least as large as the dimension of $\mathrm{End}_{FH}(T) \cong \mathrm{Hom}_{FH}(S,T)$. But this is an upper bound for the dimension of the right hand vector space. This completes the proof. □

The proposition above shows that the element γ_{X_i} for $i = 1, \ldots, s$ has a property analogous to that of a deltaword for X_i, namely that $\dim_F(S_i \gamma_{X_i}) = \dim_F(\mathrm{End}_{FH}(S_i))$ and that $S_j \gamma_{X_i}$ is zero for $j \neq i$. It follows from Lemma 3.5 that each vector in $V\gamma_{X_i}$ generates a submodule isomorphic to S_i under the action of FH. Furthermore, we can now use Algorithm 3.3 as before in order to determine a symmetry basis of V if we replace $\mathrm{Ker}((a_i)_V)$ in Algorithm 3.3 by $V\gamma_{X_i}$.

Algorithm 3.6 (Operator method for constructing a symmetry basis)

INPUT:

- A finite ordered generating system E of A

- Matrix representations X_i for all constituents S_i, $1 \leq i \leq r$, of V.

CALCULATION:

Initialize $B := []$;
For i from 1 to r:
    Determine a basis σ_i of $V\gamma_{X_i}$;
    For all z in σ_i:
        If $z \notin \langle B \rangle$:
            Append $B_E(z)$ to B;

OUTPUT:

- A symmetry basis B of V.

4 Applications

4.1 Notations and preliminary results on Ly

The aim of this section is to give a flavour of how our method of tensor condensation works in practice. In order to show the efficiency of the algorithms, we apply the method to a rather large problem: we are going to construct a new irreducible matrix representation of the sporadic simple Lyons group Ly over the field F with 5 elements. The existence of such a representation was predicted by A. Ryba, when he studied the 5-modular irreducible characters of Ly.

Let F be the field with 5 elements and G be the sporadic simple Lyons group of order 51,765,179,004,000,000, see [1]. The generators for G we choose are the standard generators a and b defined in [17], [18]. They are given by the conditions that a is an involution, that b belongs to the conjugacy class $5A$, that the order of ab is 14 and that the order of $ababab^2$ is 67.

The smallest known non-trivial F-matrix representation $D111$ of G has dimension 111. It was constructed by R. A. Parker, see [9], and the matrices for the standard generators of G are contained in R. A. Wilson's ATLAS of finite group representations, see [17]. We denote the corresponding FG-module by $V111$.

Following the conjecture of A. Ryba, which says that the new representation is contained in $V111 \otimes V111$, we are going to analyse this tensor product.

4.2 Condensing $V111 \otimes V111$

We choose the subgroup H to be a subgroup of order 67 of G generated by $h := ababab^2$. The first step is to compute a symmetry basis for $V111_{FH}$ using the deltaword method described in section 3.3.

Let $D111_{st}$ be the matrix representation of $V111$ with respect to the symmetry basis determined by the deltaword method and denote its canonical module by $V111_{st}$. Using Algorithm 2.5 we produce the matrices of $e_H a e_H$ and $e_H b e_H$ acting on the condensed module $(V111_{st} \otimes V111_{st})e_H$. The analysis of the condensed module leads to the following result.

Lemma 4.1 *The representation K of the subalgebra of $e_H FG e_H$ generated by $e_H a e_H$ and $e_H b e_H$ on the condensed module has the composition factors*

$$1a, 1b, 11a, 11a, 38a, 38b, 99a.$$

Moreover, the socle series of K is

$$1a \oplus 11a \oplus 99a, 38a, 1b, 38b, 11a.$$

Proof: Applying the standard MeatAxe program *chop*, see [12], to K, gives the above list of constituents. The MeatAxe program *socleseries* for computing the socle series, see [8], gives the explicit structure of the socle series. □

4.3 Constructing the new representation

We now take the simple submodule of K which has dimension 11 and choose a non-zero vector v_{11a} in it. We embed the vector v_{11a} into the original FG-module $V111_{st} \otimes V111_{st}$ using the program *uncondense*, implementing the algorithm described in subsection 2.6. We denote the resulting vector by $v_{517} \in F^{12321}$. We then get the following result.

Theorem 4.2 *The FG-submodule $v_{517}FG$ of $V111_{st} \otimes V111_{st}$ is a simple FG-module $S517$ of dimension 517.*

Proof: We prove the above statement by using the technique described in section 2.6 and apply standard MeatAxe routines, see [12], to determine the action of FG on the subspace. □

4.4 Running times

The following tables give the times measured on a DEC station 5000/120 using the C-implementation of the algorithms written by the second author in his diploma thesis, see [16].

PROGRAM or OPERATION	CPU TIME
STEP 1: Determination of the condensed representation	
preliminary operations on $D111_{FH}$	11 seconds
precondensation (basis- and projection- matrices)	680 seconds
determination of standard bases (using the deltaword method)	80 seconds
condensation of the tensor product	810 seconds

PROGRAM or OPERATION	CPU TIME
STEP 2: Analysis of the condensed representation	
constituents	62 seconds
peakword condensation	92 seconds
socle series	69 seconds

PROGRAM or OPERATION	CPU TIME
STEP 3: Generation of the 517-dimensional submodule	
uncondensation of the seed vector v_{11a}	1 second
constructing the invariant subspace $v_{517}FG$	80 minutes
construction of D517	25 minutes
testing irreducibility	7 minutes

4.5 Further applications

The algorithms have already been used by several people for determining the p-modular character tables of some finite simple groups. In [5], C. Jansen successfully analysed the tensor product of a module of dimension 126 with a module of dimension 770 for the sporadic simple group Co_3 over $GF(3)$ in order to determine its 3-modular character table. He also analysed successfully the tensor product of a module of dimension 286 with a module of dimension 786 for the double cover of the Suzuki sporadic simple group over $GF(3)$. G. Hiss and J. Müller, see [4], used the programs for analysing the tensor product of a module of dimension 133 with a module of dimension 1220 for the double cover of the Rudvalis simple group over $GF(5)$. They also used this result in order to determine the 5-modular character table of this group.

Finally, the authors applied the programs in order to get the socle series of the condensed projective indecomposable modules for the Mathieu group M_{23} in characteristic 2. From these it was possible to describe the socle series of the projective indecomposable modules of M_{23} in characteristic 2, see [8].

ACKNOWLEDGEMENTS

This paper is a contribution to the DFG research project "Algorithmic Number Theory and Algebra." Research of the second author is supported by the DFG through the graduate program "Mathematische Optimierung", Universität Trier. Moreover, we wish to thank Lehrstuhl D für Mathematik at the RWTH Aachen for providing the necessary computer environment without which we would not have been able to develop and test our ideas. We would also like to thank T. Breuer, C. Jansen, J. Müller and A. Ryba for fruitful discussions.

References

[1] J. H. Conway, R. T. Curtis, S. P. Norton, R. A. Parker and R. A. Wilson, *An* ATLAS *of finite groups*, Oxford University Press, 1985.

[2] W. Feit, *The representation theory of finite groups*, North-Holland Mathematical Library, 1982.

[3] J. A. Green, *Polynomial representations of GL_n*, Lecture Notes in Mathematics 830, Springer-Verlag, Berlin/Heidelberg/New York, 1980.

[4] G. Hiss and J. Müller, The 5-modular characters of the sporadic simple Rudvalis group and its covering group, *Comm. Algebra* **23** (1995), 4633–4667.

[5] C. Jansen, *Ein Atlas drei-modularer Charaktertafeln*, Ph. D. Thesis, RWTH Aachen, 1995.

[6] W. Lempken and R. Staszewski, A construction of $\widehat{3}McL$ and some representation theory in characteristic 5, *Lin. Algebra and Its Appl.* **192** (1993), 205–234.

[7] K. Lux, J. Müller and M. Ringe, Peakword condensation and submodule lattices: an application of the Meat-Axe, *J. Symbolic Comput.* **17** (1994), 529–544.

[8] K. Lux and M. Wiegelmann, An application of condensation: the PIM structures of the Mathieu group M_{23}, *in preparation*.

[9] W. Meyer, W. Neutsch and R. Parker, The minimal 5-representation of Lyons' sporadic group, *Math. Ann.* **272** (1985), 29–39.

[10] R. A. Parker, The computer calculation of modular characters (the **Meat-Axe**), in *Computational group theory (ed. M. D. Atkinson)*, pp. 267–274. Academic Press, 1984.

[11] R. A. Parker and R. A. Wilson, The computer construction of matrix representations of finite groups over finite fields, *J. Symbolic Comput.* **9** (1990), 583–590.

[12] M. Ringe, *The C-MeatAxe, Release 2.2.0*, Manual. RWTH Aachen, 1994.

[13] A. J. E. Ryba, Computer condensation of modular representations, *J. Symbolic Comput.* **9** (1990), 591–600.

[14] E. Stiefel and A. Fässler, *Gruppentheoretische Methoden und ihre Anwendungen*, Teubner, 1979.

[15] J. G. Thackray, *Modular representations of some finite groups*, Ph. D. Thesis, University of Cambridge, 1981.

[16] M. Wiegelmann, *Fixpunktkondensation von Tensorproduktmoduln*, Diplomarbeit, Lehrstuhl D für Mathematik, RWTH Aachen, 1994.

[17] R. A. Wilson, A world-wide-web ATLAS of finite group representations, http://www.mat.bham.ac.uk/atlas, 1996.

[18] R. A. Wilson, Standard generators for sporadic simple groups, *J. Algebra* **184** (1996), 505–515.

Intersections of Sylow subgroups in finite groups

V. D. Mazurov
and
V. I. Zenkov

Abstract

For every finite group G and every prime p, $i_p(G) \leq 3$, where $i_p(G)$ denotes the smallest number of Sylow p-subgroups of G whose intersection coincides with the intersection of all Sylow p-subgroups of G. For all simple groups G, $i_p(G) \leq 2$.

Introduction

Let G be a finite group and p be a prime. Denote by $O_p(G)$ the intersection of all Sylow p-subgroups of G and by $i_p(G)$ the smallest number i such that $O_p(G)$ is equal to the intersection of i Sylow p-subgroups. Obviously, $i_p(G) = 1$ if and only if G has an unique Sylow p-subgroup. J. Brodkey [2] proved that $i_p(G) \leq 2$ if a Sylow p-subgroup of G is abelian and N. Ito [5] found sufficient conditions for a finite solvable group to satisfy $i_p(G) \leq 2$.

This paper discuss some recent results in this direction. The main theorems are the following:

Theorem 1 [10] *If G is a simple non-abelian group then $i_p(G) = 2$ for every prime p dividing the order of G.*

Theorem 2 [9] *For every finite group G and every prime p, $i_p(G) \leq 3$.*

Section 2 contains a sketch of a proof of Theorem 1 which uses the Classification of Finite Simple Groups. Section 3 presents results about intersections of Sylow subgroups in arbitrary finite groups. We use the notation of the ATLAS [3].

1 Preliminary results

The following elementary results, the first of which is trivial, give a base for induction arguments in the proof of the main theorems of this paper.

Lemma 1 *Let P and Q be Sylow p-subgroups of a finite group G and $a \in Z(P), b \in Z(Q)$. If $C_G(ab)$ contains no non-trivial p-element then $P \cap Q = 1$.*

Lemma 2 *Let N be a subgroup of a finite group G and let P, Q be Sylow p-subgroups of N. Denote $V = P \cap Q$.*

(i) If $N_G(V) \le N$ then the intersection of any Sylow subgroups R and S of G containing P and Q respectively coincides with V.

(ii) If $N \le N_G(V)$ then, for any $g \in G$, the subgroup $V \cap V^g$ contains the intersection of some Sylow p-subgroup of N and some Sylow p-subgroup of N^g.

Proof: (i) follows from the inclusions $R \cap S \ge V$ and

$$N_{R \cap S}(V) = R \cap S \cap N_G(V) \le R \cap S \cap N =$$

$$= (R \cap N) \cap (S \cap N) = P \cap Q = V.$$

To prove (ii) we can suppose without loss of generality that $N \cap V^g \le P$. Let P_1 be a Sylow p-subgroup of N^g containing $Q \cap N^g$ and Q_1 be a Sylow p-subgroup of N^g such that $P_1 \cap Q_1 = V^g$. Then

$$Q \cap Q_1 = Q \cap (Q_1 \cap N^g) = (Q \cap N^g) \cap Q_1 \le P_1 \cap Q_1 = V^g.$$

Hence $Q \cap Q_1 \le Q \cap (N \cap V^g) \le Q \cap P = V$.   □

2   Sylow intersections in simple groups

By a result of J. Green (Theorem 5 of [4]) a defect group of a p-block of characters is always equal to the intersection of some two Sylow p-subgroups. Since G. Michler [7] and W. Willems [8] proved that any simple group of Lie type has a p-block of defect zero for an arbitrary prime p, Theorem 1 is true for those groups. Using the ATLAS [3] it is easy to check that a sporadic group G has no p-block of defect zero only if

$$p = 2 \text{ and } G \in \{M_{12}, M_{22}, M_{24}, J_2, HS, Suz, Ru, Co_1, Co_3, F_2\}$$

or

$$p = 3 \text{ and } G \in \{Suz, Co_3\}$$

Most of these exceptional cases can be considered using induction and the following lemma.

Lemma 3 *Let M be a maximal p-local subgroup of a finite group G which contains a Sylow p-subgroup of G and let H be a subgroup of M such that*

$N_G(H) \not\leq M$. *Suppose $O_p(M)$ is equal to the intersection of two Sylow p-subgroups of M. Then G contains two Sylow p-subgroups with trivial intersection if one of the following conditions holds:*

(i) $V = \Omega_1(O_p(M))$ is elementary abelian and H acts on V irreducibly by conjugation.

(ii) $V = O_p(M)$ is an extraspecial group of order p^{1+2m}, $\Omega_1(C_V(O^p(H))) \leq \Phi(V)$ and $O^p(H)$ has no non-trivial irreducible representation of degree $\leq m$ over a field of order p.

The ATLAS [3] shows that quartets (p, G, M, H) in Table 1 satisfy the conditions of Lemma 3.

Table 1

p	G	M	H
2	M_{22}	$2^4{:}A_6$	5
2	M_{24}	$2^4{:}A_8$	5
2	J_2	$2^{1+4}{:}A_5$	5
2	HS	$4^3{:}L_3(2)$	7
2	Suz	$2^{1+6}.U_4(2)$	$[3^4]$
3	Suz	$3^5{:}M_{11}$	11
2	Co_1	$2^{1+8}{:}O_8^+(2)$	A_9
2	Co_3	$2S_6(2)$	1
3	Co_3	$3^{1+4}{:}4S_6$	5
2	F_2	$2^{1+22}.Co_2$	$HS{:}2$

If $G = M_{12}$ then $A = \mathrm{Aut}G$ contains an involution $t \notin G$ whose centralizer in A is isomorphic to $\langle t \rangle \times A_5$. By Lemma 2(i) A contains two Sylow 2-subgroups with intersection equal to $\langle t \rangle$. Their intersections with G are Sylow 2-subgroups of G intersecting trivially.

Finally, let $G = Ru$. Denote by v an involution lying in the class $2B$ of G. Then $C_G(v) = V \times U$ where $V \simeq 2^2$, $U \simeq Sz(8)$ and $|N_G(V) : C_G(V)| = 3$. Any involution in U is equal to the square of some element of U, hence it is not a conjugate of v and belongs to the centre of some Sylow 2-subgroup of G. Let X be a subgroup of order 5 in U. Since X centralizes v, its generating elements belong to the class $5A$. The normalizer of X in U is a semi-direct product of X and a cyclic group Z of order 4, and $N_G(X) = XY \times A$, where $Y \simeq Z$, $A \simeq A_5$. It is obvious that $V \leq A \leq C_G(z)$, where z is the involution in Z. Thus, $C_G(z)$ contains A and a Sylow 2-subgroup P of $N_G(V)$, hence $V \cap O_2(C_G(z)) = 1$.

Let Q be a Sylow 2-subgroup of $N_G(V)$ intersecting P in V, let R be a Sylow 2-subgroup of $C_G(v)$ which contains P, let S be a Sylow 2-subgroup of G, which contains Q, and let u be an involution in the centre S. By Lemma

2(i),

$$O_2(C_G(z)) \cap V = O_2(C_G(z)) \cap (R \cap S) \geq O_2(C_G(z)) \cap O_2(C_G(u)).$$

Since $O_2(C_G(z)) \cap V = 1$, Lemma 2(ii) shows that G contains two Sylow 2-subgroups with trivial intersection. Thus, we reduce the proof of Theorem 1 to the case of alternating groups.

Theorem 3 *Let p be a prime.*

(i) Let $G = S_n$ be a symmetric group of degree n. Then G contains two Sylow p-subgroups with trivial intersection if and only if

$$(p,n) \notin \{(3,3),(2,2),(2,4),(2,8)\}.$$

(ii) $i_p(A_n) = 2$ if $n \geq 5$ and p divides $|A_n|$.

Step (ii) is consequence of (i). To prove (i) we use induction and Lemmas 1 and 2. The most complicated cases are $n = 8$ and $n = 16$ for $p = 2$.

Let $n = 8$. Let I be the set $\{(1,2),(3,4),(5,6),(7,8)\}$ of four independent transpositions. Consider a Sylow 2-subgroup P of G contained in $N = N_G(\langle I \rangle)$. We point out that any 2-element of N which induces by conjugation an even permutation on I is contained in $O_2(N)$. Suppose that there exists a Sylow 2-subgroup Q of G such that $P \cap Q = 1$. Since $I \not\subseteq Q$, we can suppose that $(1,3) \in Q$.

If $(2,4) \in Q$ then, up to renumbering of letters, we have $(5,7),(6,8) \in Q$. If $(2,4) \notin Q$, then we can suppose that $(2,5),(4,7),(6,8) \in Q$. Set

$$U = \langle (1,3),(2,4),(5,7),(6,8) \rangle$$

in the first case and

$$U = \langle (1,3),(2,5),(3,7),(6,8) \rangle$$

in the second case.

In the first case $a = (13)(24)(57)(68) \in U$ and a acts by conjugation on I as an even permutation. Thus, $a \in O_2(N_G(V))$, i.e. $a \in P \cap Q$. In the second case $b = (1,8)(2,7)(3,6)(5,4)$ acts by conjugation as an even permutation on I and on $\{(1,3),(2,5),(4,7),(6,8)\}$, hence

$$b \in O_2(N_G(V)) \cap O_2(N_G(U)) \leq P \cap Q.$$

Let $n = 16$, $p = 2$. Set $X = \{1,2,\ldots,6\}$, $Y = \{7,8,\ldots,16\}$.

Let P_X be a Sylow 2-subgroup of the symmetric group S_X normalizing $\langle (1,2),(3,4),(5,6) \rangle$ and centralizing $a = (1,2)(3,4)$ and let Q_X be a Sylow 2-subgroup of S_X normalizing $\langle (2,3),(4,5),(1,6) \rangle$ and centralizing

$b = (2,3)(4,5)$. Then $ab = (1,3,5,4,2)$ does not centralize any non-trivial 2-element of S_X. By Lemma 1, $P_X \cap Q_X = 1$.

Similarly, $P_Y \cap Q_Y = 1$, where P_Y is a Sylow 2-subgroup of S_Y which normalizes the group $\langle (7,8), (9,10), \ldots, (15,16) \rangle$ and centralizes the involution $(7,8)(9,10)(11,12)(13,14)$, and Q_Y is a Sylow 2-subgroup of S_Y which normalizes $\langle (8,9), (10,11), \ldots, (14,15), (7,16) \rangle$ and centralizes the involution $(8,9)(10,11)(12,13)(14,15)$.

Hence $\langle P_X, P_Y \rangle = P_X \times P_Y$ has trivial intersection with $\langle Q_X, Q_Y \rangle = Q_X \times Q_Y$.

It is obvious that $P_X \times P_Y$ is contained in the normalizer of the group $V_P = \langle (1,2), (3,4), \ldots, (15,16) \rangle$, and $Q_X \times Q_Y$ is contained in the normalizer of $V_Q = \langle (1,6), (2,3), (4,5), (7,16), (8,9), (10,11), (12,13), (14,15) \rangle$.

Let P be a Sylow 2-subgroup of $N_G(V_P)$ containing $P_X \times P_Y$ and let Q be a Sylow 2-subgroup of $N_G(V_Q)$ containing $Q_X \times Q_Y$. Then P and Q are Sylow 2-subgroups of G.

We prove now that $P \cap Q = 1$. Let $u \in P \cap Q$. If $y^u \in X$ for some $y \in Y$ then $z^u \in X$ for every $z \in Y$ such that $(y,z) \in V_P \cup V_Q$. Since the graph with vertex set Y, and edge set consisting of those transpositions in $V_P \cup V_Q$ which permute elements of Y, is connected, $Y^u \subseteq X$. This is impossible, since $|Y| > |X|$. Hence $Y^u = Y$ and $X^u = X$ for any $u \in P \cap Q$. Thus,

$$P \cap Q = (P \cap Q) \cap (S_X \times S_Y) =$$

$$= (P \cap (S_X \times S_Y)) \cap (Q \cap (S_X \times S_Y)) =$$

$$= (P_X \times P_Y) \cap (Q_X \times Q_Y) = 1.$$

Theorem 1 implies the following conjecture of R. Brauer [1]:

Corollary 1 *Let G be a finite simple non-abelian group and let p be a prime. If p^a divides $|G|$ then $|G| > p^{2a}$.*

A number-theoretical proof of Brauer's conjecture has been given by W. Kimmerle, R. Lyons, R. Sandling and D. N. Teague [6].

3 Sylow intersections in arbitrary groups

For an arbitrary finite group G the number $i_p(G)$ can be larger than 2 and the corresponding possibilities are described in the following theorem.

Theorem 4 *Let G be a finite group, p be a prime, $S = S(G)$ be the largest soluble normal subgroup of G and P be a Sylow p-subgroup of G. Assume that $P \cap P^x \neq O_p(G)$ for all $x \in G$. Then one of the following conditions holds:*

1. $(P \cap S) \cap (P \cap S)^x \neq O_p(G)$ for all $x \in G$ and, for $\overline{S} = S/O_p(G)$, either

    (a) $p = 2$ and $\overline{S}$ contains a subgroup isomorphic to $q^2{:}D_{2^n+1}$ where $q = 2^n - 1$ is a Mersenne prime, or

    (b) $p = 2$ and $\overline{S}$ contains a subgroup isomorphic to $q^2{:}\frac{1}{2}(4 \times D_{2^n+1})$ where $q = 2^n + 1$ is a Fermat prime, or

    (c) $p = 2^n - 1$ is a Mersenne prime and $\overline{S}$ contains a subgroup isomorphic to $(2^n{:}p) \wr p$, or

    (d) $p = 3$ and $\overline{S}$ contains a subgroup isomorphic to $2^{1+6}_-{:}(3 \wr 3)$.

2. $(P \cap S) \cap (P \cap S)^y = O_p(G)$ for some $y \in G$ and, for $\widetilde{G} = G/S$, either

    (a) $p = 3$ and $\widetilde{G}$ contains a normal subgroup isomorphic to K where $(O_8^+(3))^t < K \leq (O_8^+(3)\langle g \rangle)^t$, t is a natural number and g is a graph automorphism of $O_8^+(3)$ of order 3, or

    (b) $p = 2$ and $\widetilde{G}$ contains a subnormal simple subgroup isomorphic to either

        i. a group of Lie type over a field of order $2^m \pm 1$, or

        ii. one of groups $L_3(4)$, $L_n(2)$, $n \geq 3$, $O_n(2)$, $n \geq 7$, $F_4(2)$, $E_6(2)$, $E_7(2)$, $E_8(2)$, $^2F_4(2)'$.

All the cases described in Theorem 4 actually occur. In particular, $i_3(G) > 2$ for $G = O_8^+(3){:}3$ and $i_2(G) > 2$ if G is equal to $\mathrm{Aut}L_2(9)$, $\mathrm{Aut}L_2(p)$, $p = 2^m - 1$ or $(\mathrm{Aut}L_2(p)) \wr 2)$, $p = 2^m + 1$. Possibly, some groups in 2(b)(i) can be excluded, but nevertheless the following is true:

Theorem 5 *If $p = 2$ or p is a Mersenne prime then there exists a finite group G with $O_p(G) = 1$ such that $|G| < |P|^2$ where P is a Sylow p-subgroup of G. For other primes p and for any finite group G, $i_p(G) \leq 2$, and, in particular, $|G/O_p(G)| > |P/O_p(G)|^2$.*

Let P be a Sylow p-subgroup of a finite group G. Denote by $k_p(G)$ the number of orbits of P on the set of ordered pairs (P_1, P_2) of Sylow p-subgroups of G with the condition $P \cap P_1 \cap P_2 = O_p(G)$. The following theorem shows that $k_p(G) \neq 0$.

Theorem 6 *For a finite group G, one of the following conditions holds:*
(i) $k_p(G) = 1$ and G has an unique Sylow p-subgroup,
(ii) $k_p(G) = 2$, $p = 2$ and G contains exactly three Sylow 2-subgroups,
(iii) $k_p(G) \geq 3$.

Theorem 2 is a direct consequence of Theorem 6.

ACKNOWLEDGEMENTS

This work is supported by the grant 96-01-01893 of the Russian Foundation of Basic Investigations. The work of the first author is partly supported by the London Mathematical Society

References

[1] R. Brauer, On some conjectures concerning finite simple groups, in *Collected papers*, Vol. 2, pp. 171–176. MIT Press, Cambridge, Mass., 1980.

[2] J. Brodkey, A note on finite groups with an abelian Sylow group, *Proc. Amer. Math. Soc.* **14** (1963), 132–133.

[3] J. H. Conway, R. T. Curtis, S. P. Norton, R. A. Parker and R. A. Wilson, *An* ATLAS *of finite groups*, Clarendon Press, Oxford, 1985.

[4] J. Green, Blocks of modular representations, *Math. Z.* **79** (1962), 100–115.

[5] N. Ito, Über den kleinsten p-Durchschnitt auflösbarer Gruppen, *Arch. Math.* **9** (1958), 27–32.

[6] W. Kimmerle, R. Lyons, R. Sandling, D. N. Teague, Composition factors from the group ring and Artin's theorem on orders of simple groups, *Proc. London Math. Soc.* **60** (1990), 89–122.

[7] G. Michler, A finite simple group of Lie-type has p-blocks with different defects, $p \neq 2$, *J. Algebra* **104** (1986), 220–230.

[8] W. Willems, Blocks of defect zero in finite groups of Lie type, *J. Algebra* **113** (1988), 511–522.

[9] V. I. Zenkov, Intersections of nilpotent subgroups in finite groups, *Fundamental'naya i prikladnaya matematika* **1** (1996), 1–91 (In Russian).

[10] V. I. Zenkov and V. D. Mazurov, On the intersection of Sylow subgroups in finite groups, *Algebra and Logic* **35** (1996), 236–240.

Anatomy of the Monster: I

S. P. Norton

Abstract

We summarise much of what is currently known about the subgroup structure of the Fischer–Griess Monster.

Introduction

The publication of the ATLAS of Finite Groups [1] (henceforth referred to as the ATLAS) marked the end of a chapter in the history of group theory in more ways than one. The classification of finite simple groups had been completed; and, using new computational techniques, the last of the sporadic groups had been constructed, and enough progress had been made in classifying their maximal subgroups to make the rest seem like a "tidying up" job.

Since the publication of the ATLAS, these computational methods have been pushed further. The classification of the maximal subgroups for Fi_{22}, Th, Fi_{23}, J_4 and Fi'_{24} was completed by Kleidman, Linton, Parker and Wilson in [7, 8, 9, 10, 12, 13, 14]. This means that Appendix 2 of the ATLAS of Brauer Characters [11], when collated with the ATLAS itself, completes the listing for all sporadic groups except the Baby Monster and Monster. For the Baby Monster, four new maximal subgroups $((S_6 \times L_3(4){:}2){:}2, (S_6 \times S_6){:}4, L_2(31)$ and $L_2(49).2_3)$ were given in Appendix 2 of [11], and Wilson has since found four more $(M_{11}, L_3(3), L_2(17){:}2$ and $L_2(11){:}2)$; for references see [23, 24, 25]. With work on the 2-locals by Meierfrankenfeld and Shpektorov [15], an end to the Baby Monster problem is in sight.

With a lot of work also having been done on maximal subgroups of generic groups, this leaves only the Monster. Unfortunately, the degree of its smallest representation over any field, 196882, is too big for present day computational techniques to make any headway. Nor do we have any alternative means of computing with the Monster.

And yet the Monster is perhaps the most interesting of the sporadic groups. We see from page 231 of the ATLAS that it involves all but six of the sporadic groups (note that the one doubtful case, J_1, was settled in the negative by Wilson [21]) as well as many generic groups; and we shall see in section 2 that many of the relations between these groups can best be seen inside the Monster. Two properties of the Monster of particular interest are "Moonshine" and the "projective plane" formulation, mentioned respectively on pages 231 and 232 of the ATLAS. The former, and to some extent the latter, has spawned a host of papers starting with [2, 3]. Despite the emergence of links with conformal field theory and Kac–Moody algebras, Moonshine is still not fully understood. But it *has* now been proved by Ivanov and the author that the projective plane formulation provides a presentation [5, 6, 17, 18]. Incidentally, [17] has a misprint in the proof of Lemma 5—the element $(02)(39)(02)$ should be replaced by $((02)(39))^2$. With this correction the calculations that use this lemma work out correctly and no other changes to the paper are necessary.

Does this represent the start of the next chapter in the history of group theory? However this is to be answered, we feel that it is appropriate that the Proceedings of the "ATLAS ten years on" conference should contain a summary of what is known about the subgroup structure of the Monster.

Throughout the paper all notation is as in the ATLAS. All results are given without proof. We hope to present some proofs in due course in further papers in the "Anatomy of the Monster" series.

1 p-local subgroups

For odd p, complete lists of maximal p-local subgroups are shown on page 234 of the ATLAS, except for one 7-local (see Ho and Wilson [4, 22]) which had been omitted and is shown in Appendix 2 of [11]. Wilson's paper proves the classification. For $p = 2$, a classification has been obtained by Meierfrankenfeld and Shpektorov [15]; this yields no subgroups other than those shown in the ATLAS.

2   Monstralizers

We start with some results that are stated here for the Monster but which can easily be seen to be true for all groups.

Let G_1 be a subgroup of the Monster **M**. As in [16] we define the *monstralizer* of G_1 to be the centralizer $C_{\mathbf{M}}(G_1) = G_2$, say. Then G_1 is contained in the monstralizer of G_2; if $G_1 = C_{\mathbf{M}}(G_2)$ we call G_1 *closed*, and in any case we call $C_{\mathbf{M}}(G_2)$ the *closure* of G_1 and denote it by $\overline{G_1}$. Note that G_2, and hence $\overline{G_1}$, are automatically closed.

If G_1 is closed, then $Z(G_1) = Z(G_2) = G_1 \cap G_2 = H$, say; and we also have $N_{\mathbf{M}}(G_1) = N_{\mathbf{M}}(G_2) = G$, say. We now define $G_3 = G/\langle G_1, G_2 \rangle$. Then G_3 will be a subgroup of the outer automorphism groups of both G_1 and G_2. To show the pair of mutually centralizing subgroups G_1 and G_2, we may expand G as $[G_1 \circ G_2].G_3$, or, if $H = 1$, as $[G_1 \times G_2].G_3$; we call this expansion a *monstralizer pair*. The use of square brackets will not cause any confusion with the notation for commutators. If G_3 is trivial, we abbreviate this to $G \circ H$ or $G \times H$.

As we said above, all this is universally true. But it is of particular importance for the Monster. Why? Because one finds that most "large" perfect subgroups of M are closed, or have soluble extensions that are closed. This means that large subgroups of the Monster are characterized by their monstralizers, and any containment between subgroups is likely to be "explained" by a containment (the other way round) between their monstralizers. This is what we meant when we said in the introduction that many relations between simple groups could best be seen inside the Monster; and it gives a special significance to the concept of "monstralizer", which, we believe, justifies the use of such a special name.

Table 1 shows all monstralizer pairs where G_2 is the closure of a perfect group which has an element g of prime order p, $p \geq 11$. The enumeration can be done by looking through all subgroups of $C_M(g)$.

We list our monstralizer pairs in decreasing order of $o(g)$, putting G_2's containing $13B$'s before those containing $13A$'s and omitting subsequent occurrences of G_2 with g's of lower order. Within each of these sections, our list is in decreasing order of $|G_1|$, $|G_3|$ and $|G_2|$ (in that order of preference).

3   Orbits on transpositions

There are four proper perfect subgroups of the Monster whose orbit structure on transpositions (i.e. elements of class $2A$) is known. These are $2B$, $2^{1+24}Co_1$, $3Fi'_{24}$ and $2^2.{}^2E_6(2)$. The first and last were given by the author in [16] and the third in [19]. Here we consider the second.

A subgroup of type $2^{1+24}Co_1$ is determined by its central element of class $2B$, which we may call z. The 12 orbits of this group on transpositions are in one-to-one correspondence with the possible conjugacy classes of the product of z with a transposition t. Table 2 shows these orbits: the first column shows the conjugacy class of tz, the second the orbit length, and the third the structure of the orbit stabilizer, which is the centralizer of the group generated by t and z.

Table 1: Monstralizer pairs

$o(g)$	$[G_1 \circ G_2].G_3$	$o(g)$	$[G_1 \circ G_2].G_3$
71 :	$1 \times \mathbf{M}$	11 :	$[M_{12} \times L_2(11)].2$
47 :	$2 \circ 2B$		$S_6.2 \times M_{11}$
31 :	$S_3 \times Th$		$[L_2(11) \times M_{12}].2$
29 :	$[3 \circ 3Fi'_{24}].2$		$[(2 \times S_5) \circ 2M_{22}].2$
23 :	$S_4 \times 2^{11}M_{23}$		$[3^2.2A_4 \times 3^5L_2(11)].2$
	$[A_4 \times 2^{11}M_{24}].2$		$[4^2.2S_3 \times 2^{10}L_2(11)].2$
	$D_8 \circ 2^{1+22}M_{23}$		$[4^2S_3 \times 2^{10}M_{11}].2$
	$S_3 \times Fi_{23}$		$[3^2Q_8 \times 3^5M_{11}].S_3$
	$[2^2 \circ 2^{2+11+22}.M_{24}].S_3$		$[(A_4 \times S_3) \times U_5(2)].2$
	$[2^2 \circ 2^{1+23}Co_2].2$		$[A_5 \times A_{12}].2$
	$[4 \circ 2^{1+23}Co_3].2$		$[4^2.3 \times 2^{10}M_{12}.2].D_8$
	$2 \circ 2^{1+24}Co_1$		$[(2 \times S_4) \circ 2^{11}M_{22}].2$
19 :	$[A_5 \times U_3(8).3_1].2$		$[(2 \times S_4) \circ 2M_{12}].2$
	$[D_{10} \times HN].2$		$[(2 \times 5.4) \circ 2HS].2$
	$[2^2 \circ 2^2.{}^2E_6(2)].S_3$		$[2^{1+4} \circ 2^{1+20}L_2(11)].(2 \times S_3)$
17 :	$[L_2(7) \times S_4(4).2].2$		$[8.2^2 \circ 2.2^{10}M_{11}].2$
	$S_4 \times S_8(2)$		$[3^{1+2} \circ 3^{1+10}L_2(11)].D_8$
	$[7.3 \times He].2$		$[(2^2 \times S_3) \circ 2^2U_6(2)].S_3$
	$[A_4 \times O_{10}^-(2)].2$		$[(3 \times S_3) \times 3^6M_{11}].2$
	$[D_8 \circ 2F_4(2)].2$		$[(2 \times D_8) \circ 2^2.2^{10}.M_{12}].(2^2 \wr 2)$
13B :	$[13.6 \times L_3(3)].2$		$[(2 \times D_8) \circ 2^{1+21}M_{22}].2^2$
	$[2A_4 \circ (2 \times U_3(4)).2].2$		$[4.2^2 \circ 4.2^{20}M_{11}].2^2$
	$[Q_8 \circ 2.2^{12}.U_3(4).2].S_3$		$[8.2 \circ 2.2^{10}M_{12}].2^2$
	$[6 \circ 6Suz].2$		$[3^2 \circ 3^{2+5+10}M_{11}].2S_4$
	$[4 \circ 4.2^{12}(G_2(4).2)].2$		$[3^2 \circ 3^{1+11}U_5(2)].(2 \times S_3)$
	$[3 \circ 3^{1+12}2Suz].2$		$[2^3 \circ 2^3.2^{20}U_6(2)].S_4$
13A :	$3^2.2S_4 \times L_3(3).2$		$[2^3 \circ 2^{3+20+10}M_{22}].S_4$
	$[3^{1+2}.2^2 \times G_2(3)].2$		$[2^3 \circ 2^3.2^{20}M_{12}.2].S_4$
	$[3^2.D_8 \times L_4(3).2_2].2$		$[(2 \times 4) \circ (2 \times 4).2^{20}HS].D_8$
	$[2S_4 \circ (2 \times {}^2F_4(2)')].2$		$[(2 \times 4) \circ (2 \times 4).2^{20}.2M_{12}].D_8$
	$[(S_3 \times S_3) \times O_7(3)].2$		$[Q_8 \circ 2^{1+22}.3^5M_{11}].S_3$
	$S_4 \times {}^3D_4(2).3$		$[D_8 \circ 2^{1+22}McL].2$
	$[3^2.2 \times D_4(3)].S_4$		$S_3 \times 3^6.2M_{12}$
	$[(2 \times S_3) \circ (2 \times Fi_{22})].2$		

Table 2: Orbits of $2^{1+24}.Co_1$ on transpositions

tz	Orbit length	$C_{\mathbf{M}}(\langle t, z \rangle)$
$2A$	$2^4 3^3 5.7.13$	$2^{1+23} Co_2$
$2B$	$2^8 3^4 5^2 7.11.13.23$	$2^{2+8+16} O_8^+(2)$
$4A$	$2^{16} 3^7 5^3 7.13$	$2^{1+22} M_{23}$
$4B$	$2^{21} 3^5 5^3 7.11.13.23$	$(2^7 \times 2^{1+8}) S_6(2)$
$4C$	$2^{20} 3^7 5^3 7.11.13.23$	$2^{1+14+5} A_8$
$6A$	$2^{28} 3^3 5^3 7.13.23$	$2^2.U_6(2).2$
$6C$	$2^{28} 3^4 5^3 7^2 11.13.23$	$2^{3+8}.(3 \times U_4(2)).2$
$6F$	$2^{31} 3^5 5^3 7.11.13.23$	$2^{1+8} A_9$
$8B$	$2^{31} 3^7 5^3 7^2 13.23$	$2.2^{10}.M_{11}$
$10A$	$2^{35} 3^7 5.7.13.23$	$2.HS.2$
$10B$	$2^{32} 3^7 5^2 7^2 11.13.23$	$2^{1+8}(A_5 \times A_5).2$
$12C$	$2^{36} 3^5 5^3 7.11.13.23$	$2 \times S_6(2)$

4 A_5's and associated groups

The Monster has eight conjugacy classes of A_5, as can readily be seen from structure constant calculations. These are shown in Table 3.

Table 3: Classes of A_5 in $\mathbf{M}$

classes	$\overline{G}$	$C_{\mathbf{M}}(G)$	$C_{\mathbf{M}}(D_{10})$	$C_{\mathbf{M}}(A_4)$
$(2A, 3A, 5A)$	A_5	A_{12}	HN	$O_{10}^-(2)$
$(2B, 3A, 5A)$	$2 \times A_5$	$2.M_{22}.2$	$2.HS.2$	$2^{11} M_{24}$
$(2B, 3B, 5A)$	$S_6.2$	M_{11}	$2.HS.2$	$2.M_{12}.2$
$(2A, 3C, 5A)$	A_5	$U_3(8).3$	HN	$U_3(8).3$
$(2B, 3C, 5A)$	$2 \times A_5$	$2^{1+4}(A_4 \times A_5)$	$2^{1+8}(A_5 \times A_5).2$	$2^{1+4}(A_4 \times A_5)$
$(2B, 3C, 5B)$	$5^3.(4 \times A_5)$	D_{10}	$5^3.(4 \times A_5)$	$2^{1+4}(A_4 \times A_5)$
$(2B, 3B, 5B)$	Th	S_3	$5^3.(4 \times A_5)$	$2^{1+6}.3^{1+2}.4$
$(2B, 3B, 5B)$	$2B$	2	$5^3.(4 \times A_5)$	$2^{1+6}.3^{1+2}.4$

Note: In this table G is a group A_5 with conjugacy classes as shown in the first column. The others, in order, show its closure, and the monstralizers of G and its subgroups D_{10} and A_4.

The normalizers of representatives of the last two conjugacy classes of A_5 are $S_5 \times S_3$ and $A_5.4$, so we can distinguish the classes by whether the normalizers split over the centralizers. We now define the *type* of an A_5 to be T (for Thompson) or B (for Baby Monster) in these two cases, and, in the other six cases, as the sequence of letters in the first column: for example, an A_5 with monstralizer M_{11} has type BBA.

The centralizer of each of the first three types of A_5 contains various A_5's of its own, and we can ask what the conjugacy classes of these and the various diagonal A_5's are. The answers are shown in Table 4.

Table 4: Diagonal A_5's

No.	G	Orbits	H	Diagonal	Normalizer	(H, G)
1	AAA	$5 + 1^7$	AAA	BAA	$\frac{1}{2}(S_5 \wr 2 \times S_7)$	1
2	AAA	$6 + 1^6$	AAA	BBA	$\frac{1}{4}(S_5 \wr 2 \times S_6.2)$	2
3	AAA	$5^2 + 1^2$	BAA	T	$S_5 \times S_5 \times 2$	8
4	AAA	$10 + 1^2$	BBA	T	$S_5 \times S_5$	13
5	AAA	$6 + 5 + 1$	BBA	B	$A_5 \times S_5$	12
6	AAA	6^2	BAA	B	$A_5 \times S_5 \times 2$	10
7	AAA	12	AAA	B	$(A_5 \times A_5 \times 2).4$	7
8	BAA	$15 + 5 + 1^2$	AAA	T	$S_5 \times S_5 \times 2$	3
9	BAA	$15 + 5 + 1^2$	BAA	B	$\frac{1}{4}(D_8 \times (S_5 \times S_5))$	11
10	BAA	$10 + 6 + 5 + 1$	AAA	B	$S_5 \times A_5 \times 2$	6
11	BAA	$10 + 6 + 5 + 1$	BAA	B	$\frac{1}{4}(D_8 \times (S_5 \times S_5))$	9
12	BBA	$10 + 1$	AAA	B	$S_5 \times A_5$	5
13	BBA	$6 + 5$	AAA	T	$S_5 \times S_5$	4

Note: in this table the first column labels the classes of ordered pairs (G, H) of commuting A_5's and the last gives the class label for (H, G). The second, fourth and fifth columns show the types of G, H and diagonal A_5's. The third column gives the orbit structure of H in the permutation representation of (the central quotient of) $C_M(G)$ on 12, 22 or 11 points. The sixth column shows the normalizer of $G \times H$. Note that in class 7 the group is $(G \times H){:}8$ while the group in classes 9 and 11 contains $S_5 \times A_5$ and $A_5 \times S_5$ but not $S_5 \times S_5$.

In Table 5 we give a complete list of simple subgroups of the Monster containing $5A$'s, except for A_5's (for which see Table 3). Except for the group $Sz(8)$, which was eliminated by R. A. Wilson by calculating $(2, 4, 5)$ structure constants, these can be classified by working upwards from the A_5's they contain. Note that there is no canonical way of doing this; the containments shown in the table were chosen to be as close as possible to those actually used in the classification.

The first column of Table 5 shows the structure of our group G, together with further identification when there is more than one conjugacy class in the Monster of this isomorphism type; the second column shows a subgroup that can be used to define (and, normally, to prove the uniqueness of) G; the third and fourth columns show the monstralizer and closure of G.

Table 5: Simple subgroups containing $5A$-elements

G	subgroup	$C_{\mathbf{M}}(G)$	$\overline{G}$
$A_6\ (AAA)$	$A_5\ (AAA)$	$\frac{1}{2}(S_6 \wr 2)$	A_6
$A_6\ (BAA)$	$A_5\ (BAA)$	$2.M_{21}.2$	$2 \times A_6$
$A_6\ (BBA)$	$A_5\ (BBA)$	M_{11}	$S_6.2$
$L_2(11)\ (AAA)$	$A_5\ (AAA)$	M_{12}	$L_2(11)$
$L_2(11)\ (BBA)$	$A_5\ (BBA)$	$L_2(11)$	M_{12}
A_7	$A_6\ (AAA)$	$\frac{1}{2}(S_5 \wr 2)$	A_7
$L_2(16)$	$A_5\ (BAA)$	$L_3(2)$	$S_4(4).2$
$L_2(25)$	$A_5\ (BBA)$	$2S_4$	$2 \times {}^2F_4(2)'$
$M_{11}\ (AAA)$	$A_6\ (AAA)$	$S_6.2$	M_{11}
$M_{11}\ (BBA)$	$A_6\ (BBA)$	$L_2(11)$	M_{12}
A_8	A_7	$\frac{1}{2}(S_5 \times S_4)$	A_8
$U_4(2)$	$A_6\ (AAA)$	$(A_4 \times S_3 \times S_3).2$	$U_4(2)$
$U_3(4)$	$A_5\ (BCA)$	$2A_4$	$U_3(4).2$
M_{12}	$M_{11}\ (BBA)$	$L_2(11)$	M_{12}
$U_3(5)$	A_7	$5^2.4.2^2$	$U_3(5)$
A_9	A_8	$\frac{1}{2}(S_5 \times S_3)$	A_9
$S_4(4)$	$L_2(16)$	$L_3(2)$	$S_4(4).2$
$S_6(2)$	$U_4(2)$	$S_4 \times S_3$	$S_6(2)$
A_{10}	A_9	S_5	S_{10}
$L_4(3)$	$U_4(2)$	$3^2.D_8$	$L_4(3).2_2$
$U_5(2)$	$U_4(2)$	$S_3 \times A_4$	$U_5(2)$
${}^2F_4(2)'$	$L_2(25)$	$2S_4$	$2 \times {}^2F_4(2)'$
A_{11}	A_{10}	A_5	A_{12}
$O_8^+(2)$	$S_6(2)$	$\frac{1}{2}(S_4 \times S_3)$	$O_8^+(2)$
$O_8^-(2)$	$S_6(2)$	S_4	$S_8(2)$
A_{12}	A_{11}	A_5	A_{12}
He	$S_4(4)$	7.3	He
$O_7(3)$	$S_6(2)$	$S_3 \times S_3$	$O_7(3)$
$S_8(2)$	$S_4(4)$	S_4	$S_8(2)$
$O_8^+(3)$	$O_7(3)$	$3^2.2$	$O_8^+(3)$
$O_{10}^-(2)$	$S_8(2)$	A_4	$O_{10}^-(2)$
HN	A_{12}	D_{10}	HN
Fi_{23}	$O_8^+(3)$	S_3	Fi_{23}

5 (2, 3, 7) subgroups

As with the case of A_5, we calculate all twelve $(2, 3, 7)$ structure constants. Of these, five are zero, and five of the others are resolved below; the remaining two are $(2B, 3B, 7B)$ and $(2B, 3C, 7B)$, for which the respective structure constants are 61 and $242891/840 = 289\frac{131}{840}$. We have not tried to resolve the latter, but we give a partial resolution of the former below. We denote our $(2, 3, 7)$-group by L.

The six fully or partly resolved cases are as follows:

1. $(2A, 3A, 7A)$. Here the structure constant is $1/2^9 3^2 5^2 17$, and $L \cong L_3(2)$, whose monstralizer is $S_4(4).2$, and which is closed.

2. $(2B, 3A, 7A)$. Here L is either $L_3(2)$ or $2^3 L_3(2)$, accounting respectively for $1/2^9 3^3 5.7$ and $2/2^8 3^3$ of structure constant. The respective monstralizers are $2^2 M_{21}.S_3$ and $2^6.3^{1+2}.2^2$, and both groups are closed.

3. $(2B, 3C, 7A)$. Here the structure constant is $2/12$, and $L \cong 2^3 L_3(2)$, whose monstralizer is A_4, and whose closure is $2^{11} M_{24}$.

4. $(2A, 3C, 7B)$. Here the structure constant is $1/2^3 3^2 7^2$, and $L \cong L_3(2)$, whose monstralizer is $7^2.6A_4$, and which is closed.

5. $(2B, 3B, 7A)$. This case is more difficult. The group L can be $L_3(2)$, $L_2(8)$, $L_2(13)$, $^3D_4(2)$, $2Fi_{22}$, HN or Th. The case $L_3(2)$ accounts for $1/24$ of structure constant; the monstralizer is S_4 and the closure is $S_8(2)$. There are three types of $L_2(8)$ accounting respectively for $1/2160$, $3/48$ and $1/8$ of structure constant. The monstralizers are $3S_6$, $2 \times S_4$ and 2^3, and the closures are $L_2(8).3$, $2^7 S_6(2)$ and $2^{3+20} U_6(2)$. Note that in the second case the outer automorphism of $L_2(8)$ is *not* realized inside the Monster. The case $L_2(13)$ accounts for $3/108$ of structure constant and has monstralizer $3^{1+2}.2^2$ and closure $G_2(3)$. The group $^3D_4(2)$ accounts for $1/24$ of structure constant and has monstralizer S_4 and closure $^3D_4(2).3$. The remaining groups $2Fi_{22}$, HN and Th account for $9/12$, $3/10$ and $3/6$ of structure constant, have monstralizer $2 \times S_3$, D_{10} and S_3, and are closed.

6. $(2B, 3B, 7B)$. Here L closes either to $3Fi'_{24}$ (so that its monstralizer is generated by a $3A$), or to the entire Monster. In the former case L is either $L_2(13)$ (structure constant $3/3$) or $3^7 L_2(13)$ (structure constant $12/3$). This leaves a 56 of structure constant for which L has trivial centralizer. We suspect that some of this can be accounted for by groups of type $L_2(q)$ for $q \equiv \pm 1 \pmod 7$—see next section.

6 Groups of type $L_2(q)$

On page 231 of the ATLAS, there is a list of the simple groups whose order divides that of the Monster, and for each of them it is stated whether it was known to be involved, or not to be involved, in the Monster. The doubtful cases were $L_2(q)$ ($q \in Q = \{19, 27, 29, 31, 41, 49, 59, 71\}$), $Sz(8)$ and J_1. Since then Linton and Wilson have shown that $L_2(19)$, $L_2(31)$ and $L_2(49)$ are involved but that J_1 is not [12, 21, 23]. We can also remove $L_2(41)$ from the list of "possibles"—a proof is outlined below. This leaves just five doubtful cases. In section 4 we mentioned that any 5-element in any $Sz(8)$ must have class $5B$. Here is what is known about the conjugacy class structures of various $L_2(q)$'s that might be contained in the Monster.

We consider the cases where $q \in Q$. By obtaining an upper bound of 17 for the 2-rank of the Monster and Baby Monster one can show that if $L_2(q)$ is involved it must be contained. It is clear that the involutions of $L_2(q)$ must have class $2B$, as the product of any pair of $2A$'s has order at most 6. And it follows from the results of section 4 that the elements of order 5 (which exist except in the case $q = 27$) have class $5B$.

This enables us to show that $L_2(41)$ is not involved in the Monster. Otherwise the 41-normalizer 41:40 in $\mathbf{M}$ would contain elements of class $5B$, hence $40C/D$ and $8D$; but one can see inside the group $3^8 O_8^-(3).2$ that an element of order 8 normalizing a 41 acts regularly on the O_3-subgroup and so inverts some $3A$. The fusion map from $3Fi_{24}$ to $\mathbf{M}$ then yields a contradiction.

Returning to the general problem of $L_2(q)$ for $q \in Q$, one can also show that the elements of order 3 have class $3B$. If $q = 29$ or 71, the 7's have class $7B$ (as $7A$ does not normalize 29 or centralize $5B$), but if $q = 49$ they have class $7A$ (as $3B$ does not properly normalize $7B$). By building up $L_2(49)$ from $L_3(2)$ and 7^2:3 intersecting in 7:3 (with respective monstralizers S_4, $L_3(2)$, $3A_7$) one can show that any such group in the Monster centralizes a $2A$, and hence lies in the double cover of a Baby Monster inside it. As stated earlier, such an $L_2(49)$ is known to exist uniquely.

7 Matrices in the Monster

We refer again to the "projective plane" formulation for the Bimonster $\mathbf{M} \wr 2$, which is described on page 232 of the ATLAS. We know sets of words in the 26 generators which define the subgroups $2^{1+26}.(2^{24}:Co_1)$, $O_8^+(3) \wr 2$ and $2^3.^2 E_6(2)$. In [17] we obtained 26×26 matrices over $GF(2)$ which generate the group $2^{1+24} Co_1$, a double cover of a quotient group of the first of these; their images under the natural quotient map to $2^{24}:Co_1$ are the same as those of the words in the generators of $\mathbf{M} \wr 2$.

We now do the same for the two other groups. The results are shown in

Tables 6 and 7. The matrices of Table 6 are 16×16 orthogonal over $GF(3)$ and generate the natural representation of $(2.O_8^+(3)) \wr 2$, while those of Table 7 are 27×27 unitary over $GF(4)$ and generate $3.^2E_6(2)$.

These results are of considerable use when one wants to calculate within the Monster, as quite often the elements one is studying can be seen within one of the three subgroups mentioned here. It is, therefore, planned to insert these matrices into a library in the GAP system [20].

In both tables the notation for the elements of the Bimonster is as in [1, 17, 18]. We denote the elements of $GF(4)$, other than 0 and 1, by ω and $\overline{\omega}$. In both tables zeroes are denoted by dots for clarity. Several of our matrices are monomial; in such cases, for the sake of brevity, we have written the corresponding elements in "permutation" format with the rows and columns of our matrices indexed as A,B, ... ,P (in Table 6) and A,B, ... ,Z,$ (in Table 7). The "bar" in $(\overline{\alpha\beta})$, where α and β are two of our index letters, means that in the matrix the entries in the α'th row and β'th column, and in the β'th row and α'th column, are not 1, but (in Table 6) both $2 = -1$ or (in Table 7) respectively ω and $\overline{\omega}$.

Table 6: Generators for $(2.O_8^+(3)) \wr 2$

$$e_1 e_2 e_3$$

```
.  .  .  .  .  .  .  .  .  1  2  2  1  1  2  1
.  .  .  .  .  .  .  .  1  .  2  2  1  1  2  1
.  .  .  .  .  .  .  .  1  1  .  1  2  2  2  1
.  .  .  .  .  .  .  .  1  1  1  .  2  2  2  1
.  .  .  .  .  .  .  .  2  2  1  1  .  1  2  1
.  .  .  .  .  .  .  .  2  2  1  1  1  .  2  1
.  .  .  .  .  .  .  .  1  1  1  1  1  1  .  2
.  .  .  .  .  .  .  .  2  2  2  2  2  2  2  .
.  1  1  1  2  2  1  2  .  .  .  .  .  .  .  .
1  .  1  1  2  2  1  2  .  .  .  .  .  .  .  .
2  2  .  1  1  1  1  2  .  .  .  .  .  .  .  .
2  2  1  .  1  1  1  2  .  .  .  .  .  .  .  .
1  1  2  2  .  1  1  2  .  .  .  .  .  .  .  .
1  1  2  2  1  .  1  2  .  .  .  .  .  .  .  .
2  2  2  2  2  2  .  2  .  .  .  .  .  .  .  .
1  1  1  1  1  1  2  .  .  .  .  .  .  .  .  .
```

$$a_1c_1g_1$$

```
.  .  .  .  .  .  .  .  .  1  2  2  1  1  2  2
.  .  .  .  .  .  .  .  1  .  2  2  1  1  2  2
.  .  .  .  .  .  .  .  1  1  .  1  2  2  2  2
.  .  .  .  .  .  .  .  1  1  1  .  2  2  2  2
.  .  .  .  .  .  .  .  2  2  1  1  .  1  2  2
.  .  .  .  .  .  .  .  2  2  1  1  1  .  2  2
.  .  .  .  .  .  .  .  1  1  1  1  1  1  .  1
.  .  .  .  .  .  .  .  1  1  1  1  1  1  1  .
.  1  1  1  2  2  1  1  .  .  .  .  .  .  .  .
1  .  1  1  2  2  1  1  .  .  .  .  .  .  .  .
2  2  .  1  1  1  1  1  .  .  .  .  .  .  .  .
2  2  1  .  1  1  1  1  .  .  .  .  .  .  .  .
1  1  2  2  .  1  1  1  .  .  .  .  .  .  .  .
1  1  2  2  1  .  1  1  .  .  .  .  .  .  .  .
2  2  2  2  2  2  .  1  .  .  .  .  .  .  .  .
2  2  2  2  2  2  1  .  .  .  .  .  .  .  .  .
```

$$a_2c_2g_2$$

```
.  .  .  .  .  .  .  .  .  2  .  .  .  .  .  .
.  .  .  .  .  .  .  .  .  1  1  .  2  .  1  .
.  .  .  .  .  .  .  .  .  2  1  .  1  .  1  .
.  .  .  .  .  .  .  .  .  .  .  2  .  .  .  .
.  .  .  .  .  .  .  .  .  1  2  .  1  .  1  .
.  .  .  .  .  .  .  .  .  .  .  .  .  2  .  .
.  .  .  .  .  .  .  .  .  2  2  .  2  .  1  .
.  .  .  .  .  .  .  .  .  .  .  .  .  .  .  2
2  .  .  .  .  .  .  .  .  .  .  .  .  .  .  .
.  1  2  .  1  .  2  .  .  .  .  .  .  .  .  .
.  1  1  .  2  .  2  .  .  .  .  .  .  .  .  .
.  .  .  2  .  .  .  .  .  .  .  .  .  .  .  .
.  2  1  .  1  .  2  .  .  .  .  .  .  .  .  .
.  .  .  .  .  2  .  .  .  .  .  .  .  .  .  .
.  1  1  .  1  .  1  .  .  .  .  .  .  .  .  .
.  .  .  .  .  .  .  2  .  .  .  .  .  .  .  .
```

$$a_3 c_3 g_3$$

```
. . . . . . . . 1 . 1 . 2 . 1 .
. . . . . . . . . 2 . . . . . .
. . . . . . . . 2 . 1 . 1 . 1 .
. . . . . . . . . . 2 . . . . .
. . . . . . . . 1 . 2 . 1 . 1 .
. . . . . . . . . . . . 2 . . .
. . . . . . . . 2 . 2 . 2 . 1 .
. . . . . . . . . . . . . . . 2
1 . 2 . 1 . 2 . . . . . . . . .
. 2 . . . . . . . . . . . . . .
1 . 1 . 2 . 2 . . . . . . . . .
. . . 2 . . . . . . . . . . . .
2 . 1 . 1 . 2 . . . . . . . . .
. . . . . 2 . . . . . . . . . .
1 . 1 . 1 . 1 . . . . . . . . .
. . . . . . . 2 . . . . . . . .
```

$$a = (\mathrm{AI})(\mathrm{BJ})(\mathrm{CM})(\mathrm{DN})(\mathrm{EK})(\mathrm{FL})(\overline{\mathrm{GO}})(\overline{\mathrm{HP}})$$

$$f = (\mathrm{AI})(\mathrm{BJ})(\overline{\mathrm{CK}})(\overline{\mathrm{DL}})(\mathrm{EO})(\mathrm{FP})(\mathrm{GM})(\mathrm{HN})$$

$$b_1 = (\mathrm{AI})(\mathrm{BJ})(\overline{\mathrm{CK}})(\overline{\mathrm{DL}})(\mathrm{EO})(\overline{\mathrm{FP}})(\mathrm{GM})(\overline{\mathrm{HN}})$$

$$b_2 = (\mathrm{AK})(\mathrm{BL})(\mathrm{CI})(\mathrm{DJ})(\mathrm{EM})(\mathrm{FN})(\overline{\mathrm{GO}})(\overline{\mathrm{HP}})$$

$$b_3 = (\mathrm{AL})(\mathrm{BK})(\mathrm{CJ})(\mathrm{DI})(\mathrm{EM})(\mathrm{FN})(\overline{\mathrm{GO}})(\overline{\mathrm{HP}})$$

$$d_1 d_2 d_3 f_1 f_2 f_3 z_1 z_2 z_3 = (\mathrm{AI})(\mathrm{BJ})(\mathrm{CK})(\mathrm{DL})(\mathrm{EM})(\mathrm{FN})(\mathrm{GO})(\mathrm{HP})$$

S. P. Norton

Table 7: Generators for $3.^2E_6(2)$

$$b_1$$

```
1 1 1 1 1 1 ω . . . . . . . . . w̄ w̄ w̄ w̄ w̄ w̄ w̄ w̄
1 1 1 1 1 1 . ω . . . . . w̄ w̄ w̄ w̄ . . . . w̄ w̄ w̄ w̄
1 1 1 1 1 1 . . ω . . . w̄ . w̄ w̄ w̄ . w̄ w̄ w̄ . . . w̄ w̄ w̄
1 1 1 1 1 1 . . . ω . . w̄ w̄ . w̄ w̄ . w̄ w̄ w̄ . w̄ w̄ . w̄ .
1 1 1 1 1 1 . . . . ω w̄ w̄ w̄ w̄ . w̄ w̄ w̄ . w̄ w̄ . w̄ . w̄
1 1 1 1 1 1 . . . . ω w̄ w̄ w̄ w̄ . w̄ w̄ w̄ . w̄ w̄ . w̄ . .
w̄ . . . . . 1 1 1 1 1 1 . . . . w w w w w w w w w
. w̄ . . . . 1 1 1 1 1 1 w . w w w . w w w . . w w w
. . w̄ . . . 1 1 1 1 1 1 w . w w w . w w w . . w w w
. . . w̄ . . 1 1 1 1 1 1 w w w . w w w . w w . w . w .
. . . . w̄ . 1 1 1 1 1 1 w w w w . w w w . w w . w . w .
. . w w w w . . w̄ w̄ w̄ w̄ 1 1 1 1 1 1 1 1 . . . . .
. w . w w w . w̄ . w̄ w̄ w̄ 1 1 1 1 1 . . . 1 1 1 . .
w w . w w . w w̄ w̄ . w̄ w̄ 1 1 1 1 1 . 1 . 1 . 1 1 .
. w w w . w . . w̄ w̄ w̄ w̄ . 1 1 1 1 . . 1 . 1 . 1 1
. w w w w . w . . w̄ w̄ w̄ w̄ . 1 1 1 1 . . 1 . 1 . 1 1
w . . . w w w w̄ . . w̄ w̄ . w̄ . 1 1 . . 1 1 1 1 1 1 1 1 .
w . w . w w w̄ . w̄ . w̄ w̄ . w̄ 1 . . 1 . 1 1 1 1 1 . 1 .
w . w w . w w̄ . w̄ w̄ . w̄ . w̄ 1 . . 1 . 1 1 1 1 1 . 1 . 1
w . w w w . w̄ . w̄ w̄ w̄ . w̄ 1 . . 1 . 1 1 1 1 1 . 1 . 1
w w . . w w w̄ w̄ . . w̄ w̄ . . w̄ w̄ . 1 1 . 1 . 1 1 1 1 1 1
w w . w . w w̄ w̄ w̄ . . w̄ . w̄ . w̄ . 1 1 . 1 . 1 . 1 1 1 1 1
w w . w w . w̄ w̄ w̄ . w̄ w̄ . . . . 1 1 . 1 . 1 . 1 1 1 1 1 1
w w w . . w w̄ w̄ w̄ w̄ . . w̄ . 1 1 . 1 . 1 1 . 1 1 1
w w w . w . w̄ w̄ w̄ w̄ . w̄ . . 1 . 1 . 1 . 1 1 . 1 1 1 1
w w w w . . w̄ w̄ w̄ w̄ . . . . 1 1 . 1 . 1 . 1 1 1 1 1
```

$$b_2$$

```
1 1 1 1 ω ω ω . . . . . . . . . 1 1 w̄ w̄ 1 w̄ w̄ w̄ w̄ w
1 1 1 1 ω ω . ω . . . . 1 1 w̄ w̄ . 1 w̄ w̄ w̄ w̄ w̄ w
1 1 1 1 ω ω . . ω . . . 1 . 1 w̄ w̄ . 1 w̄ w̄ . . w̄ w̄ w
1 1 1 1 ω ω . . . ω . . 1 1 w̄ 1 . 1 w̄ w̄ w̄ . . w
w̄ w̄ w̄ w̄ 1 1 . . . . ω . w̄ w̄ w̄ w̄ . w w̄ w̄ w̄ . w w̄ . w .
w̄ w̄ w̄ w̄ 1 1 . . . . ω w̄ w̄ w̄ w̄ w . w̄ w̄ w̄ w . w̄ w . w .
w̄ . . . . . 1 1 1 1 ω ω . . . . w̄ w̄ w w w̄ w w w w w 1
. w̄ . . . . 1 1 1 1 ω ω . w̄ w̄ w w w . w̄ w w w . w w w 1
. . w̄ . . . 1 1 1 1 ω ω w̄ . w̄ w w w̄ . w̄ w w w . w w w 1
. . . w̄ . . 1 1 1 1 ω ω w̄ w̄ . w w w w̄ . w̄ w w w . w w w 1
. . . . w̄ . w̄ w̄ w̄ w̄ 1 1 ω ω w w 1 w w w . 1 w w . 1 w . 1 . 1
. . . . . w̄ w̄ w̄ w̄ w̄ 1 1 ω ω w w 1 . w w w 1 . w w 1 . w 1 1 . .
. 1 1 w w . . w w w̄ w̄ 1 1 1 w̄ w̄ 1 1 w̄ w̄ . . 1 w̄ w̄ .
. 1 . 1 w w . w . w w w̄ w̄ 1 1 1 w̄ w̄ 1 . 1 . 1 w̄ w̄ .
. 1 1 . w w . w w . w w w̄ w̄ 1 1 1 w̄ w̄ 1 . 1 . . w̄ w̄
. w w w w . . w̄ w̄ w̄ w̄ 1 w w w w 1 1 . 1 . 1 . 1 . 1 w̄
. w w w w̄ . . . w̄ w̄ w̄ 1 . w w w 1 1 . 1 . 1 . 1 . 1 w̄
1 . . 1 w w w . . w w̄ w̄ 1 1 . 1 1 w̄ w̄ . 1 1 w̄ w̄ . .
1 . 1 . w w w w . w . w w̄ w̄ 1 1 . 1 . 1 1 w̄ w̄ . 1 .
w . w w . w̄ w̄ . w̄ w̄ . 1 w . . . 1 . w w 1 1 . 1 . 1 . w̄
w . w w w . w̄ w̄ . w̄ w̄ 1 . . . 1 w w 1 1 . 1 . 1 . 1 . w̄
1 1 . . w w w w . . w̄ w̄ . 1 1 . 1 . 1 w̄ w̄ w̄ w̄ .
w w . w . w̄ w̄ w̄ w̄ . w̄ . 1 . w . 1 . w . 1 . w 1 1 1 . w̄
w w . w w̄ . w̄ w̄ w̄ w̄ . 1 . w . . 1 w . 1 . w 1 1 1 . w̄
w w w . . w̄ w̄ w̄ w̄ w̄ . 1 . . . w 1 . w 1 . w 1 . 1 1 1 w̄
w w w . w̄ . w̄ w̄ w̄ w̄ . 1 . . . w . 1 . w 1 . w 1 . 1 1 1 w̄
w̄ w̄ w̄ w̄ . . 1 1 1 1 . . . . . . w w . . w w . w w w w 1
```

$$b_3$$

```
1  1  ω  ω  ω  ω  ω  .  .  .  .  .  .  .  .  .  1  1  1  1  ω̄  ω̄  ω̄  ω̄  ω̄  ω̄
1  1  ω  ω  ω  ω  .  ω  .  .  .  .  1  1  1  1  .  .  .  .  ω̄  ω̄  ω̄  ω̄  ω̄  ω̄
ω̄  ω̄  1  1  1  1  .  .  ω  .  .  .  1  .  ω̄  ω̄  ω̄  .  ω̄  ω̄  ω̄  .  .  .  ω  ω  ω
ω̄  ω̄  1  1  1  1  .  .  .  ω  .  .  1  ω̄  .  ω̄  ω̄  .  ω̄  .  ω̄  .  ω  ω  .  .  ω
ω̄  ω̄  1  1  1  1  .  .  .  .  ω  .  1  ω̄  ω̄  .  ω̄  .  ω̄  ω̄  ω̄  .  ω  .  ω  .  ω
ω̄  ω̄  1  1  1  1  .  .  .  .  .  ω  1  ω̄  ω̄  ω̄  .  .  ω̄  ω̄  ω̄  .  ω  ω  .  ω  .
ω̄  .  .  .  .  .  1  1  ω  ω  ω  ω  .  .  .  .  ω̄  ω̄  ω̄  ω̄  ω̄  ω̄  ω  ω  ω  ω  ω  ω
.  ω̄  .  .  .  .  1  1  ω  ω  ω  ω  .  ω̄  ω̄  ω̄  ω̄  .  .  .  .  ω  ω  ω  ω  ω  ω
.  .  ω̄  .  .  .  ω̄  ω̄  1  1  1  1  ω̄  .  ω  ω  ω  .  ω  ω  ω  .  .  1  1  1
.  .  .  ω̄  .  .  ω̄  ω̄  1  1  1  1  ω̄  ω  .  ω  ω  .  ω  ω  ω  .  ω  1  1  .  1
.  .  .  .  ω̄  .  ω̄  ω̄  1  1  1  1  ω̄  ω  ω  .  ω  .  ω  ω  ω  .  ω  1  .  1  .  1
.  .  .  .  .  ω̄  ω̄  ω̄  1  1  1  1  ω̄  ω  ω  ω  .  ω  ω  ω  ω  .  ω̄  ω̄  ω̄  .
.  1  .  ω  ω  ω  .  ω  .  ω̄  ω̄  ω̄  ω  1  1  1  1  .  .  .  ω̄  ω̄  ω̄  .
.  1  ω  .  ω  ω  .  ω  .  ω  ω̄  ω̄  ω̄  ω  1  1  1  1  .  1  .  .  ω̄  .  ω̄  ω̄  .
.  1  ω  ω  .  ω  .  .  ω  ω̄  ω̄  ω̄  ω̄  .  ω̄  ω  1  1  1  1  .  1  .  ω̄  .  ω̄  .  ω̄
.  1  ω  ω  ω  .  .  .  ω  ω̄  ω̄  ω̄  ω̄  .  ω  1  1  1  1  .  1  .  ω̄  .  ω̄  ω̄
1  .  .  ω  ω  ω  ω  .  ω̄  ω̄  ω̄  ω̄  ω  1  .  .  1  1  1  1  ω̄  ω̄  ω̄  .
1  .  ω  .  ω  ω  ω  .  ω̄  .  ω̄  ω̄  ω̄  ω  .  1  .  1  1  1  1  ω̄  .  .  ω̄  ω̄  .
1  .  ω  ω  .  ω  ω  .  ω̄  ω̄  .  ω̄  ω̄  ω  .  .  1  .  1  1  1  1  .  ω̄  .  ω̄  .  ω̄
1  .  ω  ω  ω  .  ω  .  ω̄  ω̄  ω̄  .  ω̄  ω  .  .  .  1  1  1  1  .  ω̄  .  ω̄  .  ω̄
ω  ω  .  .  ω̄  ω̄  ω̄  ω̄  .  .  1  1  .  ω  ω  .  .  ω  ω  .  1  1  1  1  .
ω  ω  .  ω̄  .  ω̄  ω̄  ω̄  .  1  .  1  .  ω  .  ω  .  ω  .  ω  1  1  1  1  .  1
ω  ω  .  ω̄  ω̄  .  ω̄  ω̄  .  1  1  .  .  ω  .  .  ω  ω  .  ω  1  1  1  .  1  1
ω  ω  ω̄  .  .  ω̄  ω̄  ω̄  1  .  1  .  .  ω  ω  .  .  ω  ω  .  1  1  .  1  1  1
ω  ω  ω̄  .  ω̄  .  ω̄  ω̄  1  1  .  .  .  ω  .  ω  .  ω  .  ω  1  .  1  1  1  1
ω  ω  ω̄  ω̄  .  .  ω̄  ω̄  1  1  .  .  .  .  ω  ω  .  ω  ω  .  1  1  1  1  1
```

$$d_1$$

```
.  1  1  1  1  1  .  .  .  .  .  .  .  .  .  .  .  .  .  .  .  .  .  .  .  .  .  .  .
1  .  1  1  1  1  .  .  .  .  .  .  .  .  .  .  .  .  .  .  .  .  .  .  .  .  .  .  .
1  1  .  1  1  1  .  .  .  .  .  .  .  .  .  .  .  .  .  .  .  .  .  .  .  .  .  .  .
1  1  1  .  1  1  .  .  .  .  .  .  .  .  .  .  .  .  .  .  .  .  .  .  .  .  .  .  .
1  1  1  1  .  1  .  .  .  .  .  .  .  .  .  .  .  .  .  .  .  .  .  .  .  .  .  .  .
1  1  1  1  1  .  .  .  .  .  .  .  .  .  .  .  .  .  .  .  .  .  .  .  .  .  .  .  .
.  .  .  .  .  .  1  1  1  1  1  .  .  .  .  .  .  .  .  .  .  .  .  .  .  .  .  .  .
.  .  .  .  .  .  1  .  1  1  1  1  .  .  .  .  .  .  .  .  .  .  .  .  .  .  .  .  .
.  .  .  .  .  .  1  1  .  1  1  1  .  .  .  .  .  .  .  .  .  .  .  .  .  .  .  .  .
.  .  .  .  .  .  1  1  1  .  1  1  .  .  .  .  .  .  .  .  .  .  .  .  .  .  .  .  .
.  .  .  .  .  .  1  1  1  1  .  1  .  .  .  .  .  .  .  .  .  .  .  .  .  .  .  .  .
.  .  .  .  .  .  1  1  1  1  1  .  .  .  .  .  .  .  .  .  .  .  .  .  .  .  .  .  .
.  .  .  .  .  .  .  .  .  .  .  .  1  1  1  1  1  1  1  1  .  .  .  .  .  .  .  .  .
.  .  .  .  .  .  .  .  .  .  .  .  1  1  1  1  1  1  .  .  .  1  1  .  1  1  .
.  .  .  .  .  .  .  .  .  .  .  .  1  1  1  1  1  .  1  .  .  1  .  1  .  1  1  .
.  .  .  .  .  .  .  .  .  .  .  .  1  1  1  1  .  1  .  1  .  1  1  1  1  .  1
.  .  .  .  .  .  .  .  .  .  .  .  1  1  .  1  .  .  1  1  1  1  1  .  1  1  .
.  .  .  .  .  .  .  .  .  .  .  .  1  .  1  .  .  1  1  1  1  1  .  1  .  1  .  1
.  .  .  .  .  .  .  .  .  .  .  .  1  .  .  1  1  1  1  1  .  1  .  .  1  .  1
.  .  .  .  .  .  .  .  .  .  .  .  1  1  .  .  1  1  .  1  1  1  1  .  .  1
.  .  .  .  .  .  .  .  .  .  .  .  1  .  1  .  1  .  1  .  1  1  1  1  1  .  .  1
.  .  .  .  .  .  .  .  .  .  .  .  1  .  .  1  .  1  1  1  1  1  1  .  .  1
.  .  .  .  .  .  .  .  .  .  .  .  .  1  1  .  .  1  1  1  1  .  1  1  1
.  .  .  .  .  .  .  .  .  .  .  .  1  .  1  .  1  .  1  1  .  1  1  1  1  1
.  .  .  .  .  .  .  .  .  .  .  .  1  1  .  .  1  1  1  1  .  1  1  1  1  1
```

$$d_2$$

```
.   1   1   1   ω   ω   .   .   .   .   .   .   .   .   .   .   .   .   .   .   .   .   .   .
1   .   1   1   ω   ω   .   .   .   .   .   .   .   .   .   .   .   .   .   .   .   .   .   .
1   1   .   1   ω   ω   .   .   .   .   .   .   .   .   .   .   .   .   .   .   .   .   .   .
1   1   1   .   ω   ω   .   .   .   .   .   .   .   .   .   .   .   .   .   .   .   .   .   .
ω̄   ω̄   ω̄   ω̄   .   1   .   .   .   .   .   .   .   .   .   .   .   .   .   .   .   .   .   .
ω̄   ω̄   ω̄   ω̄   1   .   .   .   .   .   .   .   .   .   .   .   .   .   .   .   .   .   .   .
.   .   .   .   .   .   .   1   1   1   ω   ω   .   .   .   .   .   .   .   .   .   .   .   .
.   .   .   .   .   .   1   .   1   1   ω   ω   .   .   .   .   .   .   .   .   .   .   .   .
.   .   .   .   .   .   1   1   .   1   ω   ω   .   .   .   .   .   .   .   .   .   .   .   .
.   .   .   .   .   .   1   1   1   .   ω   ω   .   .   .   .   .   .   .   .   .   .   .   .
.   .   .   .   .   .   ω̄   ω̄   ω̄   ω̄   .   1   .   .   .   .   .   .   .   .   .   .   .   .
.   .   .   .   .   .   ω̄   ω̄   ω̄   ω̄   1   .   .   .   .   .   .   .   .   .   .   .   .   .
.   .   .   .   .   .   .   .   .   .   .   .   1   1   1   ω̄   ω̄   1   1   ω̄   ω̄   .   .   .
.   .   .   .   .   .   .   .   .   .   .   .   1   1   1   ω̄   ω̄   1   .   1   .   1   ω̄   ω̄
.   .   .   .   .   .   .   .   .   .   .   .   1   1   1   ω̄   ω̄   1   .   1   .   1   .   ω̄
.   .   .   .   .   .   .   .   .   .   .   .   ω   ω   ω   1   1   .   1   .   1   .   1   ω̄
.   .   .   .   .   .   .   .   .   .   .   .   ω   ω   ω   1   1   .   1   .   1   .   1   ω̄
.   .   .   .   .   .   .   .   .   .   .   .   1   1   .   1   1   1   ω̄   ω̄   1   .   ω̄   ω̄
.   .   .   .   .   .   .   .   .   .   .   .   1   .   1   1   1   ω̄   ω̄   1   .   ω̄   ω̄   .
.   .   .   .   .   .   .   .   .   .   .   .   ω   .   .   1   ω   ω   1   1   .   1   .   1   ω̄
.   .   .   .   .   .   .   .   .   .   .   .   ω   .   .   1   ω   ω   1   1   .   1   .   1   ω̄
.   .   .   .   .   .   .   .   .   .   .   .   .   1   1   .   1   1   .   1   ω̄   ω̄   ω̄   ω̄
.   .   .   .   .   .   .   .   .   .   .   .   ω   .   1   .   ω   .   1   ω   1   1   1   ω̄
.   .   .   .   .   .   .   .   .   .   .   .   ω   .   1   ω   .   1   ω   1   1   1   ω̄
.   .   .   .   .   .   .   .   .   .   .   .   ω   1   .   ω   1   .   ω   1   .   1   1   ω̄
.   .   .   .   .   .   .   .   .   .   .   .   ω   1   .   ω   1   .   ω   1   ω   1   1   ω̄
.   .   .   .   .   .   .   .   .   .   .   .   ω   ω   .   .   ω   ω   .   ω   ω   ω   ω   1
```

$$d_3$$

```
.   1   ω   ω   ω   ω   .   .   .   .   .   .   .   .   .   .   .   .   .   .   .   .   .   .
1   .   ω   ω   ω   ω   .   .   .   .   .   .   .   .   .   .   .   .   .   .   .   .   .   .
ω̄   ω̄   .   1   1   1   .   .   .   .   .   .   .   .   .   .   .   .   .   .   .   .   .   .
ω̄   ω̄   1   .   1   1   .   .   .   .   .   .   .   .   .   .   .   .   .   .   .   .   .   .
ω̄   ω̄   1   1   .   1   .   .   .   .   .   .   .   .   .   .   .   .   .   .   .   .   .   .
ω̄   ω̄   1   1   1   .   .   .   .   .   .   .   .   .   .   .   .   .   .   .   .   .   .   .
.   .   .   .   .   .   .   1   ω   ω   ω   ω   .   .   .   .   .   .   .   .   .   .   .   .
.   .   .   .   .   .   1   .   ω   ω   ω   ω   .   .   .   .   .   .   .   .   .   .   .   .
.   .   .   .   .   .   ω̄   ω̄   .   1   1   1   .   .   .   .   .   .   .   .   .   .   .   .
.   .   .   .   .   .   ω̄   ω̄   1   .   1   1   .   .   .   .   .   .   .   .   .   .   .   .
.   .   .   .   .   .   ω̄   ω̄   1   1   .   1   .   .   .   .   .   .   .   .   .   .   .   .
.   .   .   .   .   .   ω̄   ω̄   1   1   1   .   .   .   .   .   .   .   .   .   .   .   .   .
.   .   .   .   .   .   .   .   .   .   .   .   1   ω̄   ω̄   ω̄   ω̄   ω̄   ω̄   ω̄   ω̄   .   .   .
.   .   .   .   .   .   .   .   .   .   .   .   ω   1   1   1   1   1   .   .   ω̄   ω̄   ω̄
.   .   .   .   .   .   .   .   .   .   .   .   ω   1   1   1   1   .   1   .   ω̄   .   .   ω̄   ω̄
.   .   .   .   .   .   .   .   .   .   .   .   ω   1   1   1   1   .   1   .   ω̄   .   ω̄   ω̄
.   .   .   .   .   .   .   .   .   .   .   .   ω   1   1   1   1   .   .   1   .   ω̄   ω̄   ω̄
.   .   .   .   .   .   .   .   .   .   .   .   ω   1   .   .   1   1   1   1   ω̄   ω̄   ω̄
.   .   .   .   .   .   .   .   .   .   .   .   ω   .   1   .   1   1   1   1   ω̄   .   ω̄   ω̄
.   .   .   .   .   .   .   .   .   .   .   .   ω   .   .   1   1   1   1   1   .   ω̄   .   ω̄   ω̄
.   .   .   .   .   .   .   .   .   .   .   .   ω   ω   .   .   1   1   1   1   1   .   ω̄   .   ω̄   ω̄
.   .   .   .   .   .   .   .   .   .   .   .   .   ω   ω   .   .   ω   ω   .   .   1   1   1   1   1
.   .   .   .   .   .   .   .   .   .   .   .   .   ω   .   ω   .   ω   .   ω   .   1   1   1   1   .   1
.   .   .   .   .   .   .   .   .   .   .   .   .   ω   .   .   ω   ω   .   .   ω   ω   .   1   1   .   1   1
.   .   .   .   .   .   .   .   .   .   .   .   .   ω   ω   .   ω   .   ω   .   ω   1   .   1   1   1   1
.   .   .   .   .   .   .   .   .   .   .   .   .   ω   ω   .   .   ω   ω   .   .   ω   ω   1   1   1   1   1
```

$$a = (AG)(BH)(CI)(DJ)(EK)(FL)$$
$$c_1 = (\overline{AF})(\overline{GL})(\overline{UM})(\overline{XN})(\overline{ZO})(\overline{\$P})$$
$$c_2 = (DE)(JK)(OP)(ST)(VW)(Z\$)$$
$$c_3 = (BC)(HI)(MN)(SV)(TW)(UX)$$

References

[1] J. H. Conway, R. T. Curtis, S. P. Norton, R. A. Parker and R. A. Wilson, *An* ATLAS *of finite groups*, Oxford University Press, 1985.

[2] J. H. Conway and S. P. Norton, Monstrous moonshine, *Bull. London Math. Soc.* **11** (1979), 308–339.

[3] J. H. Conway, S. P. Norton and L. Soicher, The Bimonster, the group Y_{555}, and the projective plane of order 3, in *Computers in Algebra (ed. M. C. Tangora)*, pp. 27–50. Lecture notes in Pure and Applied Math. **111**, Marcel Dekker, 1988.

[4] C. Y. Ho, A new 7-local subgroup of the Monster, *J. Algebra* **115** (2) (1988), 513–520.

[5] A. A. Ivanov, Geometric presentations of groups with an application to the Monster, in *Proc. ICM Kyoto 1990 (ed. I. Satake)*, Vol. 2, pp. 385–395. Springer-Verlag, 1991.

[6] A. A. Ivanov, A geometric characterization of the Monster, in *Groups, combinatorics and geometry (eds. M. W. Liebeck and J. Saxl)*, pp. 46–62. LMS Lecture Note Series 165, Cambridge University Press, 1992.

[7] P. B. Kleidman and R. A. Wilson, The maximal subgroups of Fi_{22}, *Math. Proc. Cambridge Philos. Soc.* **102** (1987), 17–23.

[8] P. B. Kleidman and R. A. Wilson, Corrigendum: "The maximal subgroups of Fi_{22}", *Math. Proc. Cambridge Philos. Soc.* **103** (1988), 383.

[9] P. B. Kleidman and R. A. Wilson, The maximal subgroups of J_4, *Proc. London Math. Soc. (3)* **56** (1988), 484–510.

[10] P. B. Kleidman, R. A. Parker and R. A. Wilson, The maximal subgroups of the Fischer group Fi_{23}, *J. London Math. Soc. (2)* **39** (1989), 89–101.

[11] C. Jansen, K. Lux, R. A. Parker and R. A. Wilson, *An* ATLAS *of Brauer characters*, Clarendon Press, Oxford, 1995.

[12] S. A. Linton, The maximal subgroups of the Thompson group, *J. London Math. Soc. (2)* **39** (1989), 79–88.

[13] S. A. Linton, Correction to "The maximal subgroups of the Thompson group", *J. London Math. Soc. (2)* **43** (1991), 253–254.

[14] S. A. Linton and R. A. Wilson, The maximal subgroups of the Fischer groups Fi_{24} and Fi'_{24}, *Proc. London Math. Soc. (3)* **63** (1991), 113–164.

[15] U. Meierfrankenfeld and S. K. Shpektorov, The maximal 2-local subgroups of the Monster, Preprint.

[16] S. P. Norton, The uniqueness of the Fischer–Griess Monster, in *Finite groups—coming of age (ed. J. McKay)*, pp. 271–285. Contemp. Math. Series **45**, AMS, 1985.

[17] S. P. Norton, Presenting the Monster?, *Bull. Soc. Math. Belg. (A)* **42** (1990), 595–605.

[18] S. P. Norton, Constructing the Monster, in *Groups, combinatorics and geometry (eds. M. W. Liebeck and J. Saxl)*, pp. 63–76. LMS Lecture Note Series 165, Cambridge University Press, 1992.

[19] S. P. Norton, The Monster algebra: some new formulae, in *Moonshine, the Monster, and related topics (eds. C. Dong and G. Mason)*, pp. 433–441. Contemp. Math. Series **193**, AMS, 1996.

[20] M. Schönert *et al.*, GAP—*Groups, Algorithms, and Programming*, Lehrstuhl D für Mathematik, RWTH Aachen, 1995.

[21] R. A. Wilson, Is J_1 a subgroup of the Monster?, *Bull. London Math. Soc.* **18** (1986), 349–350.

[22] R. A. Wilson, The odd-local subgroups of the Monster, *J. Austral. Math. Soc. Series A* **44** (1988), 1–16.

[23] R. A. Wilson, Some new subgroups of the Baby Monster, *Bull. London Math. Soc.* **25** (1993), 23–28.

[24] R. A. Wilson, More on maximal subgroups of the Baby Monster, *Arch. Math. (Basel)* **61** (1993), 497–507.

[25] R. A. Wilson, The symmetric genus of the Baby Monster, *Quart. J. Math. (Oxford) (2)* **44** (1993), 513–516.

An integral meataxe

Richard A. Parker

Abstract

We describe algorithms for working with matrices of integers, performing most of the Meataxe functions on them. In particular, we can chop integral representations of groups into irreducibles, and prove irreducibility.

1 Introduction

This paper describes some algorithms sufficient to construct and work with representations of finite groups over the ordinary integers. The methods are, however, of quite general use when working with matrices of integers.

In general, the algorithms and programs closely match those used for working with matrices over finite fields, but several new problems arise.

The first problem is to find an effective replacement for Gaussian elimination—the workhorse method for finite fields. The replacement described here, called "The Module Handler", although of restricted functionality, seems to do the job, and much of the work of this paper consists of doing such jobs as "Nullspace" and "Find a $\mathbb{Z}$-basis" using only this functionality.

Another problem is to work with integers of unbounded size, and yet to prevent the size growing too much. I have some hopes that the methods described here are quite good in this respect, but have made no serious effort to analyse their behaviour.

A third "problem" that I expected is that matrices of non-zero nullity are vanishingly rare in characteristic zero. Astonishingly, in my work so far I have not encountered this problem at all. We have

Research Problem 1 *Understand why "small" combinations of group elements often have non-zero nullity.*

In section 2, the concept of a Module Handler is defined and introduced. In section 3, an implementation based on p-adic lift is described. In section 4, the main programs that use the module handler are described.

In section 5, the general method for working with representations of finite groups over $\mathbb{Z}$ is described, introducing some simple programs on the way. In section 6, the use of the fixed quadratic form is discussed. In section 7, rings other than $\mathbb{Z}$ are briefly discussed. In section 8, some thoughts on LLL are presented.

2   Free $\mathbb{Z}$-module handler

2.1   General Description

A module handler, as defined in this paper, is a set of software routines to hold an ordered basis for a free $\mathbb{Z}$-module, and to perform certain operations with that basis. The main operation is to express any given module element that is in the $\mathbb{Z}$-span of the basis in terms of that basis. Furthermore if the given element is not in the $\mathbb{Z}$-span, but some multiple of the element *is* in the $\mathbb{Z}$-span, this multiple must be found by the module handler. In practice, the multiple can often run to hundreds of digits, but the module handler described in section 3 can handle this situation readily.

A feature of the module handler is that the basis for the module is not changed by the module handler itself, a feature that is relied on by several of the algorithms. We might alternatively define (say) a "loose module handler" as a module handler that is allowed to change the basis for the $\mathbb{Z}$-module. If no subset of the given elements is a basis (a situation that often arises in practice) the loose module handler can adjust the generating set so that it is a basis, whereas the strict module handler described here can only report what linear combinations have arisen, and leave the basis change to the calling routine. In particular, therefore, the module handler itself cannot make entries grow, as it does not change them at all.

2.2   The functions of the module handler

The module handler is required to perform three functions on an ordered set of d linearly independent module elements v_i.

1. Take a new element w and

   (a) if w and v_i are linearly independent, append w to the given basis, increasing the rank by 1;

   (b) if w and v_i are linearly dependent, find the least positive integer N, and integers c_i, such that

$$N.w = \sum_{i=1}^{d} c_i.v_i.$$

2. Return the ith basis element v_i.

3. Drop the last module basis element, reducing the rank d by 1.

All algorithms start with the module handler having a module of rank d = 0.

3　A module handler using p-adic lift

How can the functions described in section 2.2 be performed?

There is no problem with the second and third functions—hand back an element and throw one away—which are essentially administrative in nature. The first one, however, requires a little thought.

In the p-adic implementation, the module handler holds a prime (in my own programs this was invariably 1949) modulo which the given basis elements are linearly independent. It can sometimes happen, of course, that the v_i and w (the new vector) are independent over $\mathbb{Z}$, but dependent modulo p. In this case a new prime must be chosen. I suspect that there is an algorithm to handle this case without changing the prime, but instead remembering which powers of p were divided by to get the independent elements—we have

Research Problem 2 *Find an algorithm that operates modulo a fixed prime even if the module has lower rank modulo that prime.*

Also held is the matrix V of module elements v_i.

To perform the calculations, two further matrices, this time of numbers mod p, must be held, referred to here as A and B. The matrices are such that

1. A is in "semi-echelon form"—the first entry of each row is 1, and below it are only zeros.

2. $A \equiv B.V \pmod{p}$

This structure enables us to find $\overline{c_i} \pmod{p}$ such that

$$w \equiv \sum_{i=1}^{d} \overline{c_i} v_i \pmod{p}$$

(if such $\overline{c_i}$ exist).

If such $\overline{c_i}$ do not exist, w can be appended to the module basis.

Otherwise we "lift" the $\overline{c_i}$ to c_i in $\mathbb{Z}$, choosing for each $\overline{c_i}$ the integer c_i nearest to zero that is congruent to $\overline{c_i} \pmod{p}$, and set

$$w_1 = (w - \sum_{i=1}^{d} c_i v_i)/p.$$

If w_1 is zero, we have found $w = \sum_{i=1}^{d} c_i v_i$ and we are done. Otherwise we proceed much as before to express w_1 (deriving w_2), and then to express w_2, and so on.

If at any stage other than the first, we are unable to express w_j (mod p) in terms of the v_i, we know our prime p is "bad" and must be discarded. We must start the process off again with a new prime.

Another possibility is that at some stage we get some $w_j = 0$. In this case w was in the Z-span of the module basis and we have found the linear combination required.

The hardest possibility is that the process goes on "indefinitely". We must therefore set a maximum number of loops—$3K$, say—and if the process loops $3K$ times, we switch to assuming that w is in the Q-span but not in the Z-span of the elements v_i, and that K is large enough that N and all the $|c_i|$ are less than p^K, and try to find the smallest positive integer N such that $N.w$ is in the Z-span of the v_i.

By keeping track (in e_i, say) of what happened during the looping, we know that

$$w \equiv \sum_{i=1}^{d} e_i v_i \quad (\text{mod } p^{3K})$$

and we seek N such that

$$N.w = \sum_{i=1}^{d} c_i v_i.$$

Hence, for each i, we have $N.e_i = c_i + p^{3K}.s_i$ for some integer s_i, so $e_i/p^{3K} = s_i/N + c_i/N.p^{3K}$. As the second term is very small, we have that s_i/N is a good approximation to e_i/p^{3K}.

It is useful to recall a lemma, true for any real number x and positive integer t, that x has at most one rational approximation a/b with $b < t$ and $(x - \frac{a}{b})^2 < 1/(4t^2)$. Applying the lemma with $t = p^K$ to e_i/p^{3K} we get that there is at most one rational a_i/b_i with $(e_i/p^{3K} - a_i/b_i)^2 < 1/(4p^{2K})$, and $b_i < p^K$.

Using continued fractions, we find an a_i/b_i that is a good enough approximation to e_i/p^{3K}, and therefore $a_i/b_i = s_i/N$, since s_i/N is the only such approximation. Hence

$$\frac{e_i}{p^{3K}} = \frac{a_i}{b_i} + \frac{c_i}{N.p^{3K}}$$

which implies $b_i e_i = a_i p^{3K} + c_i.b_i/N$, so $c_i b_i/N$ must be an integer. Hence, if $n = lcm(b_i)$, then $n.w$ is an integral linear combination of the v_i, so by minimality, $N = n$.

As we do not in fact know whether p^{3K} actually was large enough, we calculate $n = lcm(b_i)$ and attempt to express $n.w$ in terms of the v_i, for $3K$ loops. If we succeed, we have found N as desired. If we fail, it is because p^{3K} is not sufficient accuracy to determine N, so we increase K (triple it, say) and return to the looping process.

4 Algorithms that use the module handler

4.1 General Description

In this section we describe five rather similar programs that use the module handler. These programs are

- Nullspace (NS). Given a matrix M of integers, find a set of independent vectors spanning (over $\mathbb{Q}$) the space of all vectors v such that $v.M = 0$.

- Expression (CL). (This program was originally called clean by Thackray and myself many years ago). Given two matrices A and B, where A is of full row rank (all the rows are linearly independent) find a matrix X such that $B = X.A$ (if such a matrix X exists).

- Find a basis (FBA). Given a matrix A, find a matrix B whose rows are linearly independent, and which span the same $\mathbb{Z}$-module as the rows of A.

- Spin (or submodule SM). Given a matrix A and some matrices G_i, find a free $\mathbb{Z}$-basis for the smallest $\mathbb{Z}$-module containing the rows of A, and invariant under the action of all the matrices G_i. [Note—this is done in a "standard basis" way as described later].

- Purify (PU). Given a matrix X of full row rank, find a free basis for the (possibly larger) module of all integral rows in the $\mathbb{Q}$-span of the rows of X.

4.2 Nullspace (NS)

This is a simple program once a module handler is available. The given input matrix is processed one row at a time, and each row is given to the module handler, keeping track of which rows were accepted as linearly independent. If a row is not linearly independent of the previous ones, the module handler returns an N, such that $N.w = \sum_{i=1}^{d} c_i.v_i$.

By keeping track of the relationship between the input matrix and the stored rows (a subset) the linear dependence is converted to a row of the nullspace and written out.

Notice that no attempt is made by this program to find a $\mathbb{Z}$-basis for the whole nullspace. This can be obtained—if needed—by running the purify program described in section 4.6.

4.3 Expression, or clean (CL)

Given two matrices, A and B, where the rows of A are linearly independent, and the rows of B are in the $\mathbb{Z}$-span of the rows of A, find a matrix X such that $B = X.A$.

This is another simple application of the module handler. Each row of A is first given to the module handler. If a linearly dependency arises, this is, of course, an error condition.

Each row of B is then given to the module handler, and the expected result is that this row is $\sum_i c_i.v_i$ as usual, so the row c_i is written to the output file. If any other result occurs, this is again an error condition.

4.4 Free Basis (FBA)

Given a matrix A, find a matrix B whose rows are a basis for the module spanned by the rows of A. In other words, find a matrix B whose rows are linearly independent, and span the same module as that spanned by the rows of A.

This is done by presenting each row of the matrix A in turn to the module handler. If this row is independent it is appended to the module, and there is nothing more to do. If this row is in the $\mathbb{Z}$-span of the previous rows, the new row may simply be discarded and again there is nothing more to do.

The only remaining (and the hardest) case is where the new row w satisfies

$$N.w = \sum_{i=1}^{d} c_i.v_i$$

with $N > 1$.

In this case, we first look at the $GCD(N, c_i)$ for the various values of i. We need to pick a j where $GCD(N, c_j) < N$. If there is no such j, this is an error condition. See the end of this section for a discussion of what to do if there is more than one possible j.

Once a j has been selected, we calculate $GCD(N, c_j) = g = a.N + b.c_j$. For any set S of vectors, let $\langle S \rangle$ denote the $\mathbb{Z}$-span of S. Let $U = \langle v_i | i \neq j \rangle$.

We set $v'_j = b.w + a.v_j - [(b.c_i)/N]v_i$ where $[x]$ means the nearest integer to x. We also have that $GCD(a, b) = 1 = e.a + f.b$, so we set $w' = e.w - f.v_j$. We then have that $v_j = -b.w' + e.v'_j$ and $w = a.w' + f.v'_j$ modulo U so the new vectors span the same module as the old ones.

We claim that

$$\langle U, v_j, w \rangle = \langle U, v'_j, w' \rangle = H \text{ (say)}$$

and that $|H/\langle U, v'_j \rangle| = |H/\langle U, v_j \rangle|.g/N$ is less than N.

(Refer to research problem 9 for an admission that I am not satisfied with the treatment given here!)

Finally we return to the question of which j should be used. In my own implementation, the largest numerical value of j was used—the latest one in the module. This has two advantages. Firstly, it ensures that the $\mathbb{Q}$-span of the first k independent vectors is the same before the run as after, a feature that is used in the "purify" program for example. Secondly, looking ahead to "spin", it means that the disruption is carried a minimum distance back into the module, forcing as few rows as possible to be remultiplied by the generators. This is, of course only an advantage because of the poor strategy I use to remember what vectors have been multipled by the generators.

4.5 Spin

Given a matrix A, and some matrices G_i, find a free $\mathbb{Z}$-basis for the smallest $\mathbb{Z}$-module containing the rows of A, and invariant under the action of all the matrices G_i.

This is very similar to the FBA program described in the previous section, except that once the input vectors of A are processed, further vectors are obtained by multiplying the rows of the basis already obtained by the generators G_i to obtain further rows, which are included in the module until it is stable.

The interesting question here is how to keep track of which elements of the basis have been multiplied by which generators, bearing in mind that the inclusion of one single new row can (in principle) change every row of the basis, thus requiring the whole multiplication process to be restarted from the beginning. In the implementation, this is exactly the algorithm I used, since in practice the early rows are very seldom changed. I therefore kept a marker to note which row should be multiplied by which G_i next, and if at any stage the row, or an earlier one, is changed, the marker is set to start again at the beginning of the changed row.

I suspect that the implemented algorithm is not, therefore, cubic in complexity, though whether one could actually find a sequence of problems that behaved peversely enough is not clear. In any case my strategy is clearly a loony one from a complexity point of view.

Research Problem 3 *Find a strategy for doing spin that gives a bound to the complexity. Is it cubic?*

4.6   Purify (PU)

Given a matrix X of full row rank, find a free basis for the (possibly larger) module of all integral rows in the $\mathbb{Q}$-span of the rows of X.

This is done by simply appending the identity matrix to X and running the FBA procedure!

It is a (previously undescribed) feature of the FBA algorithm that, if the j is always chosen to be the largest possible value subject to $GCD(N, c_j) < N$, the v'_j will always be in the $\mathbb{Q}$-span of the v_i ($i \leq j$), as the "$-[(b.c_i)/N].v_i$" term will remove all later v_i from the expression.

It follows, therefore, that at the end of the procedure, the $\mathbb{Q}$-span of the first k rows (for all $k \leq \text{rank}(X)$) is not changed, and in particular the $\mathbb{Q}$-span of the first $\text{rank}(X)$ rows is not changed.

In the case of a matrix with few rows but many columns, this purify program is clearly a sledge-hammer to crack a nut. For example, it is clearly crazy to work with 1000×1000 matrices to purify a 1×1000 row. Much better is to find the GCD of the entries. I have not found the correct generalization to 2 or more dimensions, and we have

Research Problem 4 *Given a (rectangular) matrix M, find an integral basis for the space of all integral rows in the $\mathbb{Q}$-span of the rows of M efficient for the case where there are many columns and few rows.*

5   Representations of groups over $\mathbb{Z}$

In this section we give a description of how to work with integral representations of finite groups. The method is very similar to that used over finite fields, and (amazingly) nothing much goes wrong in the passage from finite field to integers.

The fundamental programs are essentially exactly the same.

- MU – Multiply two matrices

- AD – Add two matrices

- NS – Find a $\mathbb{Q}$-basis for the nullspace

- SP – Make the invariant subspace containing a vector, and find the action of the given generators on this subspace and on the quotient space.

- TE – Tensor two matrices together.

MU, AD and TE are simple, and should need no further explanation. SP and NS were described in the previous section. It should also be noticed that the CL program can be used to invert an invertible matrix by expressing the rows of an identity matrix in terms of the rows of the given matrix.

In my implementation, the standard base program is not needed as the spinning process (the first part of SP) is done in a basis invariant way.

What particularly astonished me about this work is how easy it seemed to be to find elements of the integral group algebra with minimal non-zero nullity. I can offer no convincing heuristics as to why this should be so, and indeed have no feeling at all even as to whether and under what conditions there should be such an abundance of group algebra elements of minimal non-zero nullity.

Given two ("random") elements X and Y and also an involution T of a group, I define $K(X, Y, T)$ to be the following set of 36 elements of the group algebra

$$
\begin{array}{ll}
X + Y + mI & -2 \leq m \leq 2 \\
X - Y + mI & 0 \leq m \leq 2 \\
X^T X + Y + Y^{-1} + mI & -3 \leq m \leq 3 \\
X^T X - Y - Y^{-1} + mI & -3 \leq m \leq 3 \\
X^T X + Y^T Y + T + mI & -3 \leq m \leq 3 \\
X^T X + Y^T Y - T + mI & -3 \leq m \leq 3
\end{array}
$$

Then looking among the $K(X, Y, T)$ for a moderate number of triples X, Y, T an element of minimum non-zero nullity is readily found.

Once such an element F has been found, we can look for the smallest invariant submodule containing a particular null-vector (or all null-vectors) of F. This submodule can only contain constituents on which F nas non-zero nullity.

As the spinning up is done in a basis-independent fashion, the basis for the resulting subspace will depend only on the null-vector chosen, and if the nullity of F is 1, the basis is absolutely standard, and the representation is absolutely irreducible. If the same F is used to find the same irreducible, the resulting matrices will be identical.

More generally, if c is the dimension of the space of matrices commuting with all the matrices of a (rationally) irreducible representation, then all nullities will be multiples of c, and if an F can be found of nullity c, any vector of the null-space will spin up to give a standard base, with respect to which the matrices will always be identical.

Given generators for a particular integral representation of a finite group, it appears to be possible to find elements of nullity c, and hence we can

1. Chop up any representation into the (rationally) irreducible modules it contains.

2. Make more (larger) representations by tensoring smaller ones.

3. Make multiple $\mathbb{Z}$-modules for the same $\mathbb{Q}$-irreducible, by spinning up the null vectors of various group algebra elements.

4. Make the quadratic form fixed by some generators. What is actually done is to calculate a matrix X that conjugates the generators to their transpose inverse. If this quadratic form is positive definite (which it is if the representation is either absolutely irreducible, or is the sum of two complex conjugate representations) the quadratic form, and hence usually the matrix entries, can be made smaller using the LLL basis reduction algorithm (see [1]).

5. Make the matrices generating the commuting algebra. This is done by spinning up various vectors in the null space.

6. Make a matrix normalizing a group by an outer automorphism. This is done by doing the automorphism (as words) to the generators, and then finding the transformation that takes them back. This is unique up to multiplication by the commuting algebra.

7. Using the quadratic form Q, the perpendicular to an invariant subspace can be found using transpose and null-space. This gives vectors that can either be purified or spun up to an invariant submodule perpendicular to the initial invariant subspace. This gives an efficient method of chopping up a new module, as the "old" stuff—the representations already known—can be assumed to be well-studied, and a suitable F known whose null vector spins up to it. The perpendicular to the old stuff is the new stuff—the representations not yet made and studied.

6   Further refinements using the quadratic form

It is a well-known fact that a complex irreducible representation fixes a Hermitian form. It thus seems sensible, when working with representations of finite groups over the integers, to keep any known quadratic forms handy. In simple cases—a complex irreducible writable over $\mathbb{Z}$—finding one is merely a question of finding the matrix that conjugates the representation to its transpose inverse.

Research Problem 5 *Given arbitrary matrices over $\mathbb{Z}$ that fix a positive definite matrix, find one.*

If the quadratic form is handy, it can be used to

Table 1: The number of elements of minimum non-zero nullity

Group	Representation		Number (out of 1800) with nullity 1
J_1	77		47
J_1	112	$56 + 56(b5)$	0(1 of nullity 2)
A_9	8		228
A_9	27		191
A_9	28		176
A_9	35a		115
A_9	35b		114
A_9	42		123
A_9	48		117
A_9	56		94
A_9	105		58
A_9	162		58
A_9	189		53
$L_2(11)$	10	$2A = -2, 3A = 1$	147
$L_2(11)$	10	$2A = 2, 3A = 1$	148
$L_2(11)$	10	$5 + 5(b11)$	0(204 of nullity 2)
$L_2(11)$	11		257
$L_2(11)$	24	$12 + 12'$	0(137 of nullity 2)
$L_3(3)$	12		224
$L_3(3)$	27		252
$L_3(3)$	39		238
$L_3(3)$	64	$16 + 16 + 16 + 16(d13)$	0(222 of nullity 4)

The table above shows that elements of nullity 1 are "common" in integral representations. 50 pairs of words in two generators T and S (where $T^2 = 1$) are chosen—$X_1 \ldots, X_{50}$ and $Y_1 \ldots, Y_{50}$. The set $K(T, X_i, Y_i)$ was then calculated, resulting in $50 \times 36 = 1800$ group algebra elements. The number of these of minimum non-zero nullity (1 unless otherwise stated) is given in the table for the generators I happened to choose for the group in question.

1. find the perpendicular to an invariant subspace using transpose and null-space. This can either be purified or spun up to an invariant sub-module perpendicular to the initial invariant subspace.

2. reduce (often drastically) the number of digits in the matrix entries. This is done by running LLL on the quadratic form.

This suggests

Research Problem 6 *(deliberately a bit vague) Does an equivalence class of representations necessarily have one with small integers in it? Why do some representations appear to need fairly large integers?*

Notice that the quadratic form can be readily carried through all the operations on representations.

In practice, it is worth looking at a fair number of rationally equivalent integral representations to find one that allows its digits to be reduced as much as possible.

7 Rings other than $\mathbb{Z}$

This section is really just starting the ball rolling on:

Research Problem 7 *What properties of $\mathbb{Z}$ are needed for the various algorithms above?*

Suppose then that D is the generalization of the integers used for entries in matrices. Even the simplest programs require that addition exists in D, and forms an abelian group with 0, and that multiplication exists, is associative and distributive on both sides over additition. This is already enough to prove that matrix multiplication is associative, and to write the simplest programs—AD, MU and TE—that require very little of the underlying ring.

It is also sensible to require that D has a 1, so that we can start to talk of invertible matrices and start to do group representation.

If the module handler is to make much sense, we need that there are no divisors of zero.

With this much, we can start to define a module handler as in section 2. A sequence of rows v_i is linearly dependent if there is a corresponding sequence of elements d_i of D such that

$$\sum_i d_i.v_i = 0$$

If the previous rows are linearly independent, but with the new row there is a linear dependence, then the new row is not multiplied by zero in the dependence.

Without serious modification, however, the module handler will be in trouble if the ring has non-principal right ideals, as then algorithms like FBA have no possible moves if presented with a linear dependence whose d_is span a non-principal ideal.

Research Problem 8 *Generalize the FBA routine to work with dependent rows if necessary to get over the problem of non-principal ideals.*

Even if we assume that all right ideals are principal, we still need an algorithm (for FBA, say) that can find the generator of an ideal. But is this enough?

Research Problem 9 *Given a linearly dependent set of rows, an algorithm for finding dependencies, and an algorithm for finding a generator of a principal ideal, what more is needed to make FBA work?*

The next step is to find an algorithm for finding linear dependencies. The p-adic lift approach taken in this paper needs a homomorphism from D into a finite field (call it reduce()), and a lift from that field back to D (call it lift()) with the convergence property:

For all d in D, the operation $d \mapsto (d - \mathrm{lift}(\mathrm{reduce}(d))/p$ reaches 0 in a finite number of steps.

Research Problem 10 *Investigate domains for the existence of primes with and without this property.*

8   The LLL algorithm

As part of the implementation, an LLL algorithm is crucial to reduce the integers in an irreducible representation before going on to tensor further. Here are a few thoughts on this part of the project.

I used a double-precision floating-point model, and was able to work happily in 189 dimensions—I did not try it on any larger examples. To achieve this, I was rather careful in that any changes made to the basis of integral vectors were not done to the model as well, but instead the relevant row of the model was recalculated from the vectors.

Research Problem 11 *For a given "accuracy" of floating point model, prove that LLL works up to a given number of dimensions.*

Here are some preliminary ideas as to what this might mean. Let us first get a mathematical abstraction a bit like floating point numbers. Let a (the accuracy) be a real number (a little) less than 1, and let B be a set of positive real numbers such that, for any positive real number x, there is a number $b \in B$ with x/b between a and $1/a$. Let F be the set $\{0, b, -b | b \in B\}$. One can then go on to define arithmetic on F, having accuracy a. Using this arithmetic to run the LLL model, it then makes sense to ask whether it works or not.

In an attempt to get yet shorter vectors, I also tried a marginal improvement. In classical LLL, at a critical stage the last two vectors in consideration may be swapped. This is usually described as choosing the shortest vector in the 2-space obtained by projecting onto the perpendicular of all previous vectors. This may be generalized in several ways to more than two dimensions, but I felt the clear favourite was to try to choose a basis for the vectors that minimized the product of the determinants of the 1-, 2-, 3- ... dimensional sublattices spanned by the first few vectors of the basis.

In a few cases a minor improvement was found, but usually the basis was not changed at all. It is a worthwhile improvement to run after LLL has finished if you are really interested in a marginal improvement.

References

[1] A. K. Lenstra, H. W. Lenstra, Jr., and L. Lovász, Factoring polynomials with rational coefficients, *Math. Ann.* **261** (1982), 515–534.

[2] R. A. Parker, The computer calculation of modular characters—The Meataxe, in *Computational group theory (ed. M. D. Atkinson)*, pp. 267–274. Academic Press, 1984.

Finite rational matrix groups: a survey

W. Plesken

Abstract

The interplay between finite rational matrix groups and integral lattices in Euclidean space is explained, in particular applications of the theory of finite rational matrix groups to the construction of modular lattices are discussed. The finite rational matrix groups of a given dimension are interrelated by the way they intersect and classification results up to dimension 31 are briefly surveyed.

1 Introduction

The investigation of finite rational matrix groups is an old topic of group theory. However, with the classification of finite simple groups on the theoretical side, and the advance of algorithmic and computational methods on the practical side, a fresh look at the subject might reveal new insights, as suggested in [20]. This has meanwhile resulted in a full classification of the maximal finite rational matrix groups up to degree 31 in [20], [17], [14], [15], [16], as well as the investigation of certain infinite series of maximal finite rational matrix groups.

As one might expect, cf. [20] Proposition (II.6), the reducible groups are not so interesting in the investigations. In this paper, therefore, I shall only deal with rational irreducible maximal finite groups, henceforth abbreviated r.i.m.f. groups. From the point of view of simply enumerating these groups, only the primitive ones play a role, i.e. those whose natural representation is not induced up (over the rationals) from a proper subgroup, cf. [20] Proposition (II.7). However, from the point of view of interrelating the r.i.m.f. groups of a given degree, the imprimitive ones are also relevant, as will become clear in section 3.

There are essentially two different sorts of problems that arise:

a) Classify the r.i.m.f. groups in a given dimension.

b) Exhibit infinite series of r.i.m.f. groups.

For both sorts of problems the ATLAS [4] plays a very helpful role: in getting rough ideas, proving non-existence of groups with certain properties, etc. In odd dimensions with few prime divisors this might sometimes be more than half of the work. However in highly composite dimensions with a high 2-power part, this yields only a small fraction of the information. This is because Clifford Theory leaves a lot of possibilities for normal subgroups. The classification results for both problems above are reported in section 4.

One of the applications of the classification results is the construction of integral lattices in Euclidean space. Namely each finite rational matrix group G gives rise to various such lattices with G acting on it as a group of automorphisms. It turns out that these lattices often have particularly good properties if G is a r.i.m.f. group. The interrelation between r.i.m.f. groups and lattices is explained in section 2 as well as the application of this towards the actual construction of r.i.m.f. groups on the computer. Particularly nice such lattices which came out in the classification up to dimension 31 are listed in section 5. These are certain k-modular lattices for certain natural numbers k, i.e. those which are isometric to a scaled version of their reciprocal lattice.

Turning to general lattices for r.i.m.f. groups again, it is interesting from the point of view of the integral representation theory, and also from the practical point of view of finding all r.i.m.f. groups containing a given finite rational matrix group it is important, to have bounds for the determinants for certain integral lattices acted upon by a finite rational matrix group. Questions of this kind were studied in [7]; new positive and also surprising negative results are reported on at the end of section 2. For instance there are r.i.m.f. groups G which do not act on any integral lattice in Euclidean space for which all prime divisors of the determinant divide the order of G.

One further aspect of r.i.m.f. groups is how they are interrelated by common irreducible subgroups. These relations can most naturally be expressed in terms of simplicial complexes. The relevant definitions are in section 3. Apart from examples very little is known about this, since most of the material of section 3 is new. In some examples the simplicial complexes are vaguely related to affine buildings and might sometimes say something about what sort of groups are generated by certain r.i.m.f. groups. Each section ends with a list of some open problems.

I should like to thank Gabriele Nebe for her efficient help via e-mail.

2   Basic structures, methods, and results

Apart from the well-known structures which are associated with a finite rational matrix group $G \leq GL_n(\mathbb{Q})$, like its natural module, its enveloping algebra, its commuting algebra $C_{\mathbb{Q}^{n \times n}}(G)$, and its centralizer $C_{GL_n(\mathbb{Q})}(G)$, there are essentially two basic data which go with a finite rational matrix group that are not so commonly known.

Definition 2.1 Let G be a finite subgroup G of $GL_n(\mathbb{Q})$ and $V = \mathbb{Q}^{1 \times n}$ the natural $\mathbb{Q}G$-module.

(i) $\mathcal{Z}(G) := \{M | M \text{ is a full } \mathbb{Z}G\text{-sublattice of } V\}$,

(ii) $\mathcal{F}(G) := \{F \in \mathbb{Q}^{n \times n} | F = F^{tr} \text{ and } gFg^{tr} = F \text{ for all } g \in G\}$

and $\mathcal{F}_{>0}(G) := \{F \in \mathcal{F}(G) | F \text{ positive definite}\}$.

$\mathcal{F}(G)$ is a $\mathbb{Q}$-vector space the dimension of which is equal to the multiplicity of the 1-character of G in the symmetric square of the natural character of G. By an old result of Burnside $\mathcal{Z}(G)$ is not empty, and by one of the standard proofs of Maschke's Theorem $\mathcal{F}_{>0}(G)$ is not empty. Clearly, the centralizer $C_{GL_n(\mathbb{Q})}(G)$ acts on both $\mathcal{Z}(G)$ and $\mathcal{F}(G)$. The classical Jordan–Zassenhaus Theorem says that the number of orbits on $\mathcal{Z}(G)$, i.e. the number of isomorphism classes of full $\mathbb{Z}G$-lattices in V, is finite. The action of the centralizer can be extended to the normalizer $N_{GL_n(\mathbb{Q})}(G)$ in both cases. In the case of $\mathcal{Z}(G)$ the orbits are in bijection with the conjugacy classes of subgroups of $GL_n(\mathbb{Z})$ which are conjugate under $GL_n(\mathbb{Q})$ to G. For the actual computation of sufficiently big portions of $\mathcal{Z}(G)$ a new implementation of the centering algorithm of [21] is available. The action of $C_{GL_n(\mathbb{Q})}(G)$ or $N_{GL_n(\mathbb{Q})}(G)$ on $\mathcal{F}(G)$ or $\mathcal{F}_{>0}(G)$ has not so intensively been studied, cf. however [6], [32], and [31]. For the actual computation of r.i.m.f. groups the following notion is important.

Definition 2.2 Let $G \leq GL_n(\mathbb{Q})$ be finite and for $(F, L) \in \mathcal{F}_{>0}(G) \times \mathcal{Z}(G)$ let

$$Aut(F, L) := \{g \in GL_n(\mathbb{Q}) | Lg = L \text{ and } gFg^{tr} = F\}$$

be the *automorphism group* of the pair (F, L).

Clearly, $G \leq Aut(F, L)$. Since orthogonal groups of positive definite quadratic forms are compact and lattice stabilizers are discrete, $Aut(F, L)$ is finite. For its electronic computation there is a very efficient implementation available, which for instance computes generators for the automorphism group of the Leech lattice in about 20 minutes on an HP9000/730, cf. [22]. The following criterion is obvious.

Remark 2.3 Let $G \leq GL_n(\mathbb{Q})$ be finite. G is maximal finite if and only if $G = Aut(F, L)$ for all $(F, L) \in \mathcal{F}_{>0}(G) \times \mathcal{Z}(G)$.

To turn this into a practical criterion, one wants to look at as few pairs (F, L) as possible. For example, consider the simplest but most common case of uniform group G, i.e. $dim\mathcal{F}(G) = 1$. Here it suffices to fix one positive definite matrix $F \in \mathcal{F}(G)$ and to let L run through a set of representatives of $N_{GL_n(\mathbb{Q})}(G)$-orbits. But there are some general ideas which might reduce the amount of work even in this easy situation.

Definition 2.4 Let $G \le GL_n(\mathbb{Q})$ be finite, $V = \mathbb{Q}^{1 \times n}$ the natural $\mathbb{Q}G$-module, and $(F, L) \in \mathcal{F}_{>0}(G) \times \mathcal{Z}(G)$.

(i) $L^{\#F} := \{x \in V \mid xFy^{tr} \in \mathbb{Z}\}$ is called the *reciprocal* or *dual lattice* of L with respect to F.

(ii) L is called integral with respect to F or F is called *integral* on L, if $L \le L^{\#F}$. In this case, $det(F, L) := |L^{\#F}/L|$ is called the *determinant* of (F, L). The isomorphism type of the discriminant group $L^{\#F}/L$ is often referred to as the *elementary divisors* of (F, L) (which of course are equal to the elementary divisors of F if $L = \mathbb{Z}^{1 \times n}$).

(iii) If (F, L) is integral, $L^{ev(F)} := \{x \in L \mid xFx^{tr} \in 2\mathbb{Z}\}$ is called the *even sublattice* of L (with respect to F).

Clearly, $L^{\#F}$ and $L^{ev(F)}$ lie again in $\mathcal{Z}(G)$. It turns out that quite a large proportion of r.i.m.f. groups G act on rather few lattices in the following sense.

Definition 2.5 A r.i.m.f. group G is called *lattice sparse*, if all lattices in $\mathcal{Z}(G)$ can be obtained by the following processes starting from any lattice in $\mathcal{Z}(G)$: multiplying with elements of $C_{GL_n(\mathbb{Q})}(G)$, taking sums and intersections, and taking duals (with respect to any $F \in \mathcal{F}_{>0}(G)$) and even sublattices.

A good example of lattice sparse groups are those automorphism groups of lattices in Euclidean space which Thompson calls utterly irreducible, cf. [27], i.e. where tensoring of the lattice with an arbitrary field always yields an irreducible module for the group. The best known examples are the Weyl group $W(E_8)$ as the automorphism group of the root lattice E_8, the covering group of the Conway group Co_1 as the automorphism group of the Leech lattice, the Thompson group Th, more precisely $C_2 \times Th$, as the automorphism group of the 248-dimensional Thompson–Smith lattice, cf. [28], [12]. Infinitely many further (tensor-indecomposable) examples have been constructed by R. Gow using the basic spin module for covering groups of S_n and Alt_n for certain n, cf. [9]. The notion of utterly irreducible has been generalized to globally irreducible in [10], cf. also [29].

Coming back to the problem of making (2.3) more practical, the following definition is useful.

Definition 2.6 In the situation of (2.4), $(F, L) \in \mathcal{F}_{>0}(G) \times \mathcal{Z}(G)$ is called *normalized*, if L is integral with respect to F such that $L^{\#F}/L$ has only elementary abelian p-Sylow subgroups of rank less then or equal to half the rank of L.

It follows from results in [33], cf. also [7], that one can restrict oneself to normalized pairs (F, L) in (2.3), to check for maximal finiteness. The next question is, whether the primes dividing the determinants can be restricted. If G is absolutely irreducible the following result follows immediately from Schur's relations.

Theorem 2.7 ([17] p. 76) *Let G be a uniform subgroup of $GL_n(\mathbb{Q})$, and let $(F, L) \in \mathcal{F}_{>0}(G) \times \mathcal{Z}(G)$ be integral and primitive, i.e. $(p^{-1}F, L)$ not integral for any prime number p. Then each prime divisor of $\det(F, L)$ divides the order of G.*

Obviously, this theorem is of basic importance if one wants to find the uniform finite r.i.m.f. supergroups of a finite rational group H, in case one has a good control over the space of invariant forms. In particular, if $dim\mathcal{F}(H) = 2$ one is in a strong position, cf. [17], Part II. For r.i.m.f. groups with a 2-dimensional space of invariant forms, the situation is not quite as good, however still good enough for practical purposes.

Theorem 2.8 ([17] p. 78) *Let G be an irreducible subgroup of $GL_n(\mathbb{Q})$ with $dim\mathcal{F}(G) = 2$. Assume that the maximal real subfield of the centre of the commuting algebra $C_{\mathbb{Q}^{n \times n}}(G)$ has an ideal class group which is generated by ideal classes which are represented by prime ideal divisors of $|G|$. Then for each $L \in \mathcal{Z}(G)$ there exists an $F \in \mathcal{F}_{>0}(G)$ integral on L such that every prime divisor of $\det(F, L)$ divides the order of G.*

This result is good enough to successfully search most 4-dimensional form spaces $\mathcal{F}(H)$ for form subspaces of r.i.m.f. supergroups of a rational matrix group H. In dimension 16 we found the first two examples of r.i.m.f groups G for which $\det(F, L)$ has prime divisors not occurring in $|G|$ for each integral pair $(F, L) \in \mathcal{F}_{>0}(G) \times \mathcal{Z}(G)$, cf. [17] p. 87. The groups in both cases were non-split extensions of cyclic groups of order 60 by a Klein four group and $dim\mathcal{Z}(G) = 4$.

Though it has not been used up to now for the classification of r.i.m.f. groups, because it was developed later, one must mention an algorithm to find the normalizer of a finite unimodular group G in $GL_n(\mathbb{Z})$ in [18]. In our context it is important that this method also yields a fundamental domain of the normalizer in the real convex hull of $\mathcal{F}_{>0}(G)$, which is a big reduction for the search of finite integral supergroups of G. This work extends work on perfect forms in [1]. It is particularly useful to find finite overgroups, in which G is normal. In this context, [17] (II.11) was very helpful, where it was proved that with a normal subgroup N of a r.i.m.f. group G a possibly bigger normal subgroup $\mathcal{B}^o(N) \geq N$, depending only on N and not on G, must be contained and normal in G.

There is an alternative to using (2.3) to find r.i.m.f. groups, particularly if the dimension gets too big for using the computer. This method leads to results only for groups close to simple groups and uses the classification of the finite simple groups in a serious way: one starts out with a faithful irreducible character of a finite group which is afforded by a rational representation. One checks with the help of [13] and [25] that the resulting matrix group is maximal finite in $GL_n(\mathbb{Q})$, cf. [20] (II.15) for an example of this procedure, where the first non-uniform r.i.m.f. group was found. In letter dated 21 June 1991, Rod Gow tells me that he can prove in this way that certain of his basic spin characters give rise to maximal finite subgroups of $PGL_n(\mathbb{C})$. There are various situations investigated by Tiep, cf. [30] and various of his papers listed in [12], where such an argument could be applied. Also the classification of the r.i.m.f. groups of degree $p - 1$ whose order is divisible by the prime p in [17], Part I, uses results in [2], which in turn use the classification of the finite simple groups, though not via the scheme set up above.

To finish this section I list some open problems.

Find new methods, computational or theoretical, to determine the minimum of a lattice in Euclidean space with a big group, say an r.i.m.f. group, acting on it. The available algorithms operate up to dimension 50 or 60 at a tolerable speed. Comparisons with the minimum of fixed sublattices of subgroups etc. have not been studied systematically.

If an r.i.m.f. group G is given in an abstract way, as described just before, investigate the minimum for some integral pair in $\mathcal{F}_{>0}(G) \times \mathcal{Z}(G)$. Virtually nothing is known on this. If the trivial 2-Brauer character does not show up in the restriction of the natural character to the $2'$-classes, the lattice has to be even. If the minimum is 2, it can be easily identified via the root lattice of a reflection group. Information on the decomposition numbers, for which [11] is now an excellent source, can be used to say something about $\mathcal{Z}(G)$, cf. [19].

Study the enveloping algebra $\overline{\mathbb{Q}G}$ (spanned by the matrices in G) of an irreducible rational matrix group G as a simple algebra with involution, where the involution is induced by the standard involution of the group ring $\mathbb{Q}G$ inverting the elements of G. Certainly a lot has been said about Schur indices, but here, of course, the story goes further. This problem was already suggested in [24].

3 Simplicial complexes of r.i.m.f. groups

There is the question of interrelating the r.i.m.f. groups of a given degree. In the following definition this will be done via common subgroups with the same commuting algebra. Other concepts are obviously possible, however this one can also be motivated from the lattice point of view: when do irreducible groups act on lattices coming from more than one r.i.m.f. group?

Definition 3.1 $\mathcal{M}_n^{irr}(\mathbb{Q})$ denotes the *full simplicial complex of r.i.m.f. groups of degree n*. The set $Max_n^{irr}(\mathbb{Q})$ of all r.i.m.f. subgroups of $GL_n(\mathbb{Q})$ is the set of vertices of $\mathcal{M}_n^{irr}(\mathbb{Q})$. The pairwise different groups $G_0, G_1, \ldots, G_k \in Max_n^{irr}(\mathbb{Q})$ form the vertices of a k-simplex of $\mathcal{M}_n^{irr}(\mathbb{Q})$, if and only if they and their intersection all have the same commuting algebra in $\mathbb{Q}^{n \times n}$, i.e.

$$C_{\mathbb{Q}^{n \times n}} \left(\bigcap_{i=0}^{k} G_i \right) = C_{\mathbb{Q}^{n \times n}}(G_i) \text{ for all } 0 \le i \le k$$

Remark 3.2 $max\{dim(S)|S \text{ is a simplex of } \mathcal{M}_n^{irr}(\mathbb{Q})\}$ is finite.

Proof: For finite irreducible $H \le GL_n(\mathbb{Q})$ let

$$Max(H) := \{G \in Max_n^{irr}(\mathbb{Q})|H \le G \text{ with } C_{\mathbb{Q}^{n \times n}}(G) = C_{\mathbb{Q}^{n \times n}}(H)\}.$$

Since $GL_n(\mathbb{Q})$ has only finitely many conjugacy classes of finite subgroups, we only have to prove that $Max(H)$ is finite. To this end note that each $G \in Max(H)$ is of the form $U(L) := \{h \in U(\overline{\mathbb{Q}H})|Lh = L\}$ with $L \in \mathcal{Z}(H)$. Here $U(\overline{\mathbb{Q}H})$ denotes the unitary group of the enveloping algebra $\overline{\mathbb{Q}H}$ of H consisting of all elements of $\overline{\mathbb{Q}H}$ inverted by the involution induced by the standard involution of $\mathbb{Q}H$. Clearly, for $L_1, L_2 \in \mathcal{Z}(H)$ one has $U(L_1) = U(L_2)$ if L_1 and L_2 are in the same orbit under the automorphism group of the natural $\mathbb{Q}H$-module, i.e. under $C_{\mathbb{Q}^{n \times n}}(H)^*$. But by the Jordan–Zassenhaus Theorem the number of these orbits is finite. □

Obviously the conjugation action of $GL_n(\mathbb{Q})$ on its subgroups, more specifically on $Max_n^{irr}(\mathbb{Q})$, induces a simplicial action on $\mathcal{M}_n^{irr}(\mathbb{Q})$, i.e. the simplices are permuted. (The maps between simplices which turn up are always assumed to be 'affinely' induced by the maps on the vertices.) This allows us to form two different quotients: either first by taking the set of orbits $Max_n^{irr}(\mathbb{Q})/GL_n(\mathbb{Q})$ and then grouping the orbits together in (possibly smaller dimensional) simplices, or in the topological sense. In the second approach one has to take a barycentric subdivision first to get a simplicial complex as quotient. This quotient of course contains a lot more information and is much harder to compute then the first quotient.

Definition 3.3 (i) The *simplicial complex* $M_n^{irr}(\mathbb{Q})$ *of r.i.m.f. groups of degree n* is the simplicial complex whose vertices are the conjugacy classes of r.i.m.f. subgroups of $GL_n(\mathbb{Q})$, i.e. $Max_n^{irr}(\mathbb{Q})/GL_n(\mathbb{Q})$. A $(k+1)$-subset $\{G_0^{GL_n(\mathbb{Q})}, \ldots, G_k^{GL_n(\mathbb{Q})}\}$ of $Max_n^{irr}(\mathbb{Q})/GL_n(\mathbb{Q})$ forms a simplex in $M_n^{irr}(\mathbb{Q})$, if the representatives $G_0, \ldots, G_k$ can be chosen to form a simplex in $\mathcal{M}_n^{irr}(\mathbb{Q})$.

(ii) The *(topological) quotient* $\mathcal{M}_n^{irr}(\mathbb{Q})/GL_n(\mathbb{Q})$ is the simplicial complex whose simplices are the $GL_n(\mathbb{Q})$ orbits on the set of simplices of the barycentric subdivision of $\mathcal{M}_n^{irr}(\mathbb{Q})$ with the identifications induced from $\mathcal{M}_n^{irr}(\mathbb{Q})$.

It is $M_n^{irr}(\mathbb{Q})$ that has ben computed in [20], [17], [15], [16] up to $n = 31$. Note that an error for $n = 8$ in [20] has been corrected in [17], Part II. The other two simplicial complexes have not been studied up to now. The main purpose of this section is to point out that these are interesting objects to study which give good insight in the interrelations between the various r.i.m.f. groups of a given degree. Obviously the orbit map $Max_n^{irr}(\mathbb{Q}) \to Max_n^{irr}(\mathbb{Q})/GL_n(\mathbb{Q})$ induces both a simplicial map of $\mathcal{M}_n^{irr}(\mathbb{Q})$ onto $M_n^{irr}(\mathbb{Q})$ and onto $\mathcal{M}_n^{irr}(\mathbb{Q})/GL_n(\mathbb{Q})$, where the first factors over the second (if everything is set up properly). One checks easily that the resulting map of $\mathcal{M}_n^{irr}(\mathbb{Q})/GL_n(\mathbb{Q})$ onto $M_n^{irr}(\mathbb{Q})$ sets up a bijection between the connected components of the two simplicial complexes. From the definition it is clear that the corresponding components need not have the same dimension, since it might well be the case that two conjugate r.i.m.f. groups have an intersection with the same commuting algebra as the two intersecting groups.

The pre-image of a component of $M_n^{irr}(\mathbb{Q})$ in $\mathcal{M}_n^{irr}(\mathbb{Q})$ obviously decomposes into components which are permuted transitively by $GL_n(\mathbb{Q})$. Therefore a key question is, what these components look like.

Proposition 3.4 *Let C be a connected component of $\mathcal{M}_n^{irr}(\mathbb{Q})$ and denote its barycentric subdivision by C'.*

(i) To each vertex v of C' a matrix group G_v is attached, namely the intersection of the r.i.m.f. groups of the vertices of the simplex of C, whose barycentre v is. All the groups G_v have equal commuting algebras in $\mathbb{Q}^{n \times n}$ and equal enveloping algebras $\overline{\mathbb{Q}G_v}$, the latter being equal as algebras with involution, where the involution is induced by the standard involution of $\mathbb{Q}G_v$. Denote this simple algebra with involution by $A(C)$ and its commuting algebra in $\mathbb{Q}^{n \times n}$ by $C(C)$.

(ii) For all vertices v of C' the matrix groups G_v have the same space $\mathcal{F}(C) := \mathcal{F}(G_v)$ and the involution of $A(C)$ is given by $\bar{a} = Fa^{tr}F^{-1}$ for all $a \in A(C)$, where $F \in \mathcal{F}(C) > 0$ is arbitrary.

(iii) The stabilizer in $GL_n(\mathbb{Q})$ of a vertex v in C' is the normalizer of G_v in $GL_n(\mathbb{Q})$.

(iv) For any finite subgroup G of $U(A(C)) := \{g \in A(C) | g\bar{g} = 1\}$ which has $C(C)$ as its commuting algebra in $\mathbb{Q}^{n \times n}$ one of two possibilities occurs: either G belongs to exactly one simplex S of C, resp. vertex $v = v(S)$ of C' such that $G \leq G_v$, and S is maximal with this property, or each r.i.m.f. group containing G has a commuting algebra properly contained in $C(C)$.

(v) Let C'_k denote the set of vertices of C' which are barycentres of k-dimensional simplices of C. Then

$$C'_k \to \{H | H \leq GL_n(\mathbb{Q})\} : v \mapsto G_v$$

is injective for the biggest k with C'_k non-empty.

Proof: (i) The vertices of $\mathcal{C}'$ are in bijection with the simplices of $\mathcal{C}$. By definition of $\mathcal{M}_n^{irr}(\mathbb{Q})$ all the G_v have the same commuting algebra, and therefore also the same enveloping algebra. The same applies to the $\mathcal{F}(G_v)$. For $g \in G_v$ one has $g^{-1} = Fg^{tr}F^{-1}$ for any $F \in \mathcal{F}_{>0}(\mathcal{C}) = \mathcal{F}_{>0}(G_v)$, thus proving that the involution is the same, as well as (ii).

(iii) Clear.

(iv) By (3.2) the number of r.i.m.f. groups with commuting algebra $C(\mathcal{C})$ containing G is finite. Their intersection is the G_v in the statement.

(v) Follows from (iv). □

That two possibilities in (iv) arise is unnatural and due to the fact that the interest lies in r.i.m.f. groups. This of course can be avoided by considering the maximal finite subgroups of $U(A(\mathcal{C}))$ with commuting algebra $C(\mathcal{C})$. In case $C(\mathcal{C}) = \mathbb{Q}$ the second possibility does not occur. Obviously the normalizers $N_{GL_n(\mathbb{Q})}(G_v)$ of the vertex groups G_v lie in the set stabilizer of $\mathcal{C}$ for any vertex of the component $\mathcal{C}$. Often they seem to generate the whole stabilizer. Some examples might give an idea of what a component might look like and how $\mathcal{M}_n^{irr}(\mathbb{Q})/GL_n(\mathbb{Q})$ compares with the much simpler $M_n^{irr}(\mathbb{Q})$.

Corollary 3.5 *Let $G \leq GL_n(\mathbb{Q})$ be absolutely irreducible and maximal finite and $Min(G)$ be the set of minimally $\mathbb{C}$-irreducible subgroups of G. The connected component $\mathcal{C}$ of $\mathcal{M}_n^{irr}(\mathbb{Q})$ containing G consists only of G, if and only if for all $H \in Min(G)$ (up to conjugacy in $GL_n(\mathbb{Q})$) and all $L \in \mathcal{Z}(H)$ (up to isomorphism) one has*

$$U(L) := \{g \in \mathbb{Q}^{n \times n} | g\overline{g} = 1, Lg = L\} \leq G.$$

Note, $U(L)$ of the last corollary is just $Aut(F, L)$ for some $F \in \mathcal{F}_{>0}(G)$. The criterion applies to all but one conjugacy class of r.i.m.f. groups up to degree 5. It is worthwhile to look at the exception in detail. Beforehand some notation and comments are needed.

Convention 3.6 The vertices of $M_n^{irr}(\mathbb{Q})$, of $\mathcal{M}_n^{irr}(\mathbb{Q})$, and also those of $\mathcal{M}_n^{irr}(\mathbb{Q})/GL_n(\mathbb{Q})$ which come from a vertex of $\mathcal{M}_n^{irr}(\mathbb{Q})$, i.e. belong to an r.i.m.f. group, are drawn with filled black circles. Those vertices of the quotient $\mathcal{M}_n^{irr}(\mathbb{Q})/GL_n(\mathbb{Q})$ which come from the barycentres of $\mathcal{M}_n^{irr}(\mathbb{Q})$, i.e. belong to proper intersections of r.i.m.f. groups, are drawn with open circles.

Clearly, in passing from $\mathcal{M}_n^{irr}(\mathbb{Q})/GL_n(\mathbb{Q})$ to $M_n^{irr}(\mathbb{Q})$, the open circles are left out. However, the latter ones determine which black vertices are grouped together into a simplex. Whereas one can attach a group G_v to each simplex of $\mathcal{M}_n^{irr}(\mathbb{Q})$, and a unique conjugacy class of groups to each vertex of $\mathcal{M}_n^{irr}(\mathbb{Q})/GL_n(\mathbb{Q})$, one can no longer do this with the simplices of $M_n^{irr}(\mathbb{Q})$.

In [17] etc. a group is chosen which is contained in suitable representatives of the groups belonging to the vertices of the simplex in $M_n^{irr}(\mathbb{Q})$.

On the face of it, there is a formal connection between the affine building of reductive p-adic groups as studied in [3] and $\mathcal{M}_n^{irr}(\mathbb{Q})$. Whereas the Bruhat–Tits building organises the relations of the maximal compact subgroups of p-adic groups as an inclusion scheme for certain of their intersections, namely the parahoric subgroups, $\mathcal{M}_n^{irr}(\mathbb{Q})$ does the same for r.i.m.f. groups and certain of their irreducible intersections. However, often the connection goes further, as the next example shows. But one should not be too optimistic, since the behaviour of r.i.m.f. groups seems to be more complex. Not only might more than one prime be involved, also various other problems in matching the vertices of a component $\mathcal{C}$ of $\mathcal{M}_n^{irr}(\mathbb{Q})$ with that of the p-adic completion of $U(A(\mathcal{C}))$ turn up. Nevertheless, one might sometimes use the buildings to say something about global properties of a component $\mathcal{C}$.

Coming to the example, let $Aut(A_n)$ denote the automorphism group of the root system A_n, and J_n resp. I_n the $(n \times n)$-all-1-matrix resp. unit matrix. Then $Aut(A_n) = Aut(I_n + J_n, \mathbb{Z}^{1 \times n}) \cong C_2 \times S_{n+1}$ is a r.i.m.f. group with the exception of $n = 8$, cf. [20].

Example 3.7 Let $G = Aut(A_4)$.

(i) The component of G in $M_4^{irr}(\mathbb{Q})$ consists of one point.

(ii) The component of G in $\mathcal{M}_4^{irr}(\mathbb{Q})/GL_4(\mathbb{Q})$ consists of a 1-simplex as follows:

$$G \bullet\!\!\!-\!\!\!-\!\!\!-\!\!\!-\!\!\!-\!\!\!-\!\!\!-\!\!\!-\!\!\!-\!\!\!-\!\!\!\circ \pm C_5{:}C_4$$

(iii) The component $\mathcal{C}$ of G in $\mathcal{M}_4^{irr}(\mathbb{Q})$ is a tree having six edges coming off each vertex. This tree can naturally be identified with the Bruhat–Tits building of $SO(\mathbb{Q}_5, I_4 + J_4)$.

(iv) The outer automorphism group induced by the normalizer of G in $GL_n(\mathbb{Q})$, resp. of $\pm C_5{:}C_4$, is trivial, resp. of order 2. These two normalizers generate the set stabilizer of $\mathcal{C}$ in $GL_n(\mathbb{Q})$, which is equal modulo scalars to the group $\Gamma O(\mathbb{Z}[\frac{1}{5}], I_4 + J_4)$ of orthogonal similitudes of $I_4 + J_4$ over $\mathbb{Z}[\frac{1}{5}]$.

Proof: $Min(G)$ consists up to conjugacy under $GL_n(\mathbb{Q})$ of just two groups $H_1(\cong C_5{:}C_4)$ and $H_2(\cong Alt_5)$. The group H_2 only gives rise to the 0-dimensional simplex G and the group H_1 defines a 1-simplex in $\mathcal{C}$, whose vertices are represented by G and G^g for some $g \in GL_n(\mathbb{Q})$ with $g(I_4 + J_4)g^{tr} = 5(I_4 + J_4)$. Moreover $G \cap G^g = \pm H$ is normalized by g and of index $6 = 1 + 5$ in both G and G^g. The rest is straightforward since each G_v lies in exactly a unique smallest parahoric subgroup of $SO(\mathbb{Q}_5, I_4 + J_4)$. $\qquad\square$

A final example might be helpful.

Example 3.8 (i) ([20]) $M_7^{irr}(\mathbb{Q})$ looks as follows:

$B_7 \bullet \!\!-\!\!-\!\!-\!\!-\!\!-\!\!-\!\!-\!\!-\!\!-\!\!-\!\!-\!\!-\!\!-\!\!-\!\!-\!\! \bullet E_7$

where the names for the root systems are also used as names for their auto-morphism groups.

(ii) $\mathcal{M}_7^{irr}(\mathbb{Q})/GL_7(\mathbb{Q})$ looks as follows:

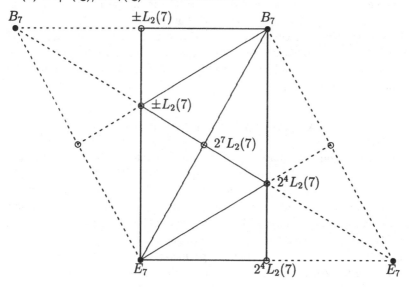

(The dotted lines together with their enclosed simplices do not belong to $\mathcal{M}_7^{irr}(\mathbb{Q})/GL_7(\mathbb{Q})$ but complement the simplices of $\mathcal{M}_7^{irr}(\mathbb{Q})/GL_7(\mathbb{Q})$ to the barycentric subdivision of a full simplex in $\mathcal{M}_7^{irr}(\mathbb{Q})$ in the preimage.)

Proof: (ii) There are two conjugacy classes of minimal $\mathbb{C}$-irreducible sub-groups of $GL_n(\mathbb{Q})$, cf. [21] part I. They give rise to two different types of 2-simplices in $\mathcal{M}_n^{irr}(\mathbb{Q})$ having vertices of types (B_7, B_7, E_7) or (B_7, E_7, E_7). The intersections of the r.i.m.f. groups one obtains by taking the stabilizer in one group of a lattice of the other group. The rest is straightforward. □

In the course of this section more open than solved problems have oc-curred already implicitly or explicitly. Nevertheless a few might be singled out. Note, the last problem of section 2 on the characterization of the simple components of rational group algebras as algebras with involutions becomes clearer now. Also all the problems on $\mathcal{M}_n^{irr}(\mathbb{Q})$ can be replaced by the corre-sponding question on the maximal finite subgroups of the unit group of such an algebra with involution.

Is it true that components of $\mathcal{M}_n^{irr}(\mathbb{Q})$ either have just one vertex or infinitely many? At the same time one might clarify the connection with the Bruhat–Tits buildings.

Is the set stabilizer of a connected component $\mathcal{C}$ of $\mathcal{M}_n^{irr}(\mathbb{Q})$ in $GL_n(\mathbb{Q})$ generated by the $N_{GL_n(\mathbb{Q})}(G_v)$ with v running through the vertices of $\mathcal{C}$? If this were the case, certainly finitely many such vertices would suffice. One can formulate sufficient conditions for this to hold. It seems to be necessary to get some deeper insight into how a vertex stabilizer can permute the vertices of the simplex in $\mathcal{M}_n^{irr}(\mathbb{Q})$ it belongs to as barycentre. Also this might give some insight into the structure of the stabilizers. For instance in example 3.7 the group acts on a tree and therefore a free product structure can be found, cf. [26]. The same game could be played with the r.i.m.f. groups themselves instead of their normalizers.

4 Classification results

As mentioned in the introduction the r.i.m.f. groups up to degree 31 have been classified including the simplicial complexes $M_n^{irr}(\mathbb{Q})$, cf. [20], [17], [14], [15], [16]. The $\mathcal{M}_n^{irr}(\mathbb{Q})/GL_n(\mathbb{Q})$, introduced in the last chapter are more complicated to compute and not yet done. Table 1 gives a statistical overview on how many conjugacy classes of r.i.m.f. groups there are in every degree, how $M_n^{irr}(\mathbb{Q})$ splits up into connected components etc.

Usually, cf. [20] (II.7) for a sufficient condition (which however is far from necessary), the wreath product $G \wr S_n$ is an r.i.m.f. group of degree nd, if G is a primitive r.i.m.f. group of degree d. The only known exception is $\{\pm 1\} \wr S_4$. Therefore, in Table 1 one has $ma(n) = \sum_{d|n} pr(d)$, unless $n = 4$. To describe the full detail of a primitive r.i.m.f. group is usually quite involved because it often involves taking various kinds of tensor products and extensions of these. Details are described in [17], Part I, while Table 2 in the next section gives an impression.

Turning to the simplicial complex $M_n^{irr}(\mathbb{Q})$, one observes that the dimensions divisible by 8 always have one component $M_n^{irr}(\mathbb{Q})$ which has more vertices than the others. It is an interesting observation how the components behave under doubling of the dimension. Roughly speaking, taking wreath products with C_2 (in the case of primitive groups) results into new components in the doubled dimension. However, different components in dimension n might give rise to the same component in dimension $2n$. This is vaguely connected to the fact, cf. [24], that modulo the Witt group $W(\mathbb{R})$ of the reals, the Witt group $W(\mathbb{Q})$ has exponent 4, cf. [20] (II.9).

Turning to infinite series of r.i.m.f. groups, $B_n (\cong C_2 \wr S_n)$ for $n \neq 4$ and $A_n (\cong C_2 \times S_{n+1})$ for $n \neq 8$ are the only known series which give a group in almost every dimension, cf. [20] (II.8) (where automorphism groups of

Table 1: Classes of r.i.m.f. groups

n	1	2	3	4	5	6	7	8
$pr(n)$	1	1	0	2	1	4	1	5
$ma(n)$	1	2	1	3	2	6	2	9
$CO(n)$	1	1^2	1^1	1^3	1^2	$1^2 2^2$	2	$1^2 7^1$
$co(n)$	1	2	1	3	2	4	1	3

n	9	10	11	12	13	14	15	16
$pr(n)$	1	5	1	11	3	9	4	22
$ma(n)$	2	8	2	19	4	12	6	31
$CO(n)$	1^2	$1^1 2^3 3^1$	2^1	$1^5 2^2 3^2 4^1$	$1^2 2^1$	$1^5 2^2 3^1$	$1^4 2^1$	$1^8 2^1 21^1$
$co(n)$	2	4	1	10	3	7	5	10

n	17	18	19	20	21	22	23	24
$pr(n)$	2	10	1	21	6	9	3	41
$ma(n)$	3	17	2	31	8	12	4	65
$CO(n)$	1^3	$1^5 2^3 3^2$	1^2	$1^9 2^2 3^2 5^1 7^1$	$1^4 4^1$	$1^5 2^2 3^1$	1^4	$1^{10} 2^2 3^1 4^2 40^1$
$co(n)$	3	10	2	15	5	8	4	16

n	25	26	27	28	29	30	31
$pr(n)$	3	11	3	23	1	17	3
$ma(n)$	5	16	5	37	2	33	4
$CO(n)$	1^5	$1^{10} 2^1 4^1$	$1^3 2^1$	$1^{16} 2^4 4^1 9^1$	1^2	$1^{12} 2^3 3^2 9^1$	4^1
$co(n)$	5	12	4	22	2	18	1

$pr(n) :=$ number of conjugacy classes of primitive r.i.m.f. subgroups of $GL_n(\mathbb{Q})$.

$ma(n) :=$ number of conjugacy classes of r.i.m.f. subgroups of $GL_n(\mathbb{Q})$.

$CO(n) :=$ splitting of $M_n^{irr,F}(\mathbb{Q})$ into connected components, e.g. $CO(n) := 1^a 2^b \cdots$ means that $M_n^{irr,F}(\mathbb{Q})$ splits up into a connected components with one vertex each, b connected components with 2 vertices each, etc.

$co(n) :=$ number of connected components of $M_n^{irr}(\mathbb{Q})$

root systems are also denoted by the name of the root system). Of course, like $\{\pm 1\}$, almost any primitive r.i.m.f. group G of degree d gives rise to an infinite series of r.i.m.f. groups $G \wr S_n \leq GL_{nd}(\mathbb{Q})$. Coming to the primitive groups, in [17], Part I, the r.i.m.f. groups of degree $p \pm 1$, which are of type $L_2(p)$, p a prime, are classified, together with the components of $M_n^{irr}(\mathbb{Q})$ to which they belong. Proving maximal finiteness of these groups can be done using Feit's results on groups of type $L_2(p)$, e.g. [7] Theorem A or [8] Theorem 3.3 (p. 347). All these groups give rise to infinite series of r.i.m.f. groups. However, the few results beyond these all seem to depend on the classification of finite simple groups, including the completeness statement of the classification of all r.i.m.f. groups of degree $p - 1$ with order divisible by p in [17], Part I, Theorem (V.10). A further series of r.i.m.f. groups is given in [16], Theorem 3.1; the groups are of degree $\frac{p^2+1}{2}$ and are isomorphic to $C_2 \times PSL_2(p^2){:}2$, where p is a prime and the outer automorphism is induced by the Frobenius automorphism. As mentioned in section 2, various other groups have not been explicitly mentioned in the literature such as certain globally irreducible groups.

Here are some open problems:

Find new infinite series of primitive r.i.m.f. groups. If possible investigate the geometry of the associated lattices, like the minimum (very difficult!) and determinant. If possible, describe its component in $M_n^{irr}(\mathbb{Q})$.

Find upper and lower bounds for the number of conjugacy classes of primitive r.i.m.f. groups in terms of the degree. Even for certain degrees like prime degrees very little is known. Of course each infinite series contributes towards a lower bound. Upper bounds are not known at all.

Find upper and lower bounds for the number of connected components of $M_n^{irr}(\mathbb{Q})$ in terms of n.

There are still some specific dimensions for which partial classifications are of interest, e.g. in dimension 32 all the r.i.m.f. groups acting on an even unimodular lattice. Or up to now no component of $M_n^{irr}(\mathbb{Q})$ has been investigated with more than two vertices, where the centre of the associated algebra with involution is of dimension bigger than 1.

5   Modular lattices

The class of l-modular lattices for some small natural numbers l has recently gained some interest, because they often give rise to rather dense lattices.

Definition 5.1 Let L be a lattice in Euclidean n-space $(\mathbb{R}^n, \langle, \rangle)$.

The *minimum* of L is defined as $min(L) := Min\{\langle l, l \rangle | 0 \neq l \in L\}$.

The *determinant* of L is defined as $det(L) := Det(\langle b_i, b_j \rangle)$ where $(b_1, \ldots, b_n)$ is a $\mathbb{Z}$-basis of L.

The *density* of L is the quotient $\frac{min(L)}{det(L)^{\frac{1}{n}}}$.

L is called *indecomposable*, if it is not the direct sum of pairwise orthogonal non-zero sublattices.

Note, the usual definition of density is a monotone function of the density defined here, cf. [5], but for comparisons the present definition will do. Clearly, for any r.i.m.f. group G of degree n and any pair $(F, L) \in \mathcal{F}_{>0}(G) \times \mathcal{Z}(G)$ one gets a lattice in Euclidean n-space, which will be indecomposable, if G is primitive. Not every lattice obtained in this way has a high density, but quite a few give very dense lattices indeed, e.g. the Conway group yields the Leech lattice. There is a class of lattices, where one can give bounds on the density.

Definition 5.2 An integral, even lattice L in Euclidean space is called *l-modular*, if the scaled version $\sqrt{l}L^{\#}$ of the reciprocal lattice $L^{\#}$ of L is isometric to L.

Obviously, l must be a natural number and the determinant of L is $l^{\frac{n}{2}}$. In case l is square-free the discriminant group $L^{\#}/L$ is isomorphic to the direct sum of $\frac{n}{2}$ copies of $\mathbb{Z}/l\mathbb{Z}$ and n is even. Quebbemann [23] proves that the theta function

$$\Theta_L(z) = \sum_{v \in L} q^{\frac{\langle v,v \rangle}{2}}, q = e^{2\pi i z}$$

is a modular form of weight $k := \frac{n}{2}$ for the Fricke group $\Gamma_*(l)$ with a certain character. In Table 2 below, l-modular lattices are called extremal, if their minimum is as big as Quebbemann's result allows, i.e. if $\Theta_L(z) = 1 + \sum_{j \geq N} a_j q^j$ with $a_N \neq 0$, then there is no Γ_*-modular form of weight k with the prescribed character having an expansion $1 + b_{N+1}q^{N+1} + \cdots$. If $l = 1$ or a prime then Quebbemann proves:

(i) k is a multiple of k_0 with $k_0 = 4$ for $l = 1$ (classical case), $k_0 = 2$ for $l \equiv 1, 2 \pmod 4$, and $k_0 = 1$ for $l \equiv 3 \pmod 4$.

(ii) In case $k_1 := \frac{24}{l+1}$ is integral, the minimum $2N$ of an extremal lattice is equal to $2 + 2[\frac{k}{k_1}]$.

In the table below, other similarity factors l also turn up. In these cases G. Nebe has checked extremality by computing a basis of the relevant space of modular forms, in case a $+$ or $-$ shows up in the second last column of the subsequent table. Finally, a lattice is called best in the table below, if it is the densest lattice in its genus, cf. [5] Chapter 15 for definition of genus. It should be noted that it is worthwhile to look at l-modular lattices for other finite rational matrix groups, not just r.i.m.f. groups. In this way, G. Nebe has found an extremal 3-modular lattice in dimension 24, which was suspected not to exist, cf. [14], [15].

Table 2: The indecomposable l-modular lattices of the r.i.m.f. groups in $GL_n(\mathbb{Q})$

n	lattice L	l	min	density	extremal	best
1	B_1	1	1	1	+	+
2	A_2	3	2	1.15	+	+
4	F_4	2	2	1.41	+	+
6	$M_{6,2}$	5	3	1.34		+
	$A_6^{(2)}$ (Craig)	7	4	1.51	+	+
8	E_8	1	2	2	+	+
	$[(SL_2(5) \overset{2}{\square} SL_2(5)):2]_8$	5	4	1.79	+	+
	$A_2 \otimes F_4$	6	4	1.63	+	+
10	$A_{10}^{(3)}$ (Craig)	11	6	1.81	+	+
12	$[6.SU_4(3).2^2]_{12}$ (Coxeter, Todd)	3	4	2.31	+	+
	$A_2 \otimes M_{6,2}$	15	6	1.55	−	−
	$[\pm 3.Alt_6.2^2]_{12}$	15	8	2.07	−	+
	$[L_2(7) \overset{2(2)}{\boxtimes} D_8]_{12}$	14	8	2.14	−	+
	$A_2 \otimes A_6^{(2)}$	21	8	1.75	−	
14	$[G_2(3)]_{14}$	3	4	2.31	+	+
	$M_{14,2}$	13	7	1.94		
16	$F_4 \tilde{\otimes} F_4$ (Barnes, Wall)	2	4	2.83	+	+
	$[(SL_2(9) \overset{2(3)}{\underset{\infty,3}{\otimes}} SL_2(9)).2]_{16}$	3	4	2.31	+	+
	$A_2 \otimes E_8$	3	4	2.31	+	+
	$[(Sp_4(3) \circ C_3) \overset{2}{\underset{\sqrt{-3}}{\boxtimes}} SL_2(3)]_{16}$	6	6	2.45	−	+
	$[2.Alt_{10}]_{16}$	5	6	2.68	+	+
	$[SL_2(5) \overset{2(2)}{\underset{\infty,2}{\boxtimes}} 2^{1+4'}.Alt_5]_{16}$	10	8	2.53		
	$A_2 \otimes [(SL_2(5) \overset{2}{\square} SL_2(5)).2]_8$	15	8	2.07	−	−
	$[SL_2(5) \overset{2(3)}{\underset{\infty,3}{\boxtimes}} SL_2(9)]_{16}$	15	10	2.58	−	+
	$[(SL_2(5) \overset{2(3)}{\underset{\infty,3}{\boxtimes}} (SL_2(3) \overset{2}{\square} C_3)]_{16}$	30	12	2.19		
	$[2.Alt_7 \overset{2(3)}{\underset{\sqrt{-7}}{\boxtimes}} \tilde{S}_3]_{16}$	21	12	2.62	−	+
	$[SL_2(7) \overset{2(3)}{\underset{\sqrt{-7}}{\boxtimes}} \tilde{S}_3]_{16}$	21	10	2.18	−	−
18	$[\pm(L_2(7) \overset{2}{\underset{\sqrt{-7}}{\boxtimes}} L_2(7)).2]_{18}$	7	6	2.27	−	
	$M_{18,2}$	17	9	2.18		
	$A_{18}^{(5)}$ (Craig)	19	10	2.29		

n	lattice L	l	min	density	extremal	best
20	$[SU_5(2) \overset{2(2)}{\circ} SL_2(3)]_{20}$	2	4	2.83	+	+
	$[2.M_{12}.2]_{20}$	2	4	2.83	+	+
	$[2.M_{22}.2]_{20}$	7	8	3.02	+	+
	$[2.L_3(4).2^2]$	7	5	1.89	−	−
	$\Lambda^3 A_6$	7	4	1.51	−	−
	$[L_2(11) \overset{2(3)}{\boxtimes} D_{12}]_{20}$	11	8	2.41	−	
	$A_2 \otimes A_{10}^{(3)}$	33	12	2.09		
22	$A_{22}^{(6)}$ (Craig)	23	12	2.50		
24	$[2.Co_1]_{24}$ (Leech)	1	4	4	+	+
	$[6.U_4(3).2 \overset{2}{\underset{\sqrt{-3}}{\boxtimes}} SL_2(3)]_{24}$	2	4	2.83	+	+
	$[6.Alt_7:2]_{24}$	2	4	2.83	+	+
	$[6.L_3(4).2 \overset{2(2)}{\otimes} D_8]_{24}$	6	8	3.27	−	+
	$[(SL_2(3) \circ C_4).2 \overset{2(3)}{\underset{\sqrt{-1}}{\boxtimes}} U_3(3)]_{24}$	6	8	3.27	−	+
	$[2.J_2 \overset{2}{\square} SL_2(5)]_{24}$	5	8	3.58	+	+
	$[L_2(7) \overset{2(2)}{\boxtimes} F_4]_{24}$	7	8	3.02	−	
	$F_4 \otimes A_6^{(2)}$	14	8	2.14		
	$[SL_2(13) \overset{2(2)}{\square} SL_2(3)]_{24}$	13	12	3.33	−	
	$[3.M_{10} \overset{2(2)}{\underset{\sqrt{-3}}{\boxtimes}} SL_2(3)]_{24}$	10	8	2.53		
	$[3.M_{10} \overset{2(2)}{\boxtimes} D_8]_{24}$	30	16	2.92		
	$A_2 \otimes [L_2(7) \overset{2(2)}{\boxtimes} D_8]_{12}$	42	16	2.47		
	$[SL_2(11) \overset{2(2)}{\underset{\sqrt{-11}}{\boxtimes}} SL_2(3)]_{24}$	22	12	2.56		
26	$[\pm S_4(5).2]_{26}$	1	3	3	−	+
	$[\pm S_6(3) \overset{2}{\square} C_3]_{26}$	3	6	3.46	+	+
	$[\pm S_4(5):2]_{26}$	5	5	2.24		
28	$F_4 \otimes E_7$	2	4	2.83	+	+
	$[\pm S_3 \overset{2(3)}{\otimes} G_2(3)]_{28}$	3	6	3.46	+	+
	$A_2 \otimes M_{14,2}$	39	14	2.42	−	−
	$[\pm L_2(13) \overset{2(3)}{\boxtimes} S_3]_{28}$	39	22	3.52		
30	$[\pm 3.U_4(3).2^2]_{30}$	3	6	3.46	+	+
	$A_2 \tilde{\otimes} A_{15}$	3	4	2.31	−	−
	$M_{30,2}$	29	15	2.79		
	$A_{30}^{(8)}$ (Craig)	31	16	2.87		

The notation for the lattices in Table 2 is explained in [17], Part I. But even without that explanation it is often easy to extract a simple non-abelian composition factor of the automorphism group of the lattice from the symbol. In the table also odd l-modular lattices are listed, which strictly speaking were excluded from the discussion above.

Among the many problems of when extremal lattices exist, I single out just one which is still open: does there exist an even unimodular lattice of dimension 72 and minimum 8?

References

[1] A.-M. Bergé, J. Martinet and F. Sigrist, Une généralisation de l'algorithme de Voronoi, *Astérisque* **209** (1992), 137–158.

[2] H. I. Blau, On real and rational representations of finite groups, *J. Algebra* **150** (1992), 57–72.

[3] F. Bruhat and J. Tits, Groupes réductives sur un corps local I. Donnée radicielles valuées, *Publ. Math. I. H. E. S.* **41** (1972), 5–251.

[4] J. H. Conway, R. T. Curtis, S. P. Norton, R. A. Parker and R. A. Wilson, ATLAS *of finite groups*, Oxford University Press, 1985.

[5] J. H. Conway, N. J. A. Sloane, *Sphere packings, lattices and groups*, Springer-Verlag, 1988.

[6] A. W. M. Dress, Induction and structure theorems for orthogonal representations of finite groups, *Ann. of Math.* **102** (1975), 291–325.

[7] W. Feit, On integral representations of finite groups, *Proc. London Math. Soc. (3)* **29** (1974), 633–683.

[8] W. Feit, *The representation theory of finite groups*, North-Holland, 1982.

[9] R. Gow, Unimodular integral lattices associated with the basic spin representations of $2A_n$ and $2S_n$, *Bull. London Math. Soc.* **21** (1989), 257–262.

[10] B. H. Gross, Group representations and lattices, *J. Amer. Math. Soc.* **3** (1990), 929–960.

[11] C. Jansen, K. Lux, R. Parker and R. Wilson, *An* ATLAS *of Brauer characters*, London Math. Soc. Monographs, New Series 11. Oxford University Press, 1995.

[12] A. I. Kostrikin and Pham Huu Tiep, *Orthogonal decompositions and integral lattices*, De Gruyter Expositiones in Math. 15, Berlin, 1994.

[13] V. Landazuri and G. M. Seitz, On the minimal degrees of projective representations of the finite Chevalley groups, *J. Algebra* **32** (1974), 418–443.

[14] G. Nebe, *Endliche rationale Matrixgruppen vom Grad 24*, Dissertation. Aachener Beiträge zur Mathematik 12, Verlag Augustinus Buchhandlung, Aachen, 1995.

[15] G. Nebe, Finite subgroups of $GL_{24}(\mathbb{Q})$, *Experimental Math.* **5** (1996), 163–195.

[16] G. Nebe, Finite subgroups of $GL_n(\mathbb{Q})$ for $25 \leq n \leq 31$, *Comm. Algebra* **24** (1996), 2341–2397.

[17] G. Nebe and W. Plesken, *Finite rational matrix groups*, AMS Memoirs 556, (1995), Part I: *Finite rational matrix groups*, 1–73; Part II: *Finite rational matrix groups of degree 16*, 74–144.

[18] J. Opgenorth, *Normalisatoren und Bravaismannigfaltigkeiten endlicher unimodularer Gruppen*, Dissertation. Aachener Beiträge zur Mathematik 16, Verlag Augustinus Buchhandlung, Aachen, 1996.

[19] W. Plesken, *Group rings over p-adic integers*, Lecture Notes in Mathematics, 1026, Springer, 1983.

[20] W. Plesken, Some applications of representation theory, in *Progress in mathematics, Vol. 95: Representation theory of finite groups and finite-dimensional algebras, (eds. G. O. Michler and C. M. Ringel)*, pp. 477–496. Birkhäuser, 1991.

[21] W. Plesken and M. Pohst, On maximal finite irreducible subgroups of $GL(n, \mathbb{Z})$. I. The five- and seven-dimensional case. II. The six-dimensional case, *Math. Comp.* **31** (1977), 536–577; III. The nine-dimensional case. IV. Remarks on even dimensions with applications to $n = 8$. V. The eight-dimensional case and a complete description of dimensions less than ten, *Math. Comp.* **34** (1980), 245–301.

[22] W. Plesken and B. Souvignier, Computing isometries of lattices, *J. Symbolic Comput.*, to appear.

[23] H.-G. Quebbemann, Modular lattices in Euclidean spaces, *J. Number Theory* **54** (1995), 190–202.

[24] W. Scharlau, *Quadratic and Hermitian forms*, Springer-Verlag, 1985.

[25] G. M. Seitz and A. E. Zalesskii, On the minimal degrees of projective representations of the finite Chevalley groups, II, *J. Algebra* **158** (1993), 233–243.

[26] J.-P. Serre, *Trees*, Springer-Verlag, 1980.

[27] J. G. Thompson, Finite groups and even lattices, *J. Algebra* **38** (1976), 523–524.

[28] J. G. Thompson, A simple subgroup of $E_8(3)$, in *Finite groups symposium (ed. N. Iwahori)*, pp. 113–116. Japan Society for the Promotion of Science, 1976.

[29] Pham Huu Tiep, Basic spin representations of $2S_n$ and $2A_n$ as globally irreducible representations, *Arch. Math.* **64** (1995), 103–112.

[30] Pham Huu Tiep, Globally irreducible representations of finite groups and integral lattices, *Geom. Dedicata* **64** (1997), 85–123.

[31] A. Turull, Bilinear forms for $SL(2, q)$, $\tilde{A}_n$ and similar groups, *Public. Mathemàtiques* **36** (1992), 1001–1010.

[32] A. Turull, Schur index two and bilinear forms, *J. Algebra* **157** (1993), 562–572.

[33] G. L. Watson, Transformations of a quadratic form which do not increase the class-number, *Proc. London Math. Soc. (3)* **12** (1962), 577–587.

Chamber graphs of some sporadic group geometries

Peter Rowley

1 Introduction

Early definitions of buildings, following the fundamental work of Tits [19], were couched in terms of simplicial complexes (see also [18, page 319]). More recently an alternative viewpoint, as expounded in [20] (or [11]), which brings chamber systems to the fore, has grown in importance.

Beginning with [1] a search commenced for geometric structures which would perform a similar service for the sporadic finite simple groups as buildings do for the finite simple groups of Lie type. That is, illuminate the internal structure of the sporadic finite simple groups and also place them in a wider context. This aim has yet (if ever?) to be realized. Nevertheless, many interesting and varied geometries have been unearthed in the past 15 or so years (see, for example, [2], [9], [10], [6]). Various aspects of these geometries have been studied—for example a great deal of effort has been expended on their point-line collinearity graphs (see, for example, [12], [13], [14], [17]). The associated chamber graph has received much less attention to date. This is surprising given that the building axioms and many of the concepts relating to buildings can be encoded in the chamber graph (of a building).

Here we survey some of the material in [15] and [16], where some tentative steps are taken in the study of chamber graphs of certain sporadic group geometries. Below we recall the definition of a geometry and a chamber system together with some related concepts, and notation sufficient for our purposes.

In section 2 we begin by parading the three geometries we shall be putting the spotlight on—the C_3-geometry for $Alt(7)$, the alternating group of degree 7, the M_{23} Petersen geometry and the M_{24} maximal 2-local geometry. There then follows an assortment of properties and observations about their chamber graphs, with particular emphasis on the geodesic closure of opposite chambers (defined in section 2.3). Section 3 is devoted exclusively to the chamber graph of the M_{24} maximal 2-local geometry, concentrating on maximal opposite sets of chambers. Much tabulated information is included which, with further study, may possibly lead to interesting configurations. Also in this section we answer in Theorem 3.1 a question raised by Conway at "The ATLAS Ten Years On" conference, after the lecture, of which this paper is an extended account.

1.1 Geometries and chamber systems

A *geometry* (over a set I) is a triple $(\Gamma, \tau, *)$ where Γ is a set, τ an onto map from Γ to I and $*$ a symmetric relation on Γ which satisfies the property that, for $x, y \in \Gamma, x * y$ implies $\tau(x) \neq \tau(y)$. Usually τ is called the *type map* and $*$ the *incidence relation* of the geometry Γ, where $(\Gamma, \tau, *)$ is abbreviated to Γ. For $x \in \Gamma$ with $\tau(x) = i$ we say x has type i, and a *flag* is a set of pairwise incident elements of Γ. The *rank* of Γ is the cardinality of I and the rank of a flag F is the cardinality of $\tau(F) = \{\tau(x) \mid x \in F\}$. For $i \in I$, put $\Gamma_i = \{x \in \Gamma \mid \tau(x) = i\}$. The automorphism group, Aut Γ, of Γ consists of all permutations of Γ which leave Γ_i invariant for each $i \in I$ and preserve the incidence relation. We say $G \leq$ Aut Γ is *flag transitive* on Γ if for any two flags F_1, F_2 of Γ of the same type, there exists $g \in G$ such that $g(F_1) = F_2$. For example, let V be an m-dimensional vector space with $m \geq 2$. Take Γ to be the set of all proper non-zero subspaces of V and $I = \{1, \ldots, m-1\}$. Then for $x, y \in \Gamma$, we define $x * y$ if either $x \subset y$ or $y \subset x$ holds (as subsets of V)—this gives us a geometry. A flag in this geometry is just a totally ordered (by inclusion) set of subspaces of V.

A *chamber system* (over the set I), denoted $(\mathbf{C}, (\mathbf{P}_i)_{i \in I})$, or $\mathbf{C}$ for short, consists of a set $\mathbf{C}$ and a system $(\mathbf{P}_i)_{i \in I}$ of partitions of $\mathbf{C}$ indexed by I. The elements of $\mathbf{C}$ are called *chambers*. Two chambers both in the same member of $\mathbf{P}_i$ for some $i \in I$ are said to be *i-adjacent*; we say two chambers are *adjacent* if they are i-adjacent for some $i \in I$. Starting with a geometry Γ (over I) we may construct a chamber system $\mathbf{C}$ (over I) as follows. Let $\mathbf{C}$ be the set of all maximal flags of Γ (a flag is maximal if it has rank equal to the rank of Γ). For $i \in I$ and $F_1, F_2 \in \mathbf{C}$ define F_1 and F_2 to be i-adjacent whenever $\tau(F_1 \cap F_2) = I\backslash\{i\}$. This yields an equivalence relation on $\mathbf{C}$ and hence a partition $\mathbf{P}_i$ of $\mathbf{C}$—we call $\mathbf{C}$ the chamber system of Γ. Observe that an automorphism of Γ induces an automorphism of $\mathbf{C}$—that is, a permutation of the chambers which respects i-adjacency for all $i \in I$. Furthermore, if G is a flag transitive group of automorphisms of Γ, then G will act transitively upon $\mathbf{C}$. In the example mentioned above, the chamber system consists of all totally ordered sets of $m-1$ (proper non-trivial) subspaces and two such would be i-adjacent if they have the same subspaces for all dimensions other than i. We remark that this chamber system is closely connected with buildings of type A_{m-1}(see [19] and [18, Section 3.3]).

1.2 Chamber graphs

Let **C** be a chamber system. The chamber graph of **C** (also denoted by **C**) has **C** as its vertex set and its edge set consists of all pairs of adjacent chambers. The chamber graphs we consider will be connected and have $G \le \mathrm{Aut}\,\mathbf{C}$ with G acting transitively on **C**. We take c_0 to be a fixed chamber of **C**, and set $B = \mathrm{Stab}_G(c_0)$. The distance function on the graph **C** will be denoted by $d(,)$, and we put

$$D_j(c_0) = \{c \in \mathbf{C} \mid d(c_0, c) = j\},$$

the jth disc (of c_0). Finally, a *geodesic* (also called a minimal gallery) from c to c', where $c, c' \in \mathbf{C}$, is a shortest path in the chamber graph **C** which starts with c and ends with c'.

2 Three chamber graphs

2.1 Three geometries

For each of $Alt(7)$, M_{23} and M_{24} we describe a geometry upon which the group acts flag transitively.

C_3 *geometry for* $Alt(7)$. Let Ω be a 7-element set. On Ω it is possible to define 30 (distinct) copies of a projective plane, upon which $Alt(7)$ has two orbits of size 15. Now let $\Gamma_0 = \Omega, \Gamma_1$ be the set of 3-element subsets of Ω and Γ_2 be one of the $Alt(7)$ orbits on projective planes. We call the elements in Γ_0, Γ_1, Γ_2, points, lines, planes respectively. A point is incident with a line if it is contained in the 3-element subset and a point is incident with all the planes, while a line is incident with a plane whenever the line is a line of the projective plane. This geometry was discovered by Neumaier [7]. The following diagram may be assigned to this rank 3 geometry (see [8, page 152]).

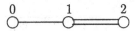

M_{23} *Petersen geometry.* This rank 4 geometry has as its diagram

where $\overset{P}{\circ\!\!-\!\!-\!\!\circ}$ is the geometry of the vertices and the edges of the Petersen graph (with a vertex and edge incident whenever the vertex is on the edge). To define this geometry we start with a 24-element set Ω upon which we have the Steiner system $S(24, 8, 5)$. Let a be a fixed element of Ω.

Now let Γ_0 consist of all octads of Ω not containing a, Γ_1 of all trios and Γ_2 of all the sextets of Ω (for the definition of octads, trios and sextets and other details concerning $S(24, 8, 5)$ see either [4] or [5]). Finally, let $\Gamma_3 = \Omega\backslash\{a\}$. An element in Γ_j and an element in Γ_k $(0 \leq j < k \leq 2)$ are defined to be incident whenever the corresponding partitions of Ω are compatible, and $b \in \Gamma_3$ is defined to be incident with $p \in \Gamma_j$ $(0 \leq j \leq 2)$ whenever $\{a, b\}$ is in the same part of the partition p (or in the octad in the case $j = 0$). Further information on this geometry is given in [6].

M_{24} maximal 2-local geometry. This geometry is particularly important as it occurs as a residue in a number of other geometries related to the sporadic simple groups (see [9]). Again we start with a 24-element set Ω and the Steiner system $S(24, 8, 5)$. Take Γ_0 to be all the octads of Ω, Γ_1 to be all the trios of Ω and Γ_2 to be all the sextets of Ω. Incidence is also given by when the corresponding partitions of Ω are compatible. So here a chamber consists of an octad, trio and sextet where the trio may be obtained from the sextet by pairing up the tetrads of the sextet and the octad is one of the octads of the trio. This geometry was first observed in [9].

To describe such chambers we use a labelling of Curtis's MOG [5]. So, for example, we describe a chamber c' by

$$c' = \begin{array}{|c c|c c|c c|} \hline 1\,1 & 3\,3 & 5\,5 \\ 1\,1 & 3\,3 & 5\,5 \\ 2\,2 & 4\,4 & 6\,6 \\ 2\,2 & 4\,4 & 6\,6 \\ \hline \end{array} \;.$$

Here, the positions of the MOG labelled by j $(1 \leq j \leq 6)$ give the tetrads of the sextet of c'. The trio of c' is understood to be specified by the pairing $12|34|56$ and the octad of c' is given by the pair 12.

For this geometry we let

$$c_0 = \begin{array}{|c|c|c|} \hline 1\,2 & 3\,4 & 5\,6 \\ 1\,2 & 3\,4 & 5\,6 \\ 1\,2 & 3\,4 & 5\,6 \\ 1\,2 & 3\,4 & 5\,6 \\ \hline \end{array} \;.$$

Note that c' and c_0 have the same octad and trio but different sextets and so they are 2-adjacent.

2.2   Disc sizes

The disc sizes of the three geometries are listed in Tables 1, 2 and 3.

Table 1: Disc sizes in the C_3-geometry for $Alt(7)$

Disc	1	2	3	4	5
Size	6	20	56	128	104

(315 chambers in total) [16]

Table 2: Disc sizes in the M_{23} Petersen geometry

Disc	1	2	3	4	5	6	7	8	9	10
Size	7	28	86	220	512	1128	2432	5152	10528	21024

11	12	13	14
38528	51840	26304	1600

(159390 chambers in total) [16]

Table 3: Disc sizes in the M_{24} maximal 2-local geometry

Disc	1	2	3	4	5	6	7	8	9	10
Size	10	44	184	544	1536	4800	10368	22272	38400	1536

(79695 chambers in total) [15]

Table 4: Sizes of the $\Delta_j(c_k)$ for the $Alt(7)$ C_3-geometry

	c_1	c_2	c_3	c_4	c_5	c_6	c_7	c_8	c_9	c_{10}	c_{11}	c_{12}	c_{13}
$|\Delta_1(c_k)|$	4	3	4	1	1	3	5	1	2	3	3	4	2
$|\Delta_2(c_k)|$	4	3	4	1	2	4	5	2	2	4	3	6	2
$|\Delta_3(c_k)|$	4	3	4	2	2	4	6	1	2	4	3	5	2
$|\Delta_4(c_k)|$	3	3	4	1	2	4	4	1	1	3	3	5	2

Remark 2.1 For the $Alt(7)$ C_3-geometry, $B \cong D_8$ has thirteen orbits (each of size 8) on $D_5(c_0)$. Representatives for each of these orbits are given in [16, Section 3] where it is also shown that the induced subgraph of $D_5(c_0)$ is connected. Quite the opposite is true for $D_{10}(c_0)$ in the M_{24} maximal 2-local geometry. There no two of the 1536 chambers in $D_{10}(c_0)$ are adjacent in **C** (that is, $D_{10}(c_0)$ is a coclique). Moreover, B acts transitively on $D_{10}(c_0)$. This is a rare phenomenon among those chamber systems studied in [16]. For instance, in the M_{23} Petersen geometry $|D_{14}(c_0)| = 1600$ and $|B| = 2^6$ and so B does not act transitively on $D_{14}(c_0)$.

2.3   Geodesic closures

Let **X** be a subset of **C**, a chamber graph. The *geodesic closure* $\overline{\mathbf{X}}$ of **X** is defined to be the set of all chambers of **C** which lie on some geodesic from c to c' where c, $c' \in \mathbf{X}$. (Note, however, that a 'geodesic closure' as here defined in not necessarily geodesically closed in the usual sense.) Let $r \in \mathbf{N}$ be maximal such that $D_r(c_0) \neq \emptyset$. Two chambers of **C** which are distance r apart will be called an *opposite pair of chambers*. In this section we look at geodesic closures of opposite pairs of chambers which should reveal certain aspects of **C**. In fact, the geodesic closure of an opposite pair of chambers in the chamber graph of a building of spherical type gives (the chambers of) an apartment. Also every apartment of the building arises in this manner (see, for example, Theorems 2.15 and 3.8 of [11]). For $c' \in D_r(c_0)$ we put

$$\Delta_j(c') = D_j(c_0) \cap \overline{\{c_0, c'\}} \quad (0 \leq j \leq r).$$

C_3 *geometry for* $Alt(7)$. Let c_0 be as in [16, Section 3] and let $c_1, \ldots, c_{13}$ be the B-orbit representatives of $D_5(c_0)$ tabulated in [16, (3.1)]. Then the sizes of the $\Delta_j(c_k)$ are as given in Table 4. In Fig. 1 we give the induced subgraph of $\overline{\{c_0, c_2\}}$. The labels 0, 1, 2 on the edge indicate the i-adjacency. So two chambers are 0-adjacent (respectively 1-, 2-adjacent) if they differ by a point (respectively a line, a plane).

M_{24} *maximal 2-local geometry.*   Let c_{10} denote the chamber

$$\begin{array}{|c|c|c|}
\hline
6\ 4 & 3\ 2 & 5\ 4 \\
3\ 3 & 5\ 1 & 5\ 6 \\
5\ 1 & 2\ 2 & 1\ 6 \\
2\ 3 & 6\ 4 & 4\ 1 \\
\hline
\end{array}.$$

Then $c_{10} \in D_{10}(c_0)$ (see the σ_0 table in (3.2) of [15]). Since $D_{10}(c_0)$ is a B-orbit we need only examine $\overline{\{c_0, c_{10}\}}$. We will write Δ_j instead of $\Delta_j(c_{10})$.

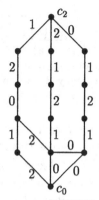

Figure 1: The induced subgraph of $\overline{\{c_0, c_2\}}$ in the $Alt(7)$ C_3-geometry

Table 5: Curtis's MOG numbering

∞	14	17	11	22	19
0	8	4	13	1	9
3	20	16	7	12	5
15	18	10	2	21	6

Altogether $\overline{\{c_0, c_{10}\}}$ has 118 chambers, the sizes of Δ_j being as follows:

j	0	1	2	3	4	5	6	7	8	9	10
$|\Delta_j|$	1	10	12	14	15	14	15	14	12	10	1

The symmetry of the Δ_j sizes is a consequence of B having one orbit on $D_{10}(c_0)$. In fact, the stabilizer in M_{24} of $\{c_0, c_{10}\}$ has order 4 and is generated by

$$x = (\infty, 3)(8, 18)(4, 10)(7, 11)(19, 21)(9, 22)(1, 5)(6, 12)$$
$$y = (\infty, 6)(0, 13)(3, 12)(15, 20)(14, 16)(8, 11)(7, 18)(2, 17)(4, 21)$$
$$(10, 19)(1, 22)(5, 9)$$

where the MOG elements are named as in [5] (see Table 5). Clearly y must stabilize $\{c_0, c_{10}\}$. In Appendix B of [15] the induced subgraph (plus i-adjacencies) of $\{c_0\} \cup \Delta_1 \cup \Delta_2 \cup \Delta_3 \cup \Delta_4 \cup \Delta_5$ is given—with the aid of y all of $\overline{\{c_0, c_{10}\}}$ may then be constructed. From this we see that in Δ_j for $2 \leq j \leq 8$ there is very little sideways adjacency (that is adjacency within Δ_j), which is very reminiscent of Coxeter complexes (see Lemma 24 of [11]). Also, particularly striking is that all those chambers in $\overline{\{c_0, c_{10}\}}$

having valency 3 (in $\overline{\{c_0, c_{10}\}}$) have adjacencies one of each type, while those chambers in $\overline{\{c_0, c_{10}\}}$ of valency 2 (in $\overline{\{c_0, c_{10}\}}$) all have adjacencies of types 0 and 2. A further inspection of $\overline{\{c_0, c_{10}\}}$ reveals that it contains two chambers of valency 10 (namely c_0 and c_{10}), two of valency 8, ten of valency 7, six of valency 5, fifty-two of valency 3 and twenty-six of valency 2. Consequently, $\text{Stab}_{M_{24}}(\overline{\{c_0, c_{10}\}}) = \langle x, y \rangle$ where x (given above) is of order 2 and $\langle x \rangle = \text{Stab}_B(c_{10})$.

In a building every pair of chambers is contained in some apartment. This is far from the case for geodesic closures (of opposite chambers) in the chamber graph of the M_{24} maximal 2-local geometry. Consulting the $\sigma_0^{(16)}$, $\sigma_0^{(192)}$, $\sigma_0^{(384)}$, $\sigma_0^{(768)}$ tables in [15] we see that $D_9(c_0)$ is the union of 16 B-orbits. Now, as the valency of c_{10} in $\overline{\{c_0, c_{10}\}}$ is 10 and $D_{10}(c_0)$ is a B-orbit, it follows that there are at least six of these B orbits such that no chamber in them can be in the geodesic closure (of opposite chambers) with c_0.

Of course, every geodesic from c_0 to c_{10} is contained in $\overline{\{c_0, c_{10}\}}$—for the record there are 74 such geodesics. Bearing in mind that $\mathbf{C}$ has 79695 chambers, $\overline{\{c_0, c_{10}\}}$ has the appearance of a very narrow corridor.

3 Maximal opposite sets of chambers

In this section $\mathbf{C}$ denotes the chamber graph of the M_{24} maximal 2-local geometry. We call a set $\mathbf{X}$ of chambers of $\mathbf{C}$ an *opposite set of chambers* if for all $c, c' \in \mathbf{X}$ with $c \neq c'$, we have $d(c, c') = 10$. A *maximal opposite set of chambers* is an opposite set of chambers of maximal possible size. We record the following fact.

Theorem 3.1 *A maximal opposite set of chambers in $\mathbf{C}$ has size 5.*

In view of the transitivity of M_{24} on pairs of chambers distance 10 apart, there is no loss in examining maximal opposite sets of chambers which contain c_0 and c_{10}. With the aid of CAYLEY [3] it was discovered that there are exactly 70 chambers in $D_{10}(c_0) \cap D_{10}(c_{10})$. Then, by hand, these 70 chambers were scrutinized and as a consequence Theorem 3.1 was observed. In fact, among the 70 chambers in $D_{10}(c_0) \cap D_{10}(c_{10})$, 52 of them are contained in at least one maximal opposite set of chambers which also contains c_0 and c_{10}. In Table 6 we list the 17 orbits of $\langle x, y \rangle$ on these 52 chambers, γ_i ($1 \leq i \leq 52$), together with the six orbits on the remaining 18. (The numbers in brackets are the orbit sizes.) The full set of 52 can be readily reconstructed from the action of x and y, which permute the subscripts of the γ_i as follows:

Table 6: Orbit representatives for $\langle x, y \rangle$ on $D_{10}(c_0) \cap D_{10}(c_{10})$

γ_1 (2)

5 6	1 5	3 4
2 1	6 2	2 2
6 4	4 1	3 5
6 3	3 1	5 4

γ_2 (4)

6 4	3 2	5 5
5 2	2 1	1 3
1 6	6 5	4 1
6 3	4 2	4 3

γ_4 (4)

3 2	4 5	4 6
1 5	6 5	2 2
4 6	1 4	2 3
3 3	6 1	1 5

γ_5 (4)

3 4	1 5	6 4
6 3	2 3	2 2
1 2	1 6	4 5
5 3	4 1	6 5

γ_6 (2)

1 2	2 3	3 6
5 3	1 5	3 4
6 5	2 4	6 4
5 4	6 2	1 1

γ_8 (2)

3 5	2 1	1 4
2 6	4 2	6 6
4 5	5 6	3 1
5 1	3 2	3 4

γ_{10} (2)

5 4	6 2	1 2
6 4	2 5	4 3
1 3	3 3	6 1
4 2	6 5	5 1

γ_{11} (4)

3 6	5 3	2 4
1 5	1 6	2 2
4 2	3 1	3 6
4 4	5 6	1 5

γ_{12} (4)

1 2	2 2	1 4
6 3	4 5	5 1
3 4	6 5	3 2
5 3	6 4	6 1

γ_{13} (4)

4 3	2 2	3 2
2 1	6 5	1 4
3 5	4 5	6 1
3 6	6 4	5 1

γ_{14} (4)

5 4	5 3	6 2
6 2	1 5	4 5
6 6	4 3	3 1
1 3	4 1	2 2

γ_{15} (2)

5 4	3 4	4 2
5 3	6 6	3 2
1 5	3 1	2 1
6 2	1 4	5 6

γ_{20} (4)

2 4	3 5	3 6
6 1	2 2	5 6
4 4	1 6	5 1
5 3	2 4	1 3

γ_{21} (4)

4 6	2 1	5 2
2 4	1 6	5 6
4 5	3 4	2 6
3 1	1 5	3 3

γ_{26} (2)

5 4	5 6	6 1
3 3	4 6	3 2
2 3	5 1	2 2
6 1	1 4	5 4

γ_{29} (2)

2 6	5 1	2 6
3 5	4 5	5 3
6 6	4 3	2 2
4 1	1 3	1 4

γ_{30} (2)

4 5	6 2	2 1
3 1	3 3	1 6
2 4	6 5	1 5
4 6	2 5	3 4

(4)

2 1	1 4	2 2
3 5	6 1	6 4
4 3	3 2	6 5
3 6	5 1	4 5

(4)

3 5	1 4	6 4
2 1	6 1	2 2
3 6	3 2	4 5
4 3	5 1	6 5

(4)

1 6	3 1	3 2
4 5	1 2	6 6
3 5	5 6	3 4
5 2	4 1	2 4

(2)

5 5	6 1	2 2
6 4	1 4	4 6
1 3	3 6	3 1
5 2	3 4	5 2

(2)

6 1	2 3	4 2
5 4	1 1	4 3
6 6	5 4	5 3
2 3	1 6	2 5

(2)

2 6	6 5	5 3
4 4	2 5	4 1
3 5	3 2	6 2
4 1	6 3	1 1

Table 7: Orbit representatives for $\langle x, y \rangle$ on the 54 maximal opposite sets

1, 12, 15	2, 4, 29	2, 41, 47	4, 35, 46	5, 10, 20	5, 10, 24
5, 20, 30	6, 13, 35	6, 21, 36	8, 11, 23	8, 11, 31	8, 23, 43
8, 25, 31	10, 20, 27	11, 13, 31	13, 26, 35	14, 18, 29	

$$x \; : \; (1,38)(2,34)(3,45)(4,50)(5,33)(6,32)(7,19)(8,9)(10,16)(11,28)$$
$$(12,42)(13,22)(14,31)(15,48)(17,46)(18,25)(20,23)(21,24)$$
$$(26,44)(27,43)(29,51)(30,41)(35,37)(36,39)(40,52)(47,49)$$
$$y \; : \; (1,38)(2,45)(3,34)(4,7)(5,49)(6)(8,9)(10,16)(11,46)(12,52)$$
$$(13,35)(14,25)(15)(17,28)(18,31)(19,50)(20,43)(21,36)(22,37)$$
$$(23,27)(24,39)(26)(29,51)(30)(32)(33,47)(40,42)(41)(44)(48).$$

In Table 7 we tabulate all the (orbits under $\langle x, y \rangle$ of) maximal opposite sets of chambers which contain c_0 and c_{10} (there we write i instead of γ_i and also omit c_0 and c_{10}). There are 54 such sets, and hence the total number of maximal opposite sets in $\mathbf{C}$ is

$$\frac{79695 \times 1536 \times 54}{2 \times 10} \; = \; 2^8.3^6.7.11.23$$
$$= \; 330,511,104.$$

With the help of x and y, a perusal of Table 7 reveals that there is quite a variation in the number of maximal opposite sets that contain $\{c_0, c_{10}, \gamma_i\}$ ($1 \le i \le 52$). Indeed, 8 of the γ_i are contained in a unique maximal opposite set, 16 are in 2, 4 are in 3, 16 are in 4, 6 are in 5 and 2 are in 6 maximal opposite sets. From this angle things look (possibly) more interesting.

ACKNOWLEDGEMENTS

I am grateful to R. T. Curtis and R. A. Wilson for permission to include the observations below on the M_{24} orbits on maximal opposite sets.

References

[1] F. Buekenhout, Diagrams for geometries and groups, *J. Combin. Theory (A)* **27** (1979), 121–151.

[2] F. Buekenhout, Diagram geometries for sporadic groups, in *Finite groups—coming of age (ed. J. McKay)*, pp. 1–32. Contemp. Math. Ser., Vol. 45, Amer. Math. Soc., 1985.

[3] J. J. Cannon, An introduction to the group theory language, Cayley, in *Computational group theory (ed. M. D. Atkinson)*, pp. 145–183. Academic Press, London/New York, 1984.

[4] J. H. Conway, Three lectures on exceptional groups, in *Finite simple groups (eds. M. B. Powell and G. Higman)*, pp. 215–247. Academic Press, New York, 1971.

[5] R. T. Curtis, A new combinatorial approach to M_{24}, *Math. Proc. Cambridge Philos. Soc.* **79** (1976), 25–42.

[6] A. A. Ivanov and S. V. Shpectorov, Geometries for sporadic groups related to the Petersen graph, II, *European J. Combin.* **10** (1989), 347–362.

[7] A. Neumaier, Some sporadic geometries related to PG(3,2), *Arch. Math. (Basel)* **42** (1984), 89–96.

[8] A. Pasini, *Diagram geometries*, Oxford University Press, 1994.

[9] M. A. Ronan and S. D. Smith, 2-local geometries for some sporadic groups, *Proc. Symp. Pure Math.* **37** (1980), 283–289.

[10] M. Ronan and G. Stroth, Minimal parabolic geometries for the sporadic groups, *European J. Combin.* **5** (1984), 59–91.

[11] M. Ronan, *Lectures on buildings*, Academic Press, 1989.

[12] P. Rowley and L. Walker, A characterization of the Co_2-minimal parabolic geometry, *Nova J. Algebra Geom.* **3** (1994), 97–157.

[13] P. Rowley and L. Walker, On the Fi_{22}-minimal parabolic geometry, *Geom. Dedicata* **61** (1996), 121–167.

[14] P. Rowley and L. Walker, The maximal 2-local geometry for J_4, I, Preprint 1994/11, Manchester Centre for Pure Mathematics.

[15] P. Rowley, The chamber graph of the M_{24} maximal 2-local geometry, Preprint 1996/8, Manchester Centre for Pure Mathematics.

[16] P. Rowley, Disc structure of certain chamber graphs, *in preparation*.

[17] Y. Segev, On the uniqueness of the Co_1 2-local geometry, *Geom. Dedicata* **25** (1988), 159–219.

[18] M. Suzuki, *Group theory 1*, Springer-Verlag, Berlin, 1982.

[19] J. Tits, *Buildings of spherical type and finite BN-pairs*, Lecture Notes 386, Springer, Berlin, 1974.

[20] J. Tits, A local approach to buildings, in *The geometric vein: Coxeter Festschrift (eds. C. Davis, B. Grünbaum, F. A. Sherk)*, pp. 519–547. Springer, New York, 1981.

Editors' note. It is natural to ask what the orbits of M_{24} are on the maximal opposite sets of chambers. It turns out that there are just five orbits, represented by $(6, 13, 35)$, $(13, 26, 35)$, $(8, 11, 31)$, $(6, 21, 36)$ and $(10, 20, 27)$. The stabilizers are respectively 2, 2, 4, 5:4 and 5:4. To see that there are at least four orbits, we introduce an invariant. For each maximal opposite set of chambers, we consider the five corresponding octads, and for each of these octads we join two points in the octad whenever they both lie in one of the other four octads. This gives us a set of five graphs, each with eight points and four edges. In the five cases listed above, the sizes of the components of these five graphs give the respective invariants:

$$
\begin{aligned}
6, 13, 35 \;&:\; (3311, 4211, 4211, 5111, 5111) \\
13, 26, 35 \;&:\; (2222, 3221, 3221, 4211, 4211) \\
8, 11, 31 \;&:\; (41111, 4211, 4211, 4211, 4211) \\
6, 21, 36 \text{ and } 10, 20, 27 \;&:\; (2222, 2222, 2222, 2222, 2222)
\end{aligned}
$$

In each case except the component of size 4 in the first graph for the third case, which is a square, all these components are paths.

We know that the stabilizer of c_0 and c_{10} is a group of order 2, generated by x, and x does not fix any of the listed maximal opposite sets of chambers. It follows at once that the stabilizer of a maximal opposite set contains no elements of order 3, and no elements acting as a transposition of S_5 on the five chambers. Therefore in all five cases the stabilizer can be no bigger than that stated above.

On the other hand, we can exhibit groups of at least the given sizes which stabilize the respective maximal opposite sets. The first two are stabilized by y only, while the third is fixed by the element

$$(\infty, 8, 3, 14)(0, 18, 15, 20)(1, 13, 21, 11)(2, 12, 7, 22)(4, 5, 10, 6)(9, 17, 19, 16)$$

of M_{24}. The case $(6, 13, 35)$ is invariant under y and the element

$$(\infty, 20, 0, 8)(1, 9, 5, 22)(2, 16, 11, 10)(3, 14, 15, 18)(4, 7, 13, 17)(6, 12, 19, 21)$$

of M_{24}, which together generate 5:4. In particular, the stabilizer is 2-transitive on the five chambers, which implies that the only other image of it which also contains c_0 and c_{10} is its image under x. It follows that it is not in the same M_{24}-orbit as $(10, 20, 27)$, whose stabilizer is also 5:4, generated by xy and the element

$$(\infty, 14, 15, 18)(0, 20, 3, 8)(1, 2, 9, 17)(4, 21, 7, 6)(5, 10, 12, 11)(13, 19, 16, 22).$$

An Atlas of sporadic group representations

Robert A. Wilson

Abstract

We describe a collection of several hundred representations of 25 of the 26 sporadic groups, and many related groups. All these representations are available on the world-wide web.

Introduction

The 'ATLAS of finite groups' [5] was originally conceived by its authors as Volume 1 of a series, as its subtitle 'Maximal subgroups and ordinary characters for simple groups' might suggest. In the event, subsequent volumes have been rather slow to appear, with Volume 2, the 'sc Atlas of Brauer characters' (or 'ABC' for short [8]), being published in 1995, just in time for this conference. Indeed, even this is only Part 1 of Volume 2, as the accidentally undeleted subtitle on page 1 proclaims, in that it only includes groups of order up to 10^9.

At this conference, several suggestions for Volume 3 have been made, most involving large quantities of data stored on computers. It seems likely that whatever Volume 3 eventually turns out to be, it will not be a big heavy book of the type hitherto associated with the word 'ATLAS'.

My own submission as a candidate for Volume 3 is a collection of explicit representations of groups. A number of these were mentioned in the 'ATLAS of finite groups' under the now notorious phrase 'Explicit matrices have been computed.' Many others have been computed since. In fact, it is difficult to know where to stop with such a collection of representations, and it (like many databases) could easily be allowed to expand to fill all the disk-space available.

For the moment, this 'ATLAS of group representations' concentrates on the sporadic simple groups, and we now have (in principle) representations of all of these and their covers and automorphism groups, with the two exceptions of the Monster[1] and the double cover of the Baby Monster. A list of the

---

[1] *Note added in proof.* This deficiency has now been corrected. A computer construction, completed on 29th May 1997, is described in [14] (see also [38]).

groups and representations that are included is given in Table 1, though it will surely be out-of-date by the time it is printed. The reader may object that there is no representation of $6 \cdot Fi_{22}$ or $6 \cdot Fi_{22}{:}2$ in the list, but this should not matter too much, as representations of $2 \cdot Fi_{22}{:}2$ and $3 \cdot Fi_{22}{:}2$ are given. (In fact, there seems to be no convenient faithful representation of $6 \cdot Fi_{22}{:}2$ to construct. Perhaps the best would be as a group of permutations on 370656 points.)

One obvious direction in which this 'ATLAS' could be extended is to include all exceptional covers of generic groups. There seems to be no serious obstacle (except lack of time and energy) to doing so, although it may be quite a challenge to construct a representation of $(2^2 \times 3) \cdot {}^2E_6(2){:}S_3$. Another possible direction is to consider characteristic 0 matrix representations of reasonably small degree—this is an obvious area for application of R. A. Parker's new integral Meat-axe [24].

Indeed, in the years since this paper was first written we have added a great deal, and set up a world-wide-web site to make these representations generally available. The URL is http://www.mat.bham.ac.uk/atlas/. The total number of representations contained here is now over 600.

The representations collected here have been obtained in various ways, which can be roughly divided into the following four categories.

1. From existing literature on hand constructions, or constructions involving limited use of a computer. For example, some representations of the Mathieu groups and Leech lattice groups, as well as J_1 and Ru.

2. From existing computer constructions. For example, representations of $3 \cdot J_3$, $3 \cdot McL$, the Fischer groups, Ly, Th, and J_4.

3. By constructing representations *ab initio*. For example, representations of $4 \cdot M_{22}$, HN, B, and new representations of $O'N$, Ly, Fi_{24} and He.

4. By constructing new representations from old ones. For example, other representations of the Mathieu groups, J_1, Ru, Suz, Fi_{22}, and others.

We now consider these various methods in more detail.

1   Existing 'hand' constructions

Here we also include some constructions which were originally computer-assisted, but which are small enough for group generators to be entered by hand. Most of these constructions are derived either from the constructions of M_{12} and M_{24} by Mathieu [17], [18], or the construction of Co_1 by Conway [2]. In particular, all the listed permutation representations of the Mathieu groups

Table 1: Available representations of sporadic groups

Group	Degree	Field		Group	Degree	Field
M_{11}	11	1		$4 \cdot M_{22}$	56	25
	12	1			16	49
	10	2			56	121
	5	3		$4 \cdot M_{22}{:}2$	32	7
	16	4		$6 \cdot M_{22}$	36	121
M_{12}	16	4		$6 \cdot M_{22}{:}2$	72	11
$M_{12}{:}2$	24	1		$12 \cdot M_{22}$	48	25
	10	2			24	121
$2 \cdot M_{12}$	24	1		$12 \cdot M_{22}{:}2$	48	11
$2 \cdot M_{12}{:}2$	48	1		J_2	6	4
	10	3		$J_2{:}2$	100	1
	12	3			14	5
J_1	266	1		$2 \cdot J_2$	6	9
	1045	1		$2 \cdot J_2{:}2$	12	3
	1463	1		M_{23}	23	1
	1540	1			11	2
	1596	1			280	23
	2926	1		HS	100	1
	4180	1			176	1
	20	2		$HS{:}2$	100	1
	56	9			352	1
	56	5			15400	1
	31	7			20	2
	7	11			22	3
	22	19			21	5
M_{22}	22	1			896	11
$M_{22}{:}2$	22	1		$2 \cdot HS$	28	5
	77	1		$2 \cdot HS{:}2$	112	3
	10	2			56	5
$2 \cdot M_{22}$	10	7		J_3	18	9
$2 \cdot M_{22}{:}2$	10	7		$J_3{:}2$	6156	1
$3 \cdot M_{22}$	6	4			36	3
	21	25		$3 \cdot J_3$	9	4
	21	7		$3 \cdot J_3{:}2$	18	2
	21	121		M_{24}	24	1
$3 \cdot M_{22}{:}2$	12	2				

Group	Degree	Field	Group	Degree	Field
McL	21	5	Co_2	2300	1
	1200	25		4600	1
$McL{:}2$	21	5		22	2
$3{\cdot}McL$	396	4		24	2
	45	25		23	3
$3{\cdot}McL{:}2$	90	5	Fi_{22}	3510	1
He	51	4	$Fi_{22}{:}2$	3510	1
	51	25		78	2
$He{:}2$	2058	1		77	3
	102	2	$2{\cdot}Fi_{22}$	28160	1
	102	5		176	3
	50	7		352	5
Ru	4060	1	$2{\cdot}Fi_{22}{:}2$	352	3
	28	2	$2{\cdot}Fi_{22}{\cdot}4$	352	5
$2{\cdot}Ru$	16240	1	$3{\cdot}Fi_{22}$	27	4
	28	5	$3{\cdot}Fi_{22}{:}2$	54	2
Suz	1782	1	HN	132	4
$Suz{:}2$	64	3	$HN{:}2$	264	2
	1782	1		133	5
$2{\cdot}Suz$	12	3	Ly	111	5
$2{\cdot}Suz{:}2$	12	3		651	3
$3{\cdot}Suz$	12	4		2480	4
$3{\cdot}Suz{:}2$	24	2	Th	248	2
	5346	1		248	3
$6{\cdot}Suz$	24	3	Fi_{23}	31671	1
$6{\cdot}Suz{:}2$	24	3		782	2
$O'N$	154	3		253	3
$O'N$	406	7	Co_1	24	2
$O'N{:}2$	154	9	$2{\cdot}Co_1$	24	3
$O'N{:}4$	154	3	J_4	112	2
$3{\cdot}O'N$	153	4	$Fi'_{24}{:}2$	306936	1
$3{\cdot}O'N{:}2$	90	7		781	3
	306	2	$3{\cdot}Fi'_{24}$	920808	1
Co_3	22	2		783	4
	22	3	$3{\cdot}Fi'_{24}{:}2$	1566	2
	276	1	B	4370	2

Note: we adopt the convention that an underlying 'field' of order 1 signifies a permutation representation.

are easily obtained in this way. Similarly, the 24-dimensional characteristic zero matrix representation of $2 \cdot Co_1$ can be written over the integers and reduced modulo any prime, and easily gives rise to all the listed matrix representations of Co_2, Co_3, Suz and its decorations, as well as those of dimension 20, 21 and 22 for HS and McL.

Other constructions of this type are the 28-dimensional representation of $2 \cdot Ru$ by Conway and Wales [4] (see also Conway's later simplification [1]), as well as the 36-dimensional representation of $3 \cdot J_3 : 2$ [3]. Easiest of all is the 7-dimensional representation of J_1 described by Janko [7]. Perhaps here we should also mention the 9-dimensional representation of $3 \cdot J_3$ over $GF(4)$, first found by Richard Parker using the Meat-axe, later tidied up by Benson and Conway into the form given in the ATLAS.

Some small permutation representations which can be obtained by hand include the representations of HS and J_2, and their automorphism groups, on 100 points, and $HS : 2$ on 352 points.

2 Existing computer constructions

Under this heading there are some famous constructions, as well as a number of unpublished ones which have often been duplicated. In addition, there are some constructions which were essentially done by hand, but which are simply too big to enter into a computer in any simple-minded manner.

Perhaps the most important original matrix construction was that of J_4 in dimension 112 over $GF(2)$ by Norton and Parker [21], pioneering a technique that has since become standard (see [25]). Other important ones to mention are Parker's construction [19] of the Lyons group in $O_{111}(5)$, and $3 \cdot O'N$ in $GL_{45}(7)$ (see [28]), as well as $3 \cdot Fi_{22}$ in $GU_{27}(2)$. The Thompson group was constructed by Smith [31] in characteristic 0, and an analogous construction in characteristic 3 was given by Linton [12]. An explicit construction of the Held group was given by Ryba [27].

There are also some important constructions of permutation representations by coset enumeration. An early example was the construction of J_3 by Higman and McKay as a permutation group on 6156 points [6]. The permutation representations of the Fischer groups can also be considered in this category, especially the representations of Fi_{24} and $3 \cdot Fi_{24}$ which were provided for us by Steve Linton, using his double-coset enumerator [13].

3 *Ab initio* constructions

The impetus to start making a systematic collection of matrix representations of the sporadic groups came in June 1991 when Klaus Lux asked me for representations of several of the large sporadic groups. Searching through my files, I found two or three of these, but did not find HN or Fi_{24}. Accordingly, I tried to construct HN, choosing the 133-dimensional orthogonal representation over $GF(5)$. This construction took me three days [30], so I next tackled the 781-dimensional orthogonal representation of Fi_{24} over $GF(3)$, which took only two days [40]. The following week I constructed the 4370-dimensional representation of the Baby Monster over $GF(2)$ (see [39]). In each case, the construction follows Parker's method [25].

Subsequently I returned to the subject at intervals when suitable interesting construction problems presented themselves. Ibrahim Suleiman and I gave the first (so far as we are aware) explicit construction of $4 \cdot M_{22}$, a group which a few years previously had been 'proved' not to exist.

While working with Christoph Jansen on computing modular character tables, we decided to tackle the very challenging problem of determining the 2-modular character table of the O'Nan group. After some exploratory calculations he suggested that it was possible that the reduction of the degree 495 characters modulo 2 might contain a degree 342 character—if so then $3 \cdot O'N$ would have an irreducible (unitary) 153-dimensional representation over $GF(4)$. It seemed clear to me that the easiest way to prove this would be to construct the representation from scratch—which we did the next day [9]. (In fact, we went on to complete the 2-modular character table [10] soon afterwards.)

By a remarkable coincidence, a very similar chain of events led us to the construction of a 154-dimensional orthogonal representation of $O'N$ over $GF(3)$. This suggested to us that we should look for other such 'surprising' representations to construct, and in fact we showed that there was only one more (with a suitable definition of 'surprising'), namely a 651-dimensional orthogonal representation of the Lyons group over $GF(3)$, a construction of which is described in [11]. A similar construction gives the 2480-dimensional unitary representation over $GF(4)$ (see [41]).

4 Standard generators

Before considering the various ways of constructing new representations from old ones, it is worth pausing briefly to discuss generators for the groups. Each representation is most conveniently stored as a list of matrices (or permutations) giving the images of certain group generators in that representation. For various reasons it is important to standardize the generators for each given group. For example, the tensor product of two representations of a group G can only be made if the same generators for G are available in both representations. A discussion of some of the issues involved in choosing such 'standard generators' can be found in [42], and some implications and applications are explored in [34], using the specific example of the group J_3.

Here we simply list in Tables 2 and 3 the defining properties of our standard generators for the groups G and G:2, where G is a sporadic simple group. For the present, we consider generators for covering groups to be standard if they map to standard generators of G or G:2 under the natural quotient map. Thus they are *not* (yet) defined up to automorphisms[2].

5 New representations from old

There are many techniques available for obtaining new representations for a group from old ones. The basic method, which was the rationale behind the original development of the Meat-axe [23], is to tensor two matrix representations (over the same field) together, and then chop up the result into irreducibles. This enables many representations in the same characteristic as the original to be constructed. Various technical refinements can be used to extend the range of this basic technique. For example, use of symmetric and exterior squares, and other higher symmetrized powers, in addition to tensor products. The ideas of condensation, exploited by Ryba [29], Lux and Wiegelmann [15], and others, can also be used here.

A matrix representation will yield a permutation representation by calculating the action of the group on an orbit of vectors (or 1-dimensional subspaces, or k-dimensional subspaces, or any other convenient objects). A permutation representation can be reduced modulo any prime and chopped up with the Meat-axe into irreducibles—in this context the condensation method really comes into its own (see for example [10], among many others). These two ideas together enable one to change characteristic—that is, given a representation of G over a field of characteristic p, obtain one over a field of characteristic q.

Some of these techniques can change the group being represented. For example, the tensor product of two faithful irreducible representations of a double cover $2 \cdot G$ will represent only G, since the central involution acts as the scalar $(-1) \times (-1) = +1$. Similarly, if an orbit of subspaces is permuted in a matrix representation, then in the resulting permutation representation the scalars act trivially. Another useful example here is the following: to obtain a representation of $12 \cdot M_{22}$, take the tensor product of a representation of $3 \cdot M_{22}$ and a representation of $4 \cdot M_{22}$.

There are two other important ways of changing the group. First, if H is a subgroup of G, and we can find words in our generators of G which give generators of H, then any representation of G can be restricted to H. Second, if we have a representation of G then we can construct a representation of

---

[2] *Note added in proof.* This deficiency has also now been corrected in most cases.

Table 2: Standard generators of sporadic simple groups

Group	Triple (a, b, ab)	Further conditions
M_{11}	$2, 4, 11$	$o((ab)^2(abab^2)^2ab^2) = 4$
M_{12}	$2B, 3B, 11$	none
J_1	$2, 3, 7$	$o(abab^2) = 19$
M_{22}	$2A, 4A, 11$	$o(abab^2) = 11 (\iff o(ab^2) = 5)$
J_2	$2B, 3B, 7$	$o([a, b]) = 12$
M_{23}	$2, 4, 23$	$o((ab)^2(abab^2)^2ab^2) = 8$
HS	$2A, 5A, 11$	none
J_3	$2A, 3A, 19$	$o([a, b]) = 9$
M_{24}	$2B, 3A, 23$	$o(ab(abab^2)^2ab^2) = 4$
McL	$2A, 5A, 11$	$o((ab)^2(abab^2)^2ab^2) = 7$
He	$2A, 7C, 17$	none
Ru	$2B, 4A, 13$	none
Suz	$2B, 3B, 13$	$o([a, b]) = 15$
$O'N$	$2A, 4A, 11$	none
Co_3	$3A, 4A, 14$	none
Co_2	$2A, 5A, 28$	none
Fi_{22}	$2A, 13, 11$	$o((ab)^2(abab^2)^2ab^2) = 12$
HN	$2A, 3B, 22$	$o([a, b]) = 5$
Ly	$2, 5A, 14$	$o(ababab^2) = 67$
Th	$2, 3A, 19$	none
Fi_{23}	$2B, 3D, 28$	none
Co_1	$2B, 3C, 40$	none
J_4	$2A, 4A, 37$	$o(abab^2) = 10$
$Fi_{24}{}'$	$2A, 3E, 29$	$o((ab)^3b) = 33$
B	$2C, 3A, 55$	$o((ab)^2(abab^2)^2ab^2) = 23$
M	$2A, 3B, 29$	none

Table 3: Standard generators of automorphism groups of sporadic groups

Group	Triple (a, b, ab)	Further conditions
$M_{12}{:}2$	$2C, 3A, 12$	$ab \in 12A(\iff o([a,b]) = 11)$
$M_{22}{:}2$	$2B, 4C, 11$	none
$J_2{:}2$	$2C, 5AB, 14$	none
$HS{:}2$	$2C, 5C, 30$	none
$J_3{:}2$	$2B, 3A, 24$	$o([a,b]) = 9$
$McL{:}2$	$2B, 3B, 22$	$o((ab)^2(abab^2)^2ab^2) = 24$
$He{:}2$	$2B, 6C, 30$	none
$Suz{:}2$	$2C, 3B, 28$	none
$O'N{:}2$	$2B, 4A, 22$	none
$Fi_{22}{:}2$	$2A, 18E, 42$	none
$HN{:}2$	$2C, 5A, 42$	none
$Fi_{24}'{:}2$	$2C, 8D, 29$	none

$G.\langle\tau\rangle$, where τ acts as an outer automorphism of G. Examples are described in [32], [9], [34], among others. Essentially, given a set $\{g_i\}$ of standard generators for G, words in the g_i are found giving images h_i of g_i under an outer automorphism τ. Then a 'standard basis' method (see [23]) is used to find explicitly a matrix (or permutation) conjugating the g_i to the h_i.

6 Future developments

There is clearly a great deal of room for further development of this 'ATLAS of group representations'. First, there are other representations of the sporadic groups which are not easy to obtain from the given ones, but which may be interesting in their own right. Second, it would be very nice to have characteristic 0 matrix representations. Of course, permutation representations can be made into characteristic 0 matrix representations, but they are often far too big to handle in this way. A number of examples exist in the literature, in varying degrees of explicitness (see for example, Conway and Wales [3], [4], Norton [20], [22], Margolin [16]), and some work has been done by Stephen Rogers [26] on integrating these and others into the 'ATLAS of group representations'.

Thirdly, there are other interesting groups which could be included. For example, as the referee pointed out, the exceptional covers of generic groups are closely related to the sporadic groups, and should be included. Since receiving the referee's report, I have made significant progress on constructing representations of these groups (see [35], [36]), but plenty remains to be done in this area.

Another way of extending the database would be to include the maximal subgroups of the sporadic groups. This could be done by including a library of procedures which, given standard generators for a group G, would produce a representative of each class of maximal subgroups of G. Some work on this has already been done by Peter Walsh [37].

All these ideas, and others, are part of our plans for the world-wide-web ATLAS of group representations [43], and are gradually being included therein.

ACKNOWLEDGEMENTS

I would like to thank the referee, and Chris Parker, for helpful comments. I would like to thank the SERC (now EPSRC) for two grants which partially supported this work. And last but not least I would like to thank Richard Parker for giving freely of his ideas, which lie behind almost everything discussed in this paper.

References

[1] J. H. Conway, A quaternionic construction for the Rudvalis group, in *Topics in group theory and computation (ed. M. P. J. Curran)*, pp. 69–81. Academic Press, 1977.

[2] J. H. Conway, A group of order 8,315,554,613,086,720,000, *Bull. London Math. Soc.* **1** (1969), 79–88.

[3] J. H. Conway and D. B.Wales, Matrix generators for J_3, *J. Algebra* **29** (1974), 474–476.

[4] J. H. Conway and D. B. Wales, The construction of the Rudvalis simple group of order 145,926,144,000, *J. Algebra* **27** (1973), 538–548.

[5] J. H. Conway, R. T. Curtis, S. P. Norton, R. A. Parker and R. A. Wilson, *An ATLAS of finite groups*, Clarendon Press, Oxford, 1985.

[6] G. Higman and J. McKay, On Janko's simple group of order 50,232,960, *Bull. London Math. Soc.* **1** (1969), 89–94.

[7] Z. Janko, A new finite simple group with Abelian Sylow 2-subgroups, and its characterization, *J. Algebra* **3** (1966), 147–186.

[8] C. Jansen, K. Lux, R. Parker and R. Wilson, *An ATLAS of Brauer characters*, Clarendon Press, Oxford, 1995.

[9] C. Jansen and R. A. Wilson, Two new constructions of the O'Nan group, *J. London Math. Soc.*, to appear.

[10] C. Jansen and R. A. Wilson, The 2-modular and 3-modular decomposition numbers for the sporadic simple O'Nan group and its triple cover, *J. London Math. Soc.*, to appear.

[11] C. Jansen and R. A. Wilson, The minimal faithful 3-modular representation for the Lyons group, *Comm. Algebra* **24** (1996), 873–879.

[12] S. A. Linton, The maximal subgroups of the Thompson group, *J. London Math. Soc.* **39** (1989), 79–88.

[13] S. A. Linton, Double coset enumeration, *J. Symbolic Comput.* **12** (1991), 415–426.

[14] S. A. Linton, R. A. Parker, P. G. Walsh and R. A. Wilson, Taming the Monster, in preparation.

[15] K. Lux and M. Wiegelmann, Tensor condensation, *these proceedings*, pp. 174–190.

[16] R. S. Margolin, A geometry for M_{24}, *J. Algebra* **156** (1993), 370–384.

[17] E. Mathieu, Mémoire sur l'étude des fonctions de plusieurs quantités, *J. Math. Pure et Appl.* **6** (1861), 241–323.

[18] E. Mathieu, Sur les fonctions cinq fois transitives de 24 quantités, *J. Math. Pure et Appl.* **18** (1873), 25–46.

[19] W. Meyer, W. Neutsch and R. Parker, The minimal 5-representation of Lyons' sporadic group, *Math. Ann.* **272** (1985), 29–39.

[20] S. P. Norton, *F and other simple groups*, Ph. D. thesis, Cambridge, 1975.

[21] S. P. Norton, The construction of J_4, in *Proceedings of the Santa Cruz conference on group theory (eds. B. Cooperstein and G. Mason)*, pp. 271–277. Amer. Math. Soc., 1980.

[22] S. P. Norton, On the Fischer group Fi_{24}, *Geom. Dedicata* **25** (1988), 483–501.

[23] R. A. Parker, The computer calculation of modular characters—the Meat-axe, in *Computational group theory (ed. M. D. Atkinson)*, pp. 267–274. Academic Press, 1984.

[24] R. A. Parker, An integral Meat-axe, *these proceedings*, pp. 215–228.

[25] R. A. Parker and R. A. Wilson, The computer construction of matrix representations of finite groups over finite fields, *J. Symbolic Comput.* **9** (1990), 583–590.

[26] S. J. F. Rogers, *Representations of finite simple groups over fields of characteristic zero*, Ph. D. thesis, Birmingham, 1997.

[27] A. J. E. Ryba, Matrix generators for the Held group, in *Computers in algebra (ed. M. C. Tangora)*, pp. 135–141. Marcel Dekker, 1988.

[28] A. J. E. Ryba, A new construction of the O'Nan simple group, *J. Algebra* **112** (1988), 173–197.

[29] A. J. E. Ryba, Condensation programs and their application to the decomposition of modular representations, *J. Symbolic Comput.* **9** (1990), 591–600.

[30] A. J. E. Ryba and R. A. Wilson, Matrix generators for the Harada–Norton group, *Experimental Math.* **3** (1994), 137–145.

[31] P. E. Smith, A simple subgroup of M? and $E_8(3)$, *Bull. London Math. Soc.* **8** (1976), 161–165.

[32] I. A. I. Suleiman and R. A. Wilson, Computer construction of matrix representations of the covering group of the Higman–Sims group, *J. Algebra* **148** (1992), 219–224.

[33] I. A. I Suleiman and R. A. Wilson, Construction of the fourfold cover of the Mathieu group M_{22}, *Experimental Math.* **2** (1993), 11–14.

[34] I. A. I. Suleiman and R. A. Wilson, Standard generators for J_3, *Experimental Math.* **4** (1995), 11–18.

[35] I. A. I. Suleiman and R. A. Wilson, Covering and automorphism groups of $U_6(2)$, *Quart. J. Math. (Oxford)*, to appear.

[36] I. A. I. Suleiman and R. A. Wilson, Construction of exceptional covers of generic groups, *Math. Proc. Cambridge Philos. Soc.*, to appear.

[37] P. G. Walsh, *Standard generators of some sporadic simple groups*, M. Phil. thesis, Birmingham, 1994.

[38] P. G. Walsh, *Computational study of the Monster and other sporadic simple groups*, Ph. D. thesis, Birmingham, 1996.

[39] R. A. Wilson, A new construction of the Baby Monster, and its applications, *Bull. London Math. Soc.* **25** (1993), 431–437.

[40] R. A. Wilson, Matrix generators for Fischer's group Fi_{24}, *Math. Proc. Cambridge Philos. Soc.* **113** (1993), 5–8.

[41] R. A. Wilson, A representation of the Lyons group in $GL_{2480}(4)$, Preprint 95/29, School of Mathematics and Statistics, University of Birmingham, 1995.

[42] R. A. Wilson, Standard generators for sporadic simple groups, *J. Algebra* **184** (1996), 505–515.

[43] R. A. Wilson *et al.*, A world-wide-web ATLAS of group representations, http://www.mat.bham.ac.uk/atlas/.

Presentations of Fischer groups

François Zara

Abstract

We survey results on presentations of the finite Fischer 3-transposition groups.

1 Introduction

Let G be a group and let D be a generating set of G such that each element of D is an involution and for each pair d and e from D, the order of the product de is 1, 2 or 3. If D is a conjugacy class of G, we say that G is a *Fischer group*, D is a *Fischer class* and (G, D) is a *Fischer pair*.

There exists a natural graph structure on D, its *Coxeter graph*, $\Gamma = \Gamma(D)$, defined by the condition that $\{d, x\}$ is an edge of Γ if and only if dx is of order 3.

Let (G, D) be a Fischer pair. If $p : G \to G/Z(G)$ is the canonical projection, then $(p(G), p(D))$ is a Fischer pair and $\Gamma(p(D))$ is isomorphic to $\Gamma(D)$.

We say that the finite group G is *reductive* if the largest normal soluble subgroup of G is central.

Fischer in [3] has proved the following theorem.

Theorem 1.1 *Let (G, D) be a finite Fischer pair. Suppose that G is a reductive group, and let $p : G \to G/Z(G)$ be the canonical projection. Then $(p(G), p(D))$ is one of the following possibilities (G_i, D_i).*

1. *Classical groups.*

   (a) $G_0 \cong S_n$ $(n \geq 5)$, D_0 *the class of transpositions;*

   (b) $G_1 \cong Sp_{2n}(2)$ $(n \geq 2)$, D_1 *the class of symplectic transvections;*

   (c) $G_2 \cong GO_{2n}^{\varepsilon}(2)$ $(n \geq 3,\ \varepsilon \in \{+, -\})$, D_2 *the class of orthogonal transvections;*

   (d) G_3 *an orthogonal group over $\mathbb{F}_3$ (see below);*

   (e) $G_4 \cong U_n(2)$ $(n \geq 4)$, D_4 *the class of unitary transvections.*

2. *Sporadic groups.*

(a) $G_5 \cong Fi_{22}$;

(b) $G_6 \cong Fi_{23}$;

(c) $G_7 \cong Fi_{24}$.

In each case D_i is uniquely determined.

3. Triality groups.

(a) $G_8 \cong D_4(2){:}S_3$;

(b) $G_9 \cong D_4(3){:}S_3$.

In each case D_i is uniquely determined.

Let E be a n-dimensional vector space over the finite field $\mathbb{F}_3$. There are two types of non-degenerate symmetric bilinear forms f over E. For one type there is an orthonormal basis, for the other type there is no such basis. Let $\varepsilon \in \{+, -\}$ be defined as follows:

$$\varepsilon \;=\; + \quad \text{if there is an orthonormal basis,}$$
$$\varepsilon \;=\; - \quad \text{if not.}$$

(Note that this definition of ε is not the same as in the ATLAS. In fact, the definitions agree for $n \equiv 0 \pmod 4$ and disagree for $n \equiv 2 \pmod 4$, while for n odd the ATLAS makes no distinction.)

Let us call $G^\varepsilon(n,3)$ the subgroup of the full orthogonal group generated by the set D_3 of orthogonal reflections associated with vectors v such that $f(v,v) = 1$, then in 1d, G_3 is isomorphic to $G^\varepsilon(n,3)/Z(G^\varepsilon(n,3))$. (In the ATLAS, there is no compact notation for these groups.)

Remark 1.2 In 1d and 1e it is really the image of the class of orthogonal reflections or the class of unitary transvections in the corresponding group.

Let us give a definition.

Definition 1.3 Let (G, D) be a Fischer pair. We say that the presentation (X, R) of G is *à la Fischer* if the following conditions are satisfied:

1. X is a subset of D; and

2. the relations in R are all of the form $(x^g y)^m = 1$, where x and y are in X, g is in G and m is 2 or 3.

The goal of this paper is to give presentations à la Fischer for the groups of Theorem 1.1, with the added condition that X is of minimal size.

In fact, as stated, this goal is unattainable. For example, if a group G has a presentation à la Fischer, then it is easy to see that $G/[G,G]$ is of order 2; but this is not true, for example, of the symplectic group. Furthermore, we cannot kill the centre with relations of the form $(gh)^m = 1$ for g, h in X and $m = 2$ or 3, because $\Gamma(p(D))$ is isomorphic to $\Gamma(D)$.

Therefore we will give presentations for some central extensions (which are Fischer groups) of the groups in Theorem 1.1. The result for the symmetric group is well known (see [7]). The classical groups have been handled by M. M. Virotte-Ducharme (see [8] and [9]). The sporadic groups have been handled by J. I. Hall and L. H. Soicher (see [6]). The triality groups have been handled by F. Zara (see [10]).

It is to be remarked that, with the exception of the three sporadic groups, everything has been done "by hand" that is, without the use of a computer.

2   Method

The preceding goal is attained by using the following results which permit us to proceed inductively.

Theorem 2.1 (Virotte-Ducharme [8]) *Let G be a group having the presentation (X, R) where X is a set of elements of order 2, of cardinality at least 3. Let b be an element of X such that:*

1. $K := \langle X - \{b\}\rangle$ *is a proper subgroup of G having a Fischer class D containing $X - \{b\}$.*

2. *For all x in $X - \{b\}$, we have $(xb)^2 = 1$ or $(xb)^3 = 1$.*

Let $B := C_K(b)$. Let us consider the following conditions:

(C.0) B acts transitively on $D - C_D(b)$;

(C.1) $\forall k \in K, \exists d, e \in D$ such that $k \in Bd \cup Bde$;

(C.2) $\exists z \in Z(K) - B$, z of order 2, such that

    (a) $(b^{zb}b)^3 = 1$ and $\forall x \in X - \{b\}$, $(xb^{zb})^2 = 1$;

    (b) $\forall k \in K, \exists d, e \in D$ such that $k \in Bd \cup Bde \cup Bz \cup Bzd$.

Then

1. *If (C.0) and (C.1) are satisfied, then $E := D \cup \{b^k \mid k \in K\}$ is a Fischer class of G and $|E| = |D| + [K : B]$.*

2. *If (C.0) and (C.2) are satisfied, then $E := D \cup \{b^k \mid k \in K\} \cup \{b^{zb}\}$ is a Fischer class of G and $|E| = |D| + [K : B] + 1$.*

This theorem gives us two pieces of information.

1. The fact that (G, E) is a Fischer pair;

2. The cardinality of the class E.

To obtain more information we can use the following result:

Proposition 2.2 (Virotte-Ducharme [8]) *Let (G, E) be a Fischer pair such that $G = \langle X \rangle$, where $X = \{x_i \mid i \in I\} \subset E$, and the elements of X satisfy certain relations R. Let L be a group generated by a set of involutions $Y = \{y_i \mid i \in I\}$ such that $x_i \mapsto y_i : X \to Y$ is a bijection and such that $\langle Y, R \rangle$ is a presentation of L. Then:*

1. *There exists a homomorphism $f : L \to G$ such that $f(y_i) = x_i$ for all i in I.*

2. *Let F be the conjugacy class of y $(y \in Y)$ in L. If E and F have the same cardinality, then the extension*

$$1 \to \operatorname{Ker} f \to L \xrightarrow{f} G \to 1$$

is central and (L, Y) is a Fischer pair.

We apply these results to find presentations of the classical Fischer groups. This is elementary but we must know all the Schur multipliers of the Fischer groups (see R. Griess [4] and [5]).

In most of the cases, the Fischer group G contains a "big" subgroup $H_{i,r}$, which is isomorphic to $W(\tilde{A}_r)/2T$ if $i = 2$ or to $W(\tilde{A}_r)/3T$ if $i = 3$, where $W(\tilde{A}_r)$ is the affine Weyl group of type A_r, T is its translation subgroup and $kT = \{t^k \mid t \in T\}$. If $i = 2$ and $r + 1$ is even, then $Z(H_{i,r})$ is of order 2 and is central in G; if $i = 3$ and $r + 1$ is a multiple of 3, then $Z(H_{i,r})$ is of order 3 and is central in G.

To find the presentations, we must start with K, whose presentation is known by induction, and then find a B which satisfies relations (C.0), (C.1) or (C.0), (C.2). To start the induction we can use the presentations which are given in the ATLAS [1].

As an example, I shall give the result for the groups $G^\varepsilon(n, 3)$.

Proposition 2.3 *Let (G, E) be a Fischer pair such that $G \cong G^\varepsilon(n, 3)$ with $n \geq 5$ and $\varepsilon \in \{+, -\}$. Let V be the natural $\mathbb{F}_3 G$-module and let $f : V \times$*

$V \to \mathbb{F}_3$ *be the symmetric invariant bilinear form such that if d is in E,* $f(v_d, v_d) = 1$ *(where v_d is a defining vector of the reflection d). Let v be a non-zero vector in V. We define $G(v) := \{g \mid g \in G, g(v) = v\}$; if v is isotropic, we define $G'(v) := \langle r_a \mid a \in V, f(a,a) = 1, f(a,v) = 0 \rangle$, where r_a denotes the reflection in the vector a, and $G''(v) := \{g \mid g \in G, g(v) \in \langle v \rangle\}$. Then:*

1. (a) *If $f(v,v) = 1$, then $G(v) \cong W(D_4){:}C_3$ if $n = 5$, and $G(v) \cong G^\varepsilon(n-1,3)$ if $n \geq 6$;*

   (b) *if $f(v,v) = -1$, then $G(v) \cong G^{-\varepsilon}(n-1,3)$;*

   (c) *if $f(v,v) = 0$, then we have $|G''(v)/G(v)| = 2$, $G'(v) = G(v)$ if $\varepsilon = +$ and $n \geq 5$, or if $\varepsilon = -$ and $n \geq 7$, and $|G(v)/G'(v)| = 3$ if $\varepsilon = -$ and $n \in \{5,6\}$. Furthermore $G'(v)$ is isomorphic to $C_3^{n-2}{:}G^{-\varepsilon}(n-2,3)$.*

2. *Except in the case $f(v,v) = 1$, $G(v)$ acts transitively on $E - (G(v) \cap E)$. If $f(v,v) = 1$, then $G(v)$ acts transitively on $E - ((G(v) \cap E) \cup \{r_v\})$.*

3. *Let g be in G. Then there exist d and e in E such that if v is not isotropic, then $g \in G(v)d \cup G(v)de$, while if v is isotropic, then $g \in G''(v)d \cup G''(v)de$.*

This shows that small dimensions create problems, and we have the same kind of problems with the unitary groups over $\mathbb{F}_4$.

3   Presentations of the classical groups

In this part we give presentations of the classical Fischer groups as obtained by the above method. It is to be remarked that for small dimensions, the results are well known and are due to Coxeter (see [2]).

3.1   Symmetric groups. $G \cong S_n$ $(n \geq 5)$

3.2 Symplectic groups over $\mathbb{F}_2$. $G \cong C_2 \times Sp_{2n}(2)$ $(n \geq 3)$

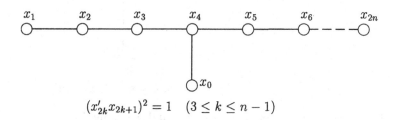

$$(x'_{2k} x_{2k+1})^2 = 1 \quad (3 \leq k \leq n-1)$$

where x'_{2k} is defined inductively in the following way: $x'_2 = x_1$, $x'_4 = x_0$ and for $k \geq 3$, x'_{2k} is the unique element of $\langle x'_{2k-4}, x'_{2k-2}, x_{2k-j} \quad (0 \leq j \leq 4) \rangle$ such that we have a graph of type $\tilde{E}_7$ as follows.

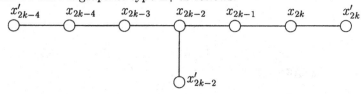

3.3 Orthogonal groups over $\mathbb{F}_2$. $G \cong GO^\varepsilon_{2n}(2)$ $(n \geq 3)$

We have several cases:

1. $\varepsilon = -$ and $n \equiv 0$ or $1 \pmod 4$, or $\varepsilon = +$ and $n \equiv 2$ or $3 \pmod 4$.

$$(x'_{2k} x_{2k+1})^2 = 1 \quad (3 \leq k \leq n-2)$$
$$(x'_{2k} x_{2n})^2 = 1 \quad (3 \leq k \leq n-2, k \text{ odd})$$

2. $\varepsilon = -$ and $n \equiv 1$ or $2 \pmod 4$, or $\varepsilon = +$ and $n \equiv 0$ or $3 \pmod 4$.

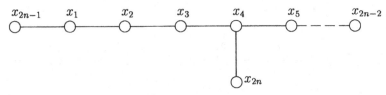

$$(x'_{2k}x_{2k+1})^2 = 1 \quad (3 \le k \le n-2)$$
$$(x'_{2k}x_{2n-1})^2 = 1 \quad (3 \le k \le n-2,\ k\ \text{even}).$$

3. $\varepsilon = -$ and $n \equiv 3 \pmod 4$, or $\varepsilon = +$ and $n \equiv 1 \pmod 4$.

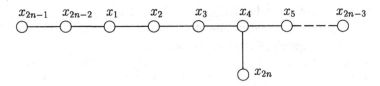

$$(x'_{2k}x_{2k+1})^2 = 1 \quad (3 \le k \le n-3)$$
$$(x'_{2k}x_{2n-2})^2 = 1 \quad (3 \le k \le n-3,\ k\ \text{even})$$
$$(yx_{2n-1})^2 = 1.$$

In each case x'_{2k} has the same meaning as in section 3.2 and y is the unique element of $\langle x_{2n-2}, x_1, x_2, \ldots, x_6, x_{2n}\rangle$ such that we have a graph of type $\tilde{E}_8$ as follows.

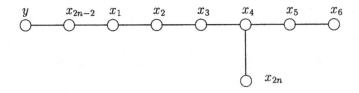

3.4 Orthogonal groups over $\mathbb{F}_3$. $G \cong G^\varepsilon(n,3)$ $(n \ge 5)$

1. $G^+(5,3)$ $(\cong C_2 \times U_4(2))$:

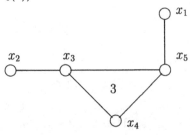

Let $x' = x_2^{x_3 x_5 x_4 x_3 x_1 x_5 x_3}$ and $x'' = x_2^{x_3 x_4 x_5 x_3 x_1 x_5 x_4 x_3}$. If $p = x_2 x_4 x_1 x' x''$ then $\langle p \rangle = Z(G^+(5,3))$.

The "3" inside the triangle means that $H = \langle x_3, x_4, x_5 \rangle$ is isomorphic to the group $H_{3,2}$, that is, we add the relation $(x_3^{x_4} x_5)^3 = 1$.

2. $G^-(6,3)$:

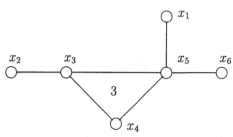

with the relation $(x_1^s x_6)^2 = 1$, where $s = (x_3 x_4 x_5)^2$.

We have $H = \langle x_3, x_4, x_5 \rangle \cong H_{3,2}$ and s generates the centre of H.

3. $C_3 \cdot G^+(7,3)$:

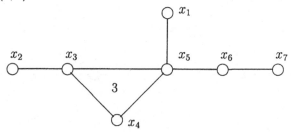

with the relations $(x_1^s x_6)^2 = 1$, $s = (x_3 x_4 x_5)^2$.

We have $H = \langle x_3, x_4, x_5 \rangle \cong H_{3,2}$ and s generates the centre of H. Let p be a generator of the centre of $\langle x_i \mid 2 \le i \le 6 \rangle$. Define $x_7' := x_7^{p x_7}$, $L := \langle x_j, x_7' \mid 3 \le j \le 7 \rangle \cong H_{3,5}$ and $\langle z \rangle = Z(L)$; then z is of order 3 and z is central in G. We have $p = x_2 x_4 x_6 e e'$, $e = x_3^{x_5 x_2 x_4 x_6 x_3 x_5 x_3}$, $e' = x_3^{x_2 e x_4 x_3}$ and $z = (x_3 x_4)(x_3 x_4)^{x_5}(x_3 x_4)^{x_5 x_6}(x_3 x_4)^{x_5 x_6 x_7}(x_3 x_4)^{x_5 x_6 x_7 x_7'}$.

4. $G^\varepsilon(n,3)$, $\varepsilon = (-1)^{n-1}$ $(n \ge 8)$:

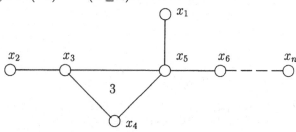

with the relations $(x_1^s x_6)^2 = 1$, $s = (x_3 x_4 x_5)^2$, $(x_8^z x_8)^2 = 1$ (z is defined above in the preceding group).

5. $G^-(5,3) \cong W(E_6)$.

6. $C_3 \cdot G^+(6,3)$:

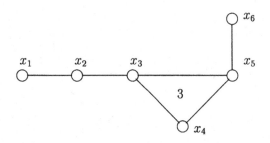

Let $m_0 := x_6 x_5 x_4 x_3 x_2 x_1$; then $Z(G) = \langle m_0^7 \rangle$ is of order 6.

7. $G^\varepsilon(n,3), \varepsilon = (-1)^n$ $(n \geq 7)$:

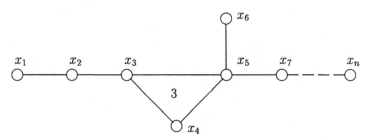

with the relations $(x_6^s x_7)^2 = 1$, $s = (x_3 x_4 x_5)^2$, $(x_7^z x_7)^2 = 1$, with $z = m_0^{14}$ (m_0 is defined above in the preceding group).

3.5 Unitary groups over $\mathbb{F}_4$. $G \cong C_2 \times SU_n(2)$ $(n \geq 5)$

1. $C_2 \times SU_5(2)$:

with the relation $(x_3^{x_2 x_4} x_5)^3 = 1$.

Let $Q := \langle x_2, x_3, x_4, x_5 \rangle$. Then Q is isomorphic to the centralizer of a transvection in $SU_5(2)$. Let $z_1 = (x_2 x_3 x_5)^2$, $z_2 = (x_3 x_5 x_4)^2$ and

$q = x_2 x_4 x_2^{z_2} x_4^{z_1}$. Then q is of order 2, and $\langle q \rangle = Z(Q)$. Define $x_0 := x_1^{q x_1}$. Then we have the following diagram:

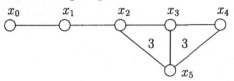

2. $C_2 \times \widetilde{U_6(2)}$

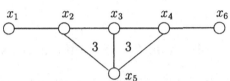

We define $b = x_1^{q x_2 x_3 x_4}$, and we add the relations $(x_3^{x_2^2 x_4} x_5)^3 = 1$ and $(x_6 b)^2 = 1$; then $G \cong C_2 \times \widetilde{U_6(2)}$ where $\widetilde{U_6(2)}$ is the non-split central extension of $SU_6(2)$ by $C_2 \times C_2$.

Let Z be the 2-part of the centre of G. We have $Z = \langle m_1, m_2 \rangle$ where m_1 and m_2 are of order 2. We now give expressions for m_1 and m_2. For $R_1 := \langle x_1, x_2, x_4, x_5, x_6, b \rangle$ we have the diagram:

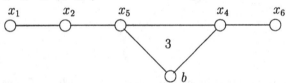

and $m_1 = m_0^{21}$, where $m_0 = x_6 x_4 b x_5 x_2 x_1$ is as defined for the group $C_3 {\cdot} G^+(6,3)$. Let $\langle p \rangle = Z(\langle x_2, x_5, x_4, x_6, b \rangle)$. We have $p = x_2 b x_6 u v$ where $u = x_2^{x_5 x_4 b x_5 x_6 x_4 x_5}$ and $v = x_2^{x_5 b x_4 x_5 x_6 x_4 b x_5}$. Let

$$f = x_2^{x_5 x_4 x_1 x_3 x_6 x_4 x_5 x_2 x_5 x_4 x_5 x_3 b},$$

and $y = f^{z_1}$ where $z_1 = (x_2 x_5 x_3)^2$ and $t = x_1^{p x_1 f y}$. Then $R_2 := \langle x_1, t, x_5, x_4, x_6, b \rangle$ is isomorphic to $C_3 {\cdot} G^+(6,3)$ and is not conjugate to R_1; we have the diagram:

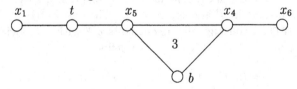

We have $\langle m_2 \rangle = Z(R_2)$ and $m_2 = (x_6 x_4 b x_5 t x_1)^{21}$.

3. $C_2 \times SU_n(2)$ $(n \geq 7)$:

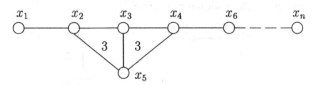

with the relations $(x_3^{x_2 x_4} x_5)^3 = 1$, $(x_6 x_1^{q x_2 x_3 x_4})^2 = 1$, $(x_7^{m_1} x_7)^2 = 1$ and $(x_7^{m_2} x_7)^2 = 1$.

4   The two triality groups

Let S be a group isomorphic to the symmetric group of degree 5 and let D be its class of transpositions, identified with the set of transpositions $\{(i,j) \mid 1 \leq i < j \leq 5\}$.

Let $G_0 := \langle S, y \mid y^2 = 1, \forall d \in D, (dy)^3 = 1 \rangle$ and $E := \{gdg^{-1} \mid g \in G_0, d \in D\}$, so that we have the following diagram:

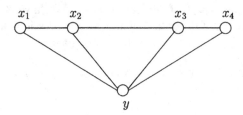

We give some properties of the group G_0. We have $C_S(y) = \{1\}$; there exists an endomorphism f of G_0 such that $f(x_i) = x_1$ $(1 \leq i \leq 4)$, $f(y) = y$; we have $f^2 = f$ so that $G_0 \cong \operatorname{Ker} f : \operatorname{Im} f$, $\operatorname{Im} f \cong S_3$, the symmetric group of degree 3 and $E = E_1 \cup E_2 \cup E_3$ with $E_i \cap E_j$ empty if $i \neq j$, where $E_1 = f^{-1}(x_1) \cap E$, $E_2 = f^{-1}(y) \cap E$ and $E_3 = f^{-1}(x_1^y) \cap E$. We have $[S, S] \subset \operatorname{Ker} f$, $\{y^s \mid s \in [S, S]\} \subset E_2$, $\{y^s \mid s \in S - [S, S]\} \subset E_3$, $\{y^{sy} \mid s \in S - [S, S]\} \subset E_1$ and $|G_0/[G_0, G_0]| = 2$. Let $G_i = \langle E_i \rangle$ $(1 \leq i \leq 3)$. In Theorem 5.2 (resp. 5.3) we identify G_i with an orthogonal group over the field $\mathbb{F}_2$ (resp. $\mathbb{F}_3$).

Put $T := \{(ijkl) \mid |\{i, j, k, l\}| = 4\}$. We say that t and t', elements of T, are *associated* if t and t' fix the same element of $\{1, 2, 3, 4, 5\}$.

Let $t = (ijkl)$ be in T and let $z = y^{ty}$. Then $N = \langle (ij), (jk), (kl), z \rangle \cong W(D_4)$ and $N \cap E_1$ contains all the elements $y^{t'y}$ with t' associated with t.

Let I be a subset of $\{1, 2, 3, 4, 5\}$. We define

$$W_I := \langle S, y^{ty} \mid t \in T, \text{ the fixed point of } t \text{ is in } I \rangle.$$

Also, $W = W_{\{1,2,3,4,5\}}$, and $W_i = W_{\{i\}}$.

We have the following results. Let t_i $(1 \leq i \leq 4)$ be pairwise non associated elements of T. Then $W = \langle S, y^{t_i y} \quad (1 \leq i \leq 4) \rangle$ and $[G_0 : W] = 180$, with G_0 acting as A_5 wr S_3 on G_0/W.

Let $z := y^{(1234)y}$; then z commutes with $(1,3)$ and $(2,4)$. Let p, $q \in (N - \{0\}) \cup \{\infty\}$. We define the following quotients of G_0:

$$G_0(p,q) := \langle S, y \mid \forall d \in D, (dy)^3 = y^2 = 1 = (z(1,5))^q = (z(2,5))^p \rangle.$$

If $W(p,q)$ is the image of W in $G_0(p,q)$, and if $(5,p) = 1$ or $(5,q) = 1$ we have $[G_0(p,q) : W(p,q)] = 3$. We can now give the results for the triality groups.

Theorem 4.1   *1. $G_0(2,3) \cong D_4(2){:}S_3$ and $G_0(3,3) \cong D_4(3){:}S_3$.*

   2. If $p = 2$, then $q = 3$;

     (a) if $|I| = 1$, then $W_I \cong W(D_5)$;

     (b) if $|I| = 2$, then $W_I \cong W(E_6)$;

     (c) if $|I| = 3$, then $W_I \cong W(E_7)$;

     (d) if $|I| = 4$, then $W_I \cong W(E_8)/Z(W(E_8)) \cong GO_8^+(2)$.

   3. If $p = q = 3$ then

     (a) if $|I| = 1$, then $W_I \cong G^+(5,3)$;

     (b) if $|I| = 2$, then $W_I \cong C_3^5{:}G^+(5,3)$;

     (c) if $|I| = 3$, then $W_I \cong G^-(7,3)$;

     (d) if $|I| = 4$, then $W_I \cong G^+(8,3)/Z(G^+(8,3))$.

In the proof of this theorem, essential use is made of Proposition 2.3.

5   The sporadic Fischer groups

Here we report on the work of Hall and Soicher (see [6]). As most of their calculations have been done by computer, we content ourselves to give the results.

1. $C_2 \times (C_6 \cdot Fi_{22})$:

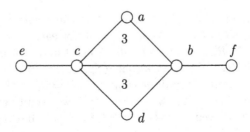

with the relations $(a^b c^d)^3 = 1$, $(ed^s)^3 = 1$ and $(fd^s)^2 = 1$, where $s = (abc)^2$.

2. $C_2 \times Fi_{23}$:

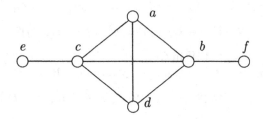

with the relations $(a^b c)^3 = (a^b d)^3 = (a^d c)^3 = (b^d c)^3 = (a^b c^d)^3 = (a^d b^c)^3 = (a^c b^d)^3 = (a^s e)^3 = (e^{dbadcb} f)^2 = 1$, where $s = (bcd)^2$.

3. Fi_{24}:

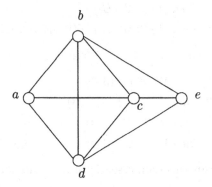

with the relations $(a^b c)^3 = (a^b d)^3 = (a^d c)^3 = (b^d c)^3 = (a^b c^d)^3 = (a^d b^c)^3 = (a^c b^d)^3 = (b^c e)^3 = (b^d e)^3 = (c^d e)^3 = (c^{db} e)^3 = (b^{dc} e)^3 = (b^{cd} e)^3 = (a^{bc} e)^3 = (a^{bd} e)^3 = (a^{cd} e)^3 = (b^{cad} e)^3 = (a^s e)^2 = (b^x b^y)^2 = (b^x c^y)^2 = (b^x d^y)^2 = (c^t e)^2 = 1$, where $s = (bcd)^2$, $x = a^s a$, $y = ee^s$ and $t = badcebadcadbebd$.

6   Conclusion

Let (G, D) be a Fischer pair. Let us define the *length of G* to be $l(G) := \inf\{|X| : X \subset D, G = \langle X \rangle\}$. We have given, for each group of Theorem 1.1, a presentation (X, R) à la Fischer with generators taken in the class D and X of cardinality $l(G)$. We remark that $l(Sp_{2n}(2)) = 2n + 1$, but, for all other classical Fischer groups G in dimension n, $l(G) = n$; as all Fischer groups of length 4 have been classified, we see that the two triality groups have length 5, and in [6], it is proved that $l(Fi_{22}) = l(Fi_{23}) = 6$, but $l(Fi_{24}) = 5$.

References

[1] J. H. Conway, R. T. Curtis, S. P. Norton, R. A. Parker and R. A. Wilson, *An* ATLAS *of finite groups*, Oxford University Press, 1985.

[2] H. S. M. Coxeter and W. O. J. Moser, *Generators and relations for discrete groups*, Springer, 1957.

[3] B. Fischer, Finite groups generated by 3-transpositions, I, *Invent. Math.* **13** (1971), 232–246.

[4] R. Griess, Schur multipliers of finite simple groups of Lie type, *Trans. Amer. Math. Soc.* **183** (1973), 355–421.

[5] R. Griess, Schur multipliers of some sporadic simple groups, *J. Algebra* **20** (1972), 320–349.

[6] J. I. Hall and L. H. Soicher, Presentations of some 3-transposition groups, *Comm. Algebra* **23** (1995), 2517–2559.

[7] E. H. Moore, Concerning the abstract groups of order $k!$ and $1/2k!$, *Proc. London Math. Soc. (1)* **28** (1897), 357–366.

[8] M. M. Virotte-Ducharme, Presentations des groupes de Fischer I, II, *Geom. Dedicata* **41** (1992), 275–335; **45** (1993), 121–162.

[9] M. M. Virotte-Ducharme, Presentations de certains couples fischeriens de type classique, *Bull. Soc. Math. France* **121** (1993), 227–270.

[10] F. Zara, Présentations des groupes de Fischer $D_4(2){:}S_3$ et $D_4(3){:}S_3$, *J. Algebra* **170** (1994), 705–734.

A brief history of the ATLAS

John H. Conway,
Robert T. Curtis
and
Robert A. Wilson

The beginnings of the ATLAS go back to around 1970 in Cambridge. In about 1969 John Conway had discovered his (then) new simple group ·1, and in the ensuing year John Thompson and Conway explored its properties in great detail. One of our aims was to compute the character table of the new group.

This process took more than a year, although most of the time spent was in two short periods at the start and end of that year. In the earlier portion, Mike Guy prepared a computer system for working with character tables—a primitive version of CAS—and in the later one we applied this to a number of starting characters that were produced rather laboriously by hand. Several of the trickiest characters were found by Nick Paterson, and these enabled us to complete the work.

Some time later, we collaborated with Don Livingstone, who used to travel from Birmingham to Cambridge and work overnight with us for the purpose, in finding a number of other character tables.

It then occurred to Conway that some kind of handbook listing "all interesting properties of all interesting groups" would be extremely useful. After taking advice from Frank Adams, he applied to the Science Research Council for a grant to support an assistant to help in producing such a work, and the ATLAS was born!

The said assistant was Conway's first student in group theory, Robert Curtis, who had just written his thesis on the subgroups of ·0, and we gamely set about collecting information about as many groups as we could.

We bought, for what then seemed the astonishingly large cost of £82, a "guard book". This was a large book with a special binding that allowed for the insertion of extra pages without bursting. Conway still has this book, although in fact we stuffed so much into it that it *did* burst at one time. Fortunately one of the chairs in the DPMMS common room had also recently burst, and he used the resulting supply of fake leather to make a new binding which was sewn on using a large bodkin. The book is now about 6" (15cm) at the binding!

Initially we had no clear idea just what information to collect, but we soon discovered that the character table of a group contains so much information that it is by far the most important thing, and from then on we concentrated mostly on character tables, although we also worked out lots of other things (such as parameters of graphs, etc.) not all of which appears in the ATLAS as published.

The character tables in the literature were printed in a wide variety of styles, and Conway and Curtis spent some time determining upon the system used in the ATLAS. We'd like to comment on this, since we're quite proud of it.

Many authors print the conjugacy classes in various kinds of order that reflects their makeup. For instance an element of order p might immediately be followed by the elements of orders $2p$, $3p$, ... that have it for their p-part. This helps the author of the table (since it conveys some information about power maps, and since (s)he will usually compute the character values of these elements at the same time), but the help it offers to the reader is offset by the greater difficulty of finding the elements (s)he wants. We decided to enumerate "in order of order", and when orders were equal, by decreasing order of centralizer.

Rather than try to specify some power map information by ordering the elements, we decided to give *complete* power maps and p-parts, and worked out a simple way of doing so.

There is also a wide variety of conventions for printing the irrational numbers that occur in character tables. Some authors print expressions such as "$(-1 + \sqrt{5})/2$", some use Greek letters explained by footnotes, and others express everything in terms of sums of roots of unity.

We wanted a system which would convey complete information without the need for footnotes, would be permanent, and would usually fit into our standard column-width of four spaces, the first of which is preferably left blank. We succeeded, by inventing special names for various "Gaussian sums" that usually turn up, and separating these from the Galois group elements "$*k$" that act on them.

This system has stood up well, and seems to have already become "the industry standard", although for modular character tables it is sometimes sorely stretched (see [2]).

Curtis had an office off the Common Room in DPMMS, which was christened "Atlantis" because that was where the ATLAS lived. A character table which had been converted into ATLAS format was said to have been "atlantified", and more generally, everything to do with the ATLAS was described as "atlantic". When a character table or other material was in reasonably good shape, it was typed onto blue paper, and photocopied for posterity. These "atlantic blue" pages were filed away in folders which were labelled with the name of the group. These folders also contained copies of all the papers we could find on the groups, and any other information we could get hold of.

Simon Norton joined in the work from a very early stage, and he and Conway worked alone for a time after Curtis took up a position in Birmingham. Then one day Clive Bach told us of someone who "was very interested in groups and computers", and so we met Richard Parker, who took over the job of "mechanizing the ATLAS". Richard later made astonishing contributions to computational group theory, and many people will be surprised to learn that he did not hold any academic position, but supported himself by doing freelance programming jobs. He would work on groups for some time, until he ran out of money, and then disappear to help some large chain of stores install a new computer system. After this he would return to Cambridge with some cash in pocket, and the cycle would begin again.

Richard designed "FORMAT X", the system by which we input known character tables into the machine. This was a tedious business, replete with accidents. The worst of these came when, after spending $14\frac{1}{2}$ hours typing in the 114 by 114 character table of $O_8^+(3)$, one of us inadvertently pressed a key that deleted it all, and had to start again from scratch!

Another of Conway's graduate students, Robert Wilson, joined the team, and took over the massive job of organizing our results. Wilson's thesis project had been the enumeration of maximal subgroups of certain groups, and he became known as "Mr Maximal Subgroups". Hence the inclusion of lists of maximal subgroups in the ATLAS.

The sequence of names is easy to remember, because (as is pointed out in the introduction to the ATLAS) their alphabetical order is also the chronological one in which they joined the work. We also noticed that we each had two initials followed by a surname of six letters, the second and fifth of which are vowels!

J. H. CONWAY

R. T. CURTIS

S. P. NORTON

R. A. PARKER

R. A. WILSON

At around the same time, Jon Thackray became involved in the computational part of the project. He abolished FORMAT X, and devised a more user-friendly format for character tables, as well as writing programs for checking the orthogonality relations between the characters, and so on.

It was understood almost from the start that Oxford University Press would be the publisher, and the benign form of Anthony Watkinson would be

Figure 1: The Conway Seal of Grudging Approval

seen in Cambridge every now and then, "prodding the work along". When it got close to publication, this role was taken over by Martin Gilchrist.

We agreed to deliver the pages in "fascicules" of about 20 groups each, and as delivery day approached, we would frantically be working on the last problems with the associated groups. The day before, we would virtually take over the Common Room, allotting one page to each seat.

Prior to this, we prepared one group at a time, in no particular order, until the material on that group reached a degree of completeness, and had been sufficiently checked, to be awarded the "Conway Seal of Grudging Approval" (see Fig. 1).

When all the groups in the fascicule had been awarded the CSGA, we were ready to print out the master copy on the Computing Service's Diabolo printer. At first, the printer was reasonably reliable, and we felt that the output was well worth the 10p per page that we paid for it. As the months went by, however, and we reached the sixth and last fascicule, the printer was wearing out, and started to live up to its name in a way which the manufacturers no doubt did not intend. It became more and more difficult to print out a whole page without at least one symbol being misprinted or left out altogether. Not all such errors were spotted in time, as the alert reader of the ATLAS will have noticed.

We photocopied all the master pages, and started trying to arrange the pieces into full A3 pages. The first few pages were relatively straightforward, but as time went by, we found ourselves forced into several uncomfortable compromises. First we conceded that certain character tables (and later other information) had to be turned through 90° in order to utilize the available space efficiently. Then we accepted the need, originally anathema, for two groups occasionally to share the same page. Later on, we even allowed the "other information" to infiltrate blank spaces in the character table itself.

The biggest headache was providing instructions to the Production Department at OUP as to how to cut and paste our small pages into the final pages, and to add in the bits that our Diabolo printer was incapable of. These included the numbering of characters by χ_i at the beginning and end, the "maps", drawing in the fusion lines in the "fus" columns of the character tables, adding pages numbers and page headings, etc. It is a tribute to them that they followed our instructions extremely accurately, and the very few errors that crept in at this stage were as likely as not due to us giving them wrong instructions.

When we got to the end, we felt there were some important things we had left out. For example, we had stripped off the partition labels from the character tables of the symmetric groups, in the interests of uniformity, but later regretted it, and reinstated the information in a separate table on page 236. Again, we had decided against including references, but Wilson felt strongly enough that we should include them, that he volunteered to trawl through the pages of Mathematical Reviews to produce a bibliography of the sporadic groups. Leonard Soicher had meanwhile been producing large numbers of presentations of sporadic groups. We managed to infiltrate several of these into the book at a very late stage, by putting them on page 232, under the Monster, and inserting forward references at the relevant places.

The pages were prepared by taking photocopies of the machine printed character tables and other information, and pasting them roughly onto large sheets of paper. When Martin Gilchrist came, he was handed these rough paste-ups together with clean copies of the things they were made up from, for the printers to make neater versions from.

Unfortunately this process, like all others, was subject to error. One of the numbers in the ATLAS is wrong merely because its minus sign has been guillotined off (you can still see about an eighth of it!). Perhaps the worst error happened because a thin strip containing five columns of the character table for the automorphism group of Fi_{22} got lost in the paste-up process.

The mention of these errors reminds us that of course we kept an error-book, and that the rate at which we discovered errors rose from about one per week to about one per day as the publication date approached! But fortunately the prediction in the ATLAS introduction was correct—there are very few errors in the actual character table entries, the above being the most serious. It was our policy throughout that we took collective and not individual responsibility for errors, so that even when an individual could be identified as the source of an error, he never was. We do still occasionally find new errors, or have them pointed out to us, but the rate has slowed to about one every six months, and we are only aware of four errors in addition to those corrected in the ABC [2]. It has to be admitted, however, that this list of corrections has itself at least one known error—on page 310, the centralizer orders of 12L, 12M and 12N should be 3456, 1296 and 576 respectively.

Then at last the final proofs of the sixth fascicule had been sent back and it was all over—or was it? We had not yet written the Introduction! This had to be substantial in order to explain all our notation, but it was always intended to be more than this, a kind of introduction to simple groups in general. The resulting tour-de-force was almost entirely the work of Conway, but perhaps the rest of us had the more difficult task—of persuading him to do it!

Perhaps a few words are in order about the binding. We had always intended the book to have a ring binding, in order that the pages would lie flat when the book was open. But when we saw the first pre-publication copies with flimsy paper covers, we asked OUP to produce a stronger binding. They did so, but unfortunately neglected to upgrade the rings to cope with the heavier covers. The result has been a source of embarrassment to us all ever since.

And finally—the ATLAS is now out of print. We hope to publish a 'reprint with corrections' in due course, but even after twelve years, the OUP Production Department still have nightmares about this book, so it may take some time to persuade them.

References

[1] J. H. Conway, R. T. Curtis, S. P. Norton, R. A. Parker and R. A. Wilson, An ATLAS of finite groups, Clarendon Press, Oxford, 1985.

[2] C. Jansen, K. Lux, R. Parker and R. Wilson, An ATLAS of Brauer characters, Clarendon Press, Oxford, 1995.

Printed in the United States
By Bookmasters